STUDENT'S SOLUTIONS MANUAL

MAX STERELYUKHIN

Simon Fraser University

CALCULUS FOR BIOLOGY AND MEDICINE

THIRD EDITION

Claudia Neuhauser

University of Minnesota

Prentice Hall
is an imprint of

Reproduced by Pearson Prentice Hall from electronic files supplied by the author.

ISBN-13: 978-0-321-64492-3
ISBN-10: 0-321-64492-1

5 6 EBM 13 12

Prentice Hall
is an imprint of

www.pearsonhighered.com

Contents

CONTENTS

CONTENTS

Chapter 1

Preview and Review

1.1 Preliminaries

Prob. 1.

(a) Walking 3 units to the right and to the left from -1 we get the numbers 2 and -4, respectively.

(b) $|x - (-1)| = 3$, so $x + 1 = \pm 3$, yielding, $x = 2$ or $x = -4$.

Prob 3.

(a) $2x - 4 = \pm 6$, so either $2x = 10$ giving the solution $x = 5$ or $2x = -2$ and the other solution is $x = -1$.

(b) $x - 3 = \pm 2$, from which we see that $x = 5$ or $x = 1$.

(c) $2x + 3 = \pm 5$, so $2x = 2$ giving $x = 1$ or $2x = -8$ and $x = -4$.

(d) The equation has no solutions, since the absolute value of a number cannot be negative, and so could never be equal to -2.

Prob. 5.

(a) We can change the absolute value for two inequalities

$$-4 \leq 5x - 2 \leq 4$$

3

which then can be solved to

$$-4 + 2 \leq 5x \leq 4 + 2$$

and dividing by 5 we get:

$$-\frac{2}{5} \leq x \leq \frac{6}{5}$$

(b) It will change to the inequalities: $1 - 3x < -8$ or $1 - 3x > 8$. Solving the first we get: $9 < 3x$ or $x > 3$, and solving the second we get $-3x > 7$. Dividing by -3 and reverting the side of the inequality we get $x < -\frac{7}{3}$.

(c) The first inequality will be $7x + 4 \geq 3$, that solves to $x \geq -\frac{1}{7}$. The second inequality will be $7x + 4 \leq -3$, and solving this we have $7x \leq -7$ or $x \leq -1$.

(d) $|6 - 5x| < 7$

$$-7 < 6 - 5x < 7$$

$$-7 - 6 < -5x < 7 - 6$$

$$-13 < -5x < 1$$

$$-\frac{1}{5} < x < \frac{13}{5}$$

Prob. 7. We use the point-slope formula to get: $y - y_0 = m(x - x_0)$, and in this case: $y - 4 = -\frac{1}{3}(x - 2)$, so multiplying both sides by 3, we get: $3(y - 4) = -(x - 2)$, or $3y - 12 = -x + 2$, and finally writing it into standard form: $x + 3y - 14 = 0$.

Prob. 9. $(x_0, y_0) = (0, -2)$ $m = -3$

$$y - y_0 = m(x - x_0)$$

$$y - (-2) = -3(x - 0)$$

$$y + 2 = -3x$$

Here $3x + y + 2 = 0$ is the Standard Form with $A = 3$, $B = 1$, $C = 2$.

Prob. 11. $(x, y_1) = (-2, -3)$

$(x_2, y_2) = (1, 4)$

First find the slope as

$$m = \frac{y_2 - y_1}{x_2 - x_1} = \frac{(4) - (-3)}{(1) - (-2)} = \frac{4 + 3}{1 + 2}$$

$$m = \frac{7}{3}$$

Second, get Standard Form with point-slope method:

$y - y_0 = m(x - x_0)$

Substitute into (x_0, y_0) either (x_1, y_1) or (x_2, y_2) - the result will be the same.

$y - 4 = \frac{7}{3}(x - 1)$

$3(y - 4) = 7(x - 1)$

$3y - 12 = 7x - 7$

$3y - 7x - 12 + 7 = 0$

$-7x + 3y - 5 = 0$ is Standard Form

or $7x - 3y + 5 = 0$.

Prob. 13. $(x_1, y_1) = (0, 4)$

$(x_2, y_2) = (3, 0)$

$m = \frac{y_2 - y_1}{x_2 - x_1} = \frac{(0) - (4)}{(3) - (0)} = -\frac{4}{3}$

$y - y_0 = m(x - x_0)$

$y - 4 = -\frac{4}{3}(x - 0)$

$3(y - 4) = -4x$

$3y - 12 = -4x$

$4x + 3y - 12 = 0$ is Standard Form.

Prob. 15. Horizontal lines are always $y = k$.

A horizontal line through $(3, \frac{3}{2})$ is $y = \frac{3}{2}$. The Standard Form is $2y - 3 = 0$

Prob. 17. Vertical lines are always $x = h$. A vertical line through $(-1, \frac{7}{2})$ is $x = -1$. Standard form is $x + 1 = 0$.

Prob. 19. Here $m = 3$ and the y-intercept $(0, 2)$. Use slope-intercept form $y = mx + b$.

Here $y = 3x + 2$. Standard form is $3x - y + 2 = 0$.

Prob. 21. Here $m = \frac{1}{2}$ and y-intercept $(0, 2)$.

$y = mx + b$

$y = \frac{1}{2}x + 2$ is slope-intercept form

$x - 2y + 4$ is Standard form.

Prob. 23. Here $m = -2$ and x-intercept $(1, 0)$

$y - y_0 = m(x - x_0)$

$y - 0 = -2(x - 1)$

$y = -2x + 2$ slope-intercept form

$2x + y - 2 = 0$ Standard form

Prob. 25. Here $m = -\frac{1}{4}$ and x-intercept $(3, 0)$

$y - y_0 = m(x - x_0)$

$y - 0 = -\frac{1}{4}(x - 3)$

$y = -\frac{1}{4}x + \frac{3}{4}$ slope-intercept form and $x + 4y - 3 = 0$ is Standard form.

Prob. 27. Line through $(2, -3)$ parallel to $x + 2y - 4 = 0$

$x + 2y - 4 = 0$ is Standard form.

Change to slope-intercept form to find slope of the given line

$x + 2y - 4 = 0$

$2y - 4 = -x$

$2y = -x + 4$

$\frac{2y}{2} = \frac{-x+4}{2}$

$y = -\frac{1}{2}x + 2$, slope is $m = -\frac{1}{2}$.

Thus the line we want has $m = -\frac{1}{2}$, point is $(2, -3)$

$y - y_0 = m(x - x_0)$

$y - (-3) = -\frac{1}{2}(x - 2)$

$y + 3 = -\frac{1}{2}(x - 2)$

$y + 3 = -\frac{1}{2}x + 1$

$2(y + 3) = (-\frac{1}{2}x + 1)2$

$2y + 6 = -x + 2$

$x + 2y + 4 = 0$ is Standard form of the line parallel to $x + 2y - 4 = 0$ passing through $(2, -3)$.

Prob. 29. Line passing through $(-1, -1)$ parallel to line passing through $(0, 1)$ and $(3, 0)$.

First find slope $m = \frac{y_2 - y_1}{x_2 - x_1}$

$m = \frac{(0)-(1)}{(3)-(0)} = -\frac{1}{3}$

Second use point-slope form

$y - y_0 = m(x - x_0)$

$y - (-1) = -\frac{1}{3}(x - (-1))$

$y + 1 = -\frac{1}{3}(x + 1)$

$-3(y+1) = x+1$

$-3y - 3 = x+1$

$x + 3y + 4 = 0$ Standard form.

Prob. 31. Line through $(1,4)$ perpendicular to $2y - 5x + 7 = 0$

$2y - 5x + 7 = 0$

$2y = 5x - 7$

$\frac{2y}{2} = \frac{5x-7}{2}$

$y = \frac{5}{2}x - \frac{7}{2}, \; m = \frac{5}{2}$

$m_\perp = -\frac{1}{m}$ (recall $m_1 \cdot m_2 = -1$)

$m_\perp = \frac{-1}{\left(\frac{5}{2}\right)} = -\frac{2}{5}$

use pt-slope form with $(1,4)$

$y - y_0 = m_\perp(x - x_0)$

$y - 4 = -\frac{2}{5}(x - 1)$

$5(y - 4) = -2(x - 1)$

$5y - 20 = -2x + 2$

$2x - 2 + 5y - 20 = 0$

$2x + 5y - 22 = 0$

Prob. 33. Line through $(5, -1)$ perpendicular to line passing through $(-2, 1)$ and $(1, -2)$

$m = \frac{y_2 - y_1}{x_2 - x_1} = \frac{(-2) - (1)}{(1) - (-2)} = -\frac{3}{3}$

$m = -1$

$m_\perp = -\frac{1}{m} = \frac{-1}{-1} = 1$

use pt-slope form with $(5, -1)$

$y - y_0 = m_\perp(x - x_0)$

$y - (-1) = 1(x - 5)$

$y + 1 = x - 5$

$y = x - 6$

$x - y - 6 = 0$

Prob. 35. So the line is horizontal through $(4, 2)$, and the equation is $y = 2$, in standard form the equation is $y - 2 = 0$.

Prob. 37. The line is vertical with equation $x = -1$, so in standard form the equation is $x + 1 = 0$.

Prob. 39. The line is vertical with equation $x = 1$, so in standard form the equation is $x - 1 = 0$.

Prob. 41. The line is horizontal with equation $y = 3$, so in standard form the equation is $y - 3 = 0$.

Prob. 43. $y = 30.5x$ is in the form of $y = mx$ meaning the two quantities x and y are linearly related. Thus y is proportional to x, and m is the constant of proportionality.

(a) To use this relationship, recall that 1 foot $= 30.5$ cm. Therefore let x be the variable for feet, and y be the variable for centimeters.

(b) Convert into centimeters.

    (i) $y = 30.5x$, $x = 6$ft

       $y = 30.5(6) = 183$cm

    (ii) $y = 30.5x$, $x = 3$ft 2in. Note that 2 inches $= \frac{2}{12}$ feet ≈ 0.167ft

       Let $x = 3.167$ feet (approx.)

       $y = 30.5(3.167) = 96.58$cm

    (iii) $y = 30.5x$, $x = 1$ft 7in.

       Note 7in $= \frac{7}{12}$ feet ≈ 0.583ft.

       Let $x = 1.583$ft (approx.)

       $y = 30.5(1.583) = 48.29$cm.

(c) Convert into feet

    (i) $y = 30.5x$, $y = 173$cm

       $173 = 30.5x$

       $x = 5.67$ feet

    (ii) $y = 30.5x$, $y = 75$cm

       $75 = 30.5x$

       $x = 2.459$ feet

(iii) $y = 48$cm

$$y = 30.5x$$

$$48 = 30.5x$$

$$x = 1.574 \text{ feet}$$

Prob. 45. Distance = rate · time (Recall $y = m \cdot x$)

time = 15 mins = $\frac{1}{4}$ hour = 0.25 hrs

distance = 10 mi

Constant of proportionality is miles per hour or "mph"

distance = speed · time

10 mi = speed · 0.25 hrs

$\frac{10mi}{0.25hrs}$ = speed (the constant of proportionality)

Thus speed = 40 $\frac{miles}{hour}$ (or "mph")

Prob. 47. 1 foot = 0.305 meters so 3.279 ft = 1 meter. Now we can convert $1\text{m}^2 = (1\text{m})(1\text{m}) = (3.279\text{ft})(3.279\text{ft})$ $= 10.75\text{ft}^2$ (approx.)

Prob. 49. 1 liter = 33.81 ounces

(a) $y = mx$, $m = \frac{1liter}{33.81ounces}$

y (liters) = $m \cdot x$ (ounces)

$y = \frac{1}{33.81} \cdot x$

(b) $x = 12$ ounces

$y = \frac{1}{33.8}(12) = 0.355$ (approx.)

So 12 ounces is 0.355 liters.

Prob. 51. 1 mile = 1.609 kilometers

(a) $\left(55\frac{miles}{hour}\right) \cdot \left(\frac{1.609kilometers}{1mile}\right) = 88.5\frac{kilometers}{hour}$

(b) $\left(130\frac{kilometers}{hour}\right) \cdot \left(\frac{1mile}{1.609kilometers}\right) = 80.8\frac{miles}{hour}$

Prob. 53.

(a) Denoting the two scales by K and C, they relate by equation

$$K = mC + b$$

and the slope of the line m is equal to 1 because 1 K denotes the same temperature difference as $1°C$, so

$$K = C + b$$

and substituting the two points $(-273.15, 0)$ we get $b = 273.15$ so

$$K = C + 273.15$$

(b) Nitrogen boils at $77.4K$ and oxygen at $90.2K$

So the boiling points will be given in Celsius by the equation

$$C = K - 273.15$$

so they are $77.4 - 273.15 = -195.75°$ for nitrogen and $90.2 - 273.15 = -182.95°$ for oxygen.

Now, $F = \frac{9}{5}C + 32$, and the boiling points in ° F are

Nitrogen: $F = \frac{9}{5}(-195.75) + 32 = -320.35°F$

Oxygen: $F = \frac{9}{5}(-182.95) + 32 = -297.31°F$

and the nitrogen will end up being distilled first.

Prob. 55. $r^2 = (x - x_0)^2 + (y - y_0)^2$

$r = 3, \quad (x_0, y_0) = (-1, 4)$

$9 = (x - (-1))^2 + (y - 4)^2$

$9 = (x + 1)^2 + (y - 4)^2$

Prob. 57. $r^2 = (x - x_0)^2 + (y - y_0)^2$

(a) $r = 3, \quad (x_0, y_0) = (2, 5)$

$3^2 = (x - 2)^2 + (y - 5)^2$

$9 = (x - 2)^2 + (y - 5)^2$

(b) where does the circle intersect the y-axis?

When $x = 0$ the circle is on the y-axis.

$$9 = (x - 2)^2 + (y - 5)^2$$

$$9 = (0 - 2)^2 + (y - 5)^2$$

$$9 = 4 + (y - 5)^2$$

$$5 = (y - 5)^2$$

$$\sqrt{5} = y - 5$$

$$5 \pm \sqrt{5} = y$$

(c) Does the circle intersect the x-axis? If $y = 0$ the circle would be on the x-axis.

$$9 = (x - 2)^2 + (y - 5)^2$$

$$9 = (x - 2)^2 + (0 - 5)^2$$

$$9 = (x - 2)^2 + 25$$

$$-16 = (x - 2)^2$$

Since the square of a real number cannot be negative, the circle does not intersect the x-axis.

Prob. 59. Find center and radius.

$$(x - 2)^2 + y^2 = 16$$
$$(x - 2)^2 + (y - 0)^2 = 4^2$$
center $(x_0, y_0) = (2, 0), \quad r = 4$

Prob. 61. $0 = x^2 + y^2 - 4x + 2y - 11$

$$0 = (x^2 - 4x + 4) + (y^2 + 2y + 1) - 11 - 5$$
$$0 = (x - 2)^2 + (y + 1)^2 - 16$$
$$16 = (x - 2)^2 + (y + 1)^2$$
$$16 = (x - 2)^2 + (y - (-1))^2$$
$r = 4$ and center is $(x_0, y_0) = (2, -1)$

Prob. 63.

(a) Convert to radian measure

$$(75°)(\frac{\pi}{180°}) = \frac{5}{12}\pi radians.$$

(b) Convert to degree measure

$$(\frac{17}{12}\pi)(\frac{180°}{\pi}) = \frac{3060°}{12} = 255°$$

Prob. 65.

(a)

$$\sin\left(\frac{-5}{4}\pi\right) = \frac{\sqrt{2}}{2}$$

(b)

$$\cos\left(\frac{5}{6}\pi\right) = \frac{-\sqrt{3}}{2}$$

(c)

$$\tan\left(\frac{\pi}{3}\right) = \sqrt{3}$$

Prob. 67.

(a) Find values of $\alpha \in [0, 2\pi)$

$$\sin\alpha = -\frac{1}{2}\sqrt{3}$$

$$\pi + \frac{\pi}{3} = \frac{4}{3}\pi$$

and

$$2\pi - \frac{\pi}{3} - \frac{5}{3}\pi$$

$$\alpha = \frac{4}{3}\pi \text{ and } \frac{5}{3}\pi$$

(b) Find values of $\alpha \in [0, 2\pi)$

$$\tan\alpha = \sqrt{3}$$

$\frac{\pi}{3}$ and $\pi + \frac{\pi}{3}$

$$\alpha = \frac{\pi}{3} \text{ and } \frac{4}{3}\pi$$

Prob. 69.

$$\sin^2\theta + \cos^2\theta = 1$$

$$\frac{\sin^2\theta + \cos^2\theta}{\cos^2\theta} = \frac{1}{\cos^2\theta} \quad \texttt{Divide both sides by } \cos^2\theta$$

$$\frac{\sin^2\theta}{\cos^2\theta} + \frac{\cos^2\theta}{\cos^2\theta} = \frac{1}{\cos^2\theta} \quad \texttt{Distribute}$$

$$\left(\texttt{Recall} \quad \frac{\sin\theta}{\cos\theta} = \tan\theta, \quad \texttt{and} \quad \frac{1}{\cos\theta} = \sec\theta\right)$$

$$\tan^2\theta + 1 = \sec^2\theta \quad \texttt{make substitution}$$

Prob. 71.

$$2\cos\theta\sin\theta = \sin\theta, \quad [0, 2\pi)$$

$$\frac{2\cos\theta\sin\theta}{\sin\theta} = \frac{\sin\theta}{\sin\theta} \quad \texttt{as long as } \sin\theta \neq 0, \texttt{ or } \theta \neq 0, \ \theta \neq \pi$$

$$2\cos\theta = 1$$

$$\cos\theta = \tfrac{1}{2}$$

$$\theta = \frac{\pi}{3} \quad \texttt{and} \quad 2\pi - \frac{\pi}{3}$$

Hence

$$\theta = 0, \pi, \frac{\pi}{3} \quad \texttt{and} \quad \frac{5}{3}\pi$$

Prob. 73.

(a) $4^3 4^{-2/3}$ Recall

$$a^r a^s \quad = a^{r+s}$$

$$4^{3+(-2/3)} \quad = 4^{7/3} = 16\sqrt[3]{4}$$

(b)

$$\frac{3^2 3^{1/2}}{3^{-1/2}} \quad = \frac{3^{5/2}}{3^{-1/2}} = 3^{5/2} \cdot 3^{1/2}$$

$$= 3^{5/2+1/2} = 3^{6/2} = 3^3 = 27$$

(c)

$$\frac{5^k 5^{2k-1}}{5^{1-k}} \quad = 5^k \cdot 5^{2k-1} \cdot 5^{k-1}$$

$$5^{k+2k-1+k-1} \quad = 5^{4k-2}$$

Prob. 75.

(a)

$$\log_4 x \quad = -2$$

$$x \quad = 4^{-2} = \frac{1}{4^2} = \frac{1}{16}$$

(b)

$$\log_{1/3} x \quad = -3$$

$$x \quad = \left(\frac{1}{3}\right)^{-3} = \frac{1}{\left(\frac{1}{3}\right)^3} = \frac{1}{\left(\frac{1}{27}\right)} = 27$$

(c)

$$\log_{10} x \quad = -2$$

$$x \quad = 10^{-2} = \frac{1}{10^2} = \frac{1}{100}$$

Prob. 77.

(a)

$$\log_{1/2} 32 = x$$

$$32 = \left(\tfrac{1}{2}\right)^x = \tfrac{1^x}{2^x} = \tfrac{1}{2^x}$$

$$\text{So } 2^x = \tfrac{1}{32}$$

$$2^x = 2^{-5}$$

$$x = -5$$

(b)

$$\log_{1/3} 81 = x$$

$$81 = \left(\tfrac{1}{3}\right)^x = \tfrac{1^x}{3^x}$$

$$3^x = \tfrac{1}{81} = \tfrac{1}{3^4} = 3^{-4}$$

$$x = -4$$

(c)

$$\log_{10} 0.001 = x$$

$$10^x = 0.001 = \tfrac{1}{1000} = \tfrac{1}{10^3}$$

$$10^x = 10^{-3}$$

$$x = -3$$

Prob. 79.

(a) $-\ln\left(\tfrac{1}{3}\right) = \ln\left[\left(\tfrac{1}{3}\right)^{-1}\right] = \ln 3$

(b) $\log_4(x^2 - 4) = \log_4[(x-2)(x+2)] = \log_4(x-2) + \log_4(x+2)$

(c) $\log_2 4^{(3x-1)} = (3x-1)\log_2(2^2) = (3x-1)(2 \cdot \log_2 2) = (3x-1)(2 \cdot 1) = 6x - 2$

Prob. 81.

(a)

$$e^{3x-1} = 2$$

$$\ln e^{(3x-1)} = \ln 2$$

$$(3x-1)(\ln e) = \ln 2 \qquad \texttt{Recall } \ln e = 1$$

$$3x - 1 = \ln 2$$

$$3x = 1 + \ln 2$$

$$x = \frac{1 + \ln 2}{3}$$

(b)

$$e^{-2x} = 10$$

$$\ln e^{(-2x)} = \ln 10$$

$$(-2x)(\ln e) = \ln 10$$

$$-2x = \ln 10$$

$$x = \frac{\ln 10}{-2} = -\frac{1}{2} \cdot \ln 10$$

$$\texttt{or } x = \ln 10^{-1/2} = \ln\left(\frac{1}{\sqrt{10}}\right)$$

(c)

$$e^{x^2-1} = 10$$

$$\ln e^{(x^2-1)} = \ln 10$$

$$(x^2 - 1)(\ln e) = \ln 10 \qquad (\ln e = 1)$$

$$x^2 - 1 = \ln 10$$

$$x^2 = 1 + \ln 10$$

$$x = \pm\sqrt{1 + \ln 10}$$

Prob. 83.

(a)

$$\ln(x-3) = 5$$
$$e^{\ln(x-3)} = e^5, \quad \texttt{Recall } e^{\ln x} = x$$
$$x - 3 = e^5$$
$$x = e^5 + 3$$

(b)

$$\ln(x+2) + \ln(x-2) = 1$$
$$\ln(x+2)(x-2) = 1$$
$$e^{\ln[(x+2)(x-2)]} = e$$
$$(x+2)(x-2) = e$$
$$x^2 - 4 = e$$
$$x^2 = 4 + e$$
$$x = \sqrt{4+e}$$

(c)

$$\log_3 x^2 - \log_3 2x = 2$$
$$\log_3 \frac{x^2}{2x} = 2$$
$$\frac{x^2}{2x} = 3^2$$
$$\frac{x}{2} = 9$$
$$x = 18$$

Prob. 85. $3 - 2i - (-2 + 5i) = 3 - 2i + 2 - 5i = 5 - 7i$

Prob. 87. $(4 - 2i) + (9 + 4i) = 4 + 9 - 2i + 4i = 13 + 2i$

Prob. 89. $3(5 + 3i) = 15 + 9i$

Prob. 91. $(6 - i)(6 + i) = 36 - i^2 = 36 - (-1) = 37$

Prob. 93.

$$z \;\; = 3 - 2i$$

$$\text{conjugate } \overline{z} \;\; = 3 + 2i$$

Prob. 95. $\overline{z + v} = \overline{3 - 2i} + \overline{3 + 5i} = 3 + 2i + 3 - 5i = 6 - 3i$

Prob. 97.

$$\overline{vw} = \overline{(3 + 5i)(1 - i)} = \overline{8 + 2i} = 8 - 2i$$

Prob. 99. Let

$$z \qquad\qquad = a + bi$$

$$z + \overline{z} \qquad = a + bi + \overline{a + bi}$$

$$\qquad\qquad = a + bi + a - bi = 2a$$

and

$$z - \overline{z} \quad = (a + bi) - \overline{(a + bi)}$$

$$\qquad\qquad = a + bi - (a - bi)$$

$$\qquad\qquad = a + bi - a + bi = 2bi$$

Prob. 101. $2x^2 - 3x + 2 = 0$. Here $a = 2$, $b = -3$, $c = 2$

Recall

$$x_{1,2} \qquad = \frac{-b \pm \sqrt{b^2 - 4ac}}{2a}$$

$$x_{1,2} \quad = \frac{-(-3) \pm \sqrt{(-3)^2 - 4(2)(2)}}{2(2)}$$

$$x_{1,2} \qquad = \frac{3 \pm \sqrt{9 - 16}}{4}$$

$$x_{1,2} \quad = \frac{3 \pm \sqrt{-7}}{4} = \frac{3 \pm \sqrt{7i^2}}{4}$$

Recall $i^2 = -1$, and $\sqrt{ab} = \sqrt{a} \cdot \sqrt{b}$

$$x_{1,2} \qquad = \frac{3 \pm \sqrt{7} \cdot \sqrt{i^2}}{4} = \frac{3 \pm \sqrt{7}i}{4}$$

$$x_1 \; = \frac{3 + \sqrt{7}i}{4} \quad \textbf{and} \quad x_2 = \frac{3 - \sqrt{7}i}{4}$$

Prob. 103. Here $a = -1, b = 1$, and $c = 2$

$$x_{1,2} = \frac{-b \pm \sqrt{b^2 - 4ac}}{2a}$$

$$x_{1,2} = \frac{-(1) \pm \sqrt{(1)^2 - 4(-1)(2)}}{2(-1)}$$

$$x_{1,2} = \frac{-1 \pm \sqrt{1+8}}{-2}$$

$$x_{1,2} = \frac{-1 \pm \sqrt{9}}{-2} = \frac{-1 \pm 3}{-2}$$

$$x_1 = \frac{-1+3}{-2} = \frac{2}{-2} = -1$$

$$x_2 = \frac{-1-3}{-2} = \frac{-4}{-2} = 2$$

Prob. 105. $4x^2 - 3x + 1 = 0$

$$a = 4, \quad b = -3, \quad c = 1$$

$$x_{1,2} = \frac{-b \pm \sqrt{b^2 - 4ac}}{2a}$$

$$x_{1,2} = \frac{-(-3) \pm \sqrt{(-3)^2 - 4(4)(1)}}{2(4)}$$

$$x_{1,2} = \frac{3 \pm \sqrt{9 - 16}}{8}$$

$$x_{1,2} = \frac{3 \pm \sqrt{-7}}{8} = \frac{3 \pm \sqrt{7i^2}}{8} \quad \text{Note } i^2 = -1$$

$$x_{1,2} = \frac{3 \pm \sqrt{7}i}{8}$$

$$x_1 = \frac{3 + \sqrt{7}i}{8}$$

$$x_2 = \frac{3 - \sqrt{7}i}{8}$$

Prob. 107. $3x^2 - 4x - 7 = 0$. Compute the discriminant, $b^2 - 4ac$

$$a = 3, \quad b = -4, \quad c = -7$$

$$(-4)^2 - 4(3)(-7) = 16 + 84 = 100 > 0$$

Hence, two real solutions. These are -1 and $\frac{7}{3}$.

Prob. 109. $-x^2 + 2x - 1 = 0$. Compute the discriminant, $b^2 - 4ac$

$$a = -1, \quad b = 2, \quad c = -1$$

$$(2)^2 - 4(-1)(-1)$$

$$4 - 4 = 0$$

Two identical real solutions, that is, only one real solution. This solution is 1.

Prob. 111. $3x^2 - 5x + 6 = 0$

$$a = 3, \quad b = -5, \quad c = 6$$

Compute the discriminant

$$b^2 - 4ac$$

$$(-5)^2 - 4(3)(6)$$

$$25 - 72 = -47 < 0$$

Hence, two complex solutions which are conjugates of each other. These are $\frac{5+\sqrt{47}i}{6}$ and $\frac{5-\sqrt{47}i}{6}$.

Prob. 113. Let $z = a + bi$. Conjugate $\bar{z} = a - bi$

$$\overline{(\bar{z})} = \overline{(a - bi)} = a + bi = z.$$

Hence $z = \overline{(\bar{z})}$.

Prob. 115. Let

$$z = a + bi$$

$$w = c + di.$$

Does $\overline{zw} = \bar{z} \cdot \bar{w}$?

First,

$$\bar{z} \cdot \bar{w} = \overline{(a + bi)(c + di)} = \overline{(ac - bd) + (ad + bc)i}$$

$$= (ac - bd) - (ad + bc)i$$

On the other hand,

$$\overline{z} \cdot \overline{w} \quad = \overline{(a+bi)} \cdot \overline{(c+di)}$$
$$= (a-bi) \cdot (c-di)$$
$$= ac - adi - bci + bdi^2$$
$$= ac - (ad+bc)i - bd$$
$$= (ac-bd) - (ad+bc)i$$

So $\overline{zw} = \overline{z} \cdot \overline{w}$ is true.

1.2   Elementary Functions

Prob. 1. $f(x) = x^2$, $x \in \mathbb{R}$. Range $[0, +\infty)$

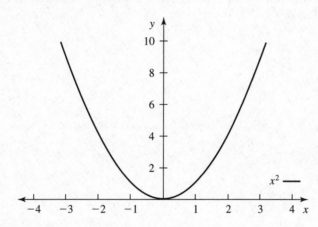

Prob. 3. $f(x) = x^2$, $-1 < x \le 0$. Range $[0, 1)$

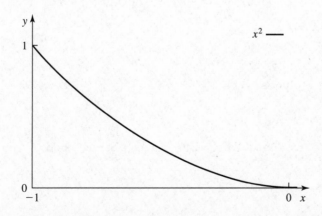

Prob. 5.

(a) Factor $\frac{x^2-1}{x-1} = \frac{(x+1)(x-1)}{x-1} = x+1$ (if $x \neq 1$)

$$f(x) = \frac{x^2-1}{x-1}, \quad x \neq 1$$

(b) No the functions are not equal, they have different domains. The function f is not defined at the point $x = 1$.

Prob. 7. The function is odd, as the tests below indicate.

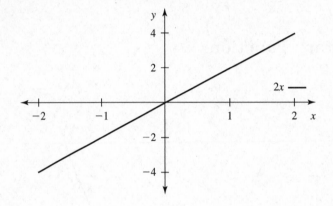

Test for "Even";

$$\text{Does } f(x) \qquad\qquad = f(x)?$$
$$\text{Ask if } ; 2x \qquad\qquad = 2(-x)?$$
$$\text{Does } 2x \;\; = -2x? \quad \text{no, not "even"}$$

Test for "Odd";

$$\text{Does } f(x) \qquad\qquad = -f(-x)$$
$$\text{Ask if } ; 2x \qquad\qquad = -2(-x)?$$
$$\text{Does } 2x \;\; = 2x? \quad \text{yes, hence "odd"}$$

Prob. 9. The function is even, as the tests below indicate.

$f(x) = |3x|$

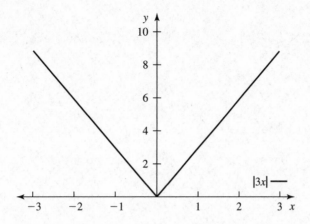

Test for "Even";

$$\text{Does } f(x) \qquad = f(-x)?$$

$$\text{Ask if } ; |3x| \qquad = |3(-x)|?$$

$$\text{Does } |3x| \ = |-3x|? \quad \text{yes, hence "even"}$$

Test for "Odd";

$$\text{Does } f(x) \qquad = -f(-x)?$$

$$\text{Ask if } ; |3x| \qquad = -|3(-x)|?$$

$$\text{Does } |3x| \ = -|3x|? \quad \text{no, not "odd"}$$

Prob. 11. The function is even, as the tests below indicate.

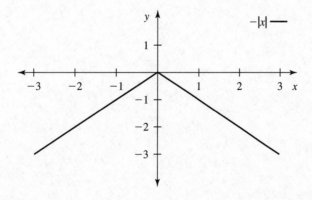

Test for "Even";

$$\text{Does } f(x) \qquad = f(-x)?$$

$$\text{Does } -|x| \quad = -|-x|? \quad \text{yes, hence ``even''}$$

Test for "Odd";

$$\text{Does } f(x) \qquad = -f(-x)?$$

$$\text{Does } -|x| \quad = |-x|? \quad \text{no, not ``odd''}$$

Prob. 13.

(a) $(f \circ g)(x) = f[g(x)]$

$$f(x) \qquad\qquad = x^2$$

$$f(g(x)) \quad = f(3+x) = (3+x)^2$$

$$(f \circ g)(x) \qquad = (3+x)^2$$

(b) $(g \circ f)(x) = g[f(x)]$

$$g(x) \qquad\qquad = 3 + x$$

$$g(f(x)) \quad = g(x^2) = 3 + (x^2)$$

$$(g \circ f)(x) \qquad = 3 + x^2$$

Prob. 15.

(a)

$$(f \circ g)(x) \qquad = f(g(x))$$

$$f(g(x)) \quad = f(2x) = 1 - (2x)^2$$

$$(f \circ g)(x) \qquad = 1 - 4x^2$$

Domain $x \geq 0$

(b)

$$(g \circ f)(x) \quad = g(f(x))$$

$$g(x) \quad = 2x$$

$$g(f(x)) \quad = g(1 - x^2) = 2(1 - x^2)$$

$$(g \circ f)(x) \quad = 2 - 2x^2$$

Here we require that $f(x) \geq 0$, or that $1 - x^2 \geq 0$. This means the domain is $-1 \leq x \leq 1$.

Prob. 17.

$$f(x) \quad = 3x^2 \quad x \geq 3$$

$$g(x) \quad = \sqrt{x} \quad x \geq 0$$

$$(f \circ g)(x) \quad = f(g(x))$$

$$= f(\sqrt{x}) = 3(\sqrt{x})^2$$

$$(f \circ g)(x) \quad = f(g(x)) = 3x$$

Domain is $\{x \in \mathbb{R} : x \geq 0 \text{ and } g(x) \geq 3\} = \{x \in \mathbb{R} : \sqrt{x} \geq 3\} = \{x \in \mathbb{R} : x \geq 9\}$.

Prob. 19.

$$f(x) \quad = x^2 \quad x \geq 0$$

$$g(x) \quad = \sqrt{x} \quad x \geq 0$$

Note $f \circ g \neq g \circ f$ in general. Here however, we show that $f \circ g = g \circ f$.

$$f \circ g \quad = f(g(x)) = f(\sqrt{x}) = (\sqrt{x})^2$$

$$f \circ g \quad = x$$

and

$$g \circ f \quad = g(f(x)) = g(x^2) = \sqrt{x^2}$$

$$g \circ f \quad = x$$

In addition, both compositions have the same domain, namely $\{x \in \mathbb{R} : x \geq 0\}$.

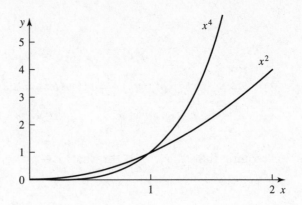

Prob. 21.

$f(x) > g(x)$ when $0 < x < 1$

$f(x) < g(x)$ when $x > 1$

Prob. 23. $y = x^n, \ x \geq 0$

$y_1 = x^1$

$y_2 = x^2$

$y_3 = x^3$

$y_4 = x^4$

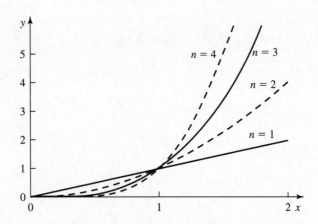

The curves intersect at the points $(0,0)$ and $(1,1)$.

Prob. 25.

(a) $f(x) = x^2$

$g(x) = x^3 \qquad x \geq 0$

(b) We can start with the inequalities $0 \leq x \leq 1$, and multiply both sides by x^2 which is positive, so it will not change the direction of the inequalities that become $0 \leq x^3 \leq x^2$

(c) In the same way, we start with $x \geq 1$, and multiply both sides by the positive value x^2, which then becomes $x^3 \geq x^2$

Prob. 27.

(a) Show $y = x^2, x \in \mathbb{R}$ is even.

$f(x) = x^2, f(-x) = (-x)^2$

Since $x^2 = (-x)^2$, f(x) is an even function.

(b) Show $y = x^3$, $x \in \mathbb{R}$ is odd.

$f(x) = x^3, -f(-x) = -(-x)^3$

Does $f(x) = -f(-x)$?

$x^3 = -(-x^3) = -1 \cdot (-1 \cdot x^3)$

$x^3 = x^3$ hence the function is odd.

Prob. 29. $A + B \rightarrow AB$, $a = [A] = 3$, $b = [B] = 4$

(a) When $x = 1$, we have $R = 9$. Substituting this into the equation $R(x) = k(a-x)(b-x)$, we get

$9 = k(a-1)(b-1)$

$$9 = k(3 - 1)(4 - 1)$$

$$9 = k \cdot 6$$

$$k = \frac{9}{6} = \frac{3}{2}$$

Thus $R(x) = \frac{3}{2}(3 - x)(4 - x)$.

(b) $R(x) = \frac{3}{2}(3 - x)(4 - x)$

Since the reaction rate cannot be negative, we must find the roots of $R(x)$. As $R(x)$ is already factored, the roots are 3 and 4. Thus the domain must be the interval $[0, 3]$.

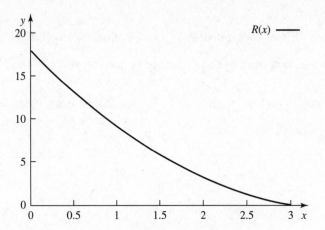

Prob. 31. Constant speed of beetle $= 1\frac{\text{meter}}{\text{hour}}$

$$\text{rate} \cdot \text{time} = \text{distance}.$$

In one hour

$$\left(1\ \frac{\text{meter}}{\text{hour}}\right)(1\ \text{hour}) = 1\ \text{meter}.$$

In two hours

$$\left(1\ \frac{\text{meter}}{\text{hour}}\right)(2\ \text{hours}) = 2\ \text{meters}.$$

In three hours

$$\left(1\ \frac{\text{meter}}{\text{hour}}\right)(3\ \text{hours}) = 3\ \text{meters}.$$

Now $\text{distance} = \text{rate} \cdot \text{time}$,

Let $\text{distance} = y$

$\text{rate} = m$

time $= x$

Then $y = mx$, which is a polynomial of degree 1; i.e., a line with slope m.

Prob. 33. $f(x) = \frac{1}{1-x}$

Domain $(-\infty, 1) \cup (1, +\infty)$

Range $(-\infty, 0) \cup (0, +\infty)$

Prob. 35. $f(x) = \frac{x-2}{x^2-9}$

Domain is $\{x : x \neq -3, 3\}$ or $(-\infty, -3) \cup (-3, +3) \cup (3, +\infty)$

Range $(-\infty, +\infty)$

Prob. 37.

$$y = \frac{1}{x} \quad y = \frac{1}{x^2}, \quad x > 0$$

The curves intersect at the point $(1, 1)$, when $\frac{1}{x^2} = \frac{1}{x}$

$$\frac{1}{x} > \frac{1}{x^2}, \text{ if } x > 1 \quad \text{and} \quad \frac{1}{x} < \frac{1}{x^2}, \text{ if } 0 < x < 1$$

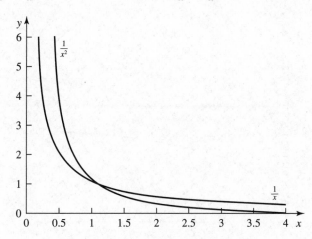

Prob. 39. (a)

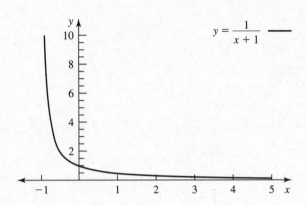

(b) Range of $f(x)$ is $(0, \infty)$,   (c) $x = -\frac{1}{2}$,

(d) $f(x) = a$ has exactly one solution, since the graph is one-to-one.

Prob. 41. (a)

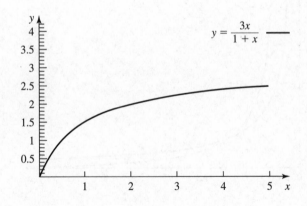

(b) Range of $f(x)$ is $[0, 3)$,   (c) $x = 2$,

(d) For each horizontal line $y = a$, where $0 \le a < 3$, there is only one intersection with the graph, so there is only one possible solution for $f(x) = a$. This solution is given by the equation

$$a = \frac{3x}{1 + x}$$

$$a(1 + x) = 3x$$

$$a = (3 - a)x$$

so

$$x = \frac{a}{3 - a}.$$

Since $0 \le a < 3$, we see that $x \ge 0$, which is in the domain.

Prob. 43. The percentage increase is

$$\frac{r(0.2) - r(0.1)}{r(0.1)} \times 100\% = \frac{5/6 - 5/11}{5/11} \times 100\% \approx 0.8333 \times 100\% = 83.33\%$$

The percentage increase is

$$\frac{r(20) - r(10)}{r(10)} \times 100\% = \frac{100/21 - 50/11}{50/11} \times 100\% \approx 0.0476 \times 100\% = 4.76\%$$

Prob. 45. (a)

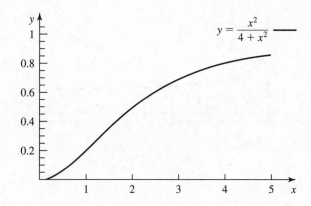

(b) range of $f(x)$ is $[0, 1)$,   (c) $f(x)$ approaches 1, from below.

Prob. 47.

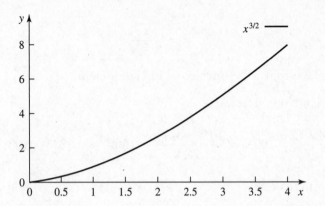

Prob. 49.

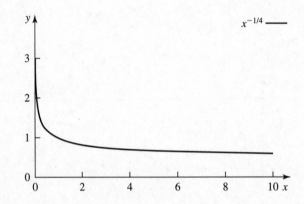

Prob. 51. (a)

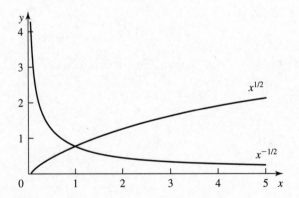

(b) We have

$$0 < x \le 1 \quad \Rightarrow \quad 0 < \sqrt{x}\sqrt{x} \le 1 \quad \Rightarrow \quad 0 < \sqrt{x} \le \frac{1}{\sqrt{x}},$$

since $\sqrt{x}$ is positive.

The last inequality can be written as $x^{1/2} \le x^{-1/2}$.

(c) Here,

$$x \ge 1 \quad \Rightarrow \quad \sqrt{x}\sqrt{x} \ge 1 \quad \Rightarrow \quad \sqrt{x} \ge \frac{1}{\sqrt{x}},$$

since $\sqrt{x}$ is positive.

The last inequality can be written as $x^{1/2} \ge x^{-1/2}$.

Prob. 53. The leaf area increases.

Prob. 55. As wood density increases.

Prob. 57. (a) 1,2,4,8,16   (b)

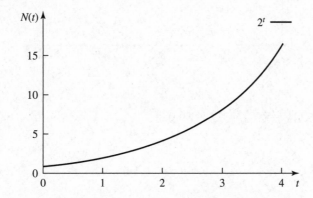

Prob. 59. Since $\lambda = \frac{\ln 2}{T_h} = \frac{\ln 2}{5730}$, the amount left after 2000 years is

$$20 \exp\left[-\frac{\ln 2}{5730}2000\right] = 20 \cdot 2^{-200/573} \approx 15.70 \texttt{ micrograms}.$$

Prob. 61. $\lambda = \frac{\ln 2}{7 \, \text{days}}$

Prob. 63. (a) $W(t) = (300\,\text{micrograms})\exp\left[-\frac{\ln 2}{140\,\text{days}}t\right] = 300\exp\left[-\frac{\ln 2}{140}t\right]$

(b) 20% of 300 micrograms is 60 micrograms. We set $W(t) = 60$ and solve for t.

$60 = 300\exp\left[-\frac{\ln 2}{140}t\right] \quad \Rightarrow \quad 5 = \exp\left[\frac{\ln 2}{140}t\right] \quad \Rightarrow \quad t = \frac{\ln 5}{\ln 2}140 \approx 325\,\text{days}$

(c)

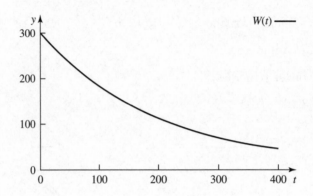

Prob. 65. $\frac{W(t)}{W(0)} = \exp\left[-\frac{\ln 2}{5730}15{,}000\right] \approx 16.3\%$

Prob. 67.

(a) The population with $r = 3$ grows faster.

(b) $\frac{250}{200} = \frac{N(t+1)}{N(t)} = \frac{N_0 e^{r(t+1)}}{N_0 e^{rt}} = \frac{e^{rt}e^r}{e^{rt}} = e^r$. Thus we get $e^r = 1.25$, or $r = \ln 1.25$.

Prob. 69. (a) yes   (b) no   (c) yes   (d) yes   (e) no   (f) yes

Prob. 71.

(a) f is strictly increasing, so it is one to one. To find its inverse, we start with

$$y = x^2 + 1$$

and solving for x, we have $y - 1 = x^2$ or $x = \sqrt{y-1}$. So the inverse function is $f^{-1}(x) = \sqrt{x-1}$, with domain $\{x \geq 1\}$.

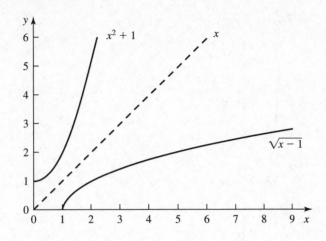

(b)

Prob. 73.

(a) f is strictly increasing, so it is one to one. To find its inverse, we start with

$$y = 1/x^3$$

and solving for x, we have $1/y = x^3$ or $x = \sqrt[3]{1/y}$. So the inverse function is $f^{-1}(x) = \sqrt[3]{\frac{1}{x}}$, with domain $\{x > 0\}$.

(b)

Prob. 75. $f^{-1}(x) = \log_3 x$, $x > 0$

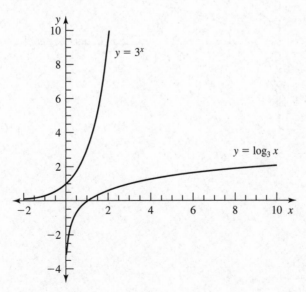

Prob. 77. $f^{-1}(x) = \log_{1/4} x,\ x > 0$

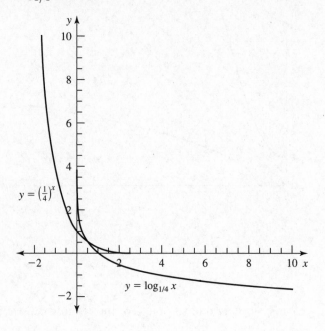

Prob. 79. $f^{-1}(x) = \log_2(x),\ x \geq 1$

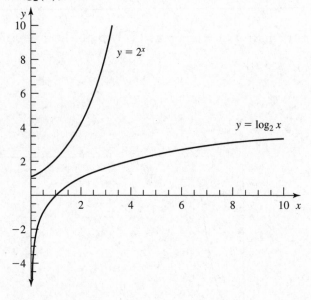

Prob. 81. (a) $2^{5\log_2 x} = 2^{\log_2 x^5} = x^5$

(b) $3^{4\log_3 x} = 3^{\log_3 x^4} = x^4$

(c) $5^{5\log_{1/5} x} = 5^{-\log_{1/5} x^{-5}} = \left(\frac{1}{5}\right)^{\log_{1/5} x^{-5}} = x^{-5}$

(d) $4^{-2\log_2 x} = 2^{-4\log_2 x} = 2^{\log_2 x^{-4}} = x^{-4}$

(e) $2^{3\log_{1/2} x} = 2^{-\log_{1/2} x^{-3}} = \left(\frac{1}{2}\right)^{\log_{1/2} x^{-3}} = x^{-3}$

(f) $4^{-\log_{1/2} x} = 2^{-2\log_{1/2} x} = 2^{-\log_{1/2} x^2} = \left(\frac{1}{2}\right)^{\log_{1/2} x^2} = x^2$

Prob. 83. (a) $5\ln x$,   (b) $6\ln x$,   (c) $\ln(x-1)$,   (d) $-4\ln x$

Prob. 85. (a) $e^{x\ln 3}$,   (b) $e^{(x^2-1)\ln 4}$,   (c) $e^{-(x+1)\ln 2}$,   (d) $e^{(-4x+1)\ln 3}$

Prob. 87. We have $y = (1/2)^x = e^{x\ln(1/2)} = e^{-x\ln 2}$, and so $\mu = \ln 2$.

Prob. 89. $K = -\frac{3}{4}\ln\left(1 - \frac{4}{3}\cdot\frac{47}{300}\right) \approx 0.1757$

Prob. 91. Same period; $2\sin x$ has twice the amplitude of $\sin x$.

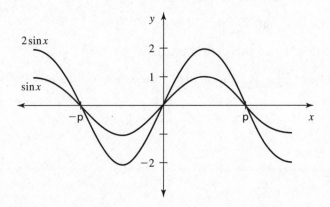

Prob. 93. Same period; $2\cos x$ has twice the amplitude of $\cos x$.

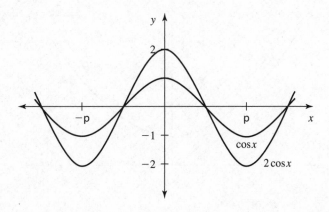

Prob. 95. Same period; $y = 2 \tan x$ is vertically stretched by a factor of 2.

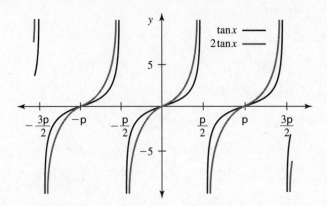

Prob. 97. Amplitude is 3, and the period is $\frac{\pi}{2}$.

Prob. 99. Amplitude is 4, and the period is 1.

Prob. 101. Amplitude is 4, and the period is 8π.

Prob. 103. Amplitude is 3 and the period is 10.

Prob. 105. Since $\sec x = \frac{1}{\cos x}$, it is undefined for $\cos x = 0$, meaning:

$$x = \frac{\pi}{2} + \pi n = \frac{\pi}{2}(1 + 2n), \quad n \in Z.$$

1.3 Graphing

Prob. 1.

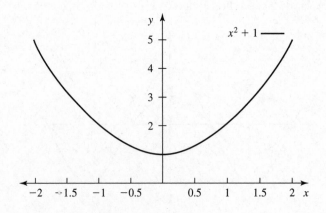

Prob. 3.

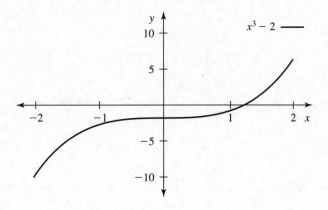

Prob. 5.

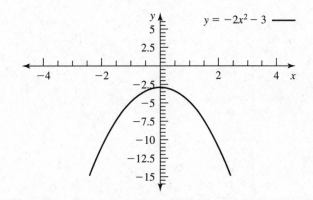

Prob. 7.

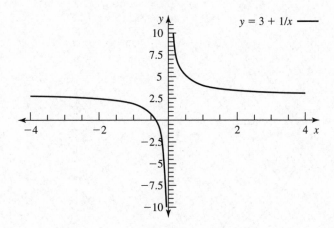

Prob. 9.

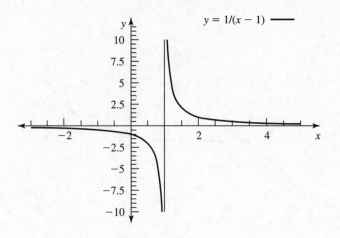

Prob. 11.

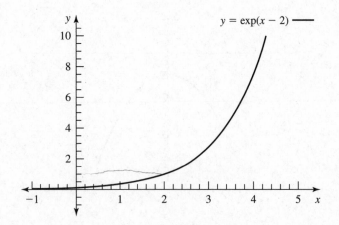

Prob. 13.

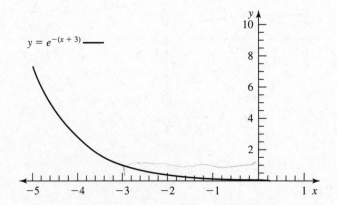

Prob. 15.

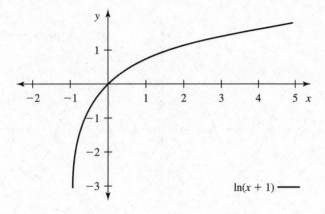

$\ln(x + 1)$ ———

Prob. 17.

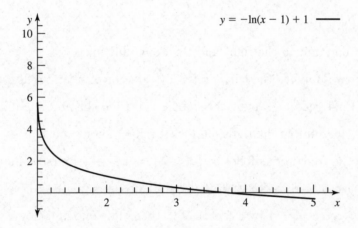

$y = -\ln(x - 1) + 1$ ———

Prob. 19.

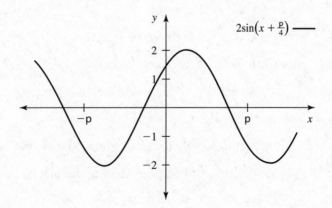

$2\sin\left(x + \frac{p}{4}\right)$ ———

Prob. 21.

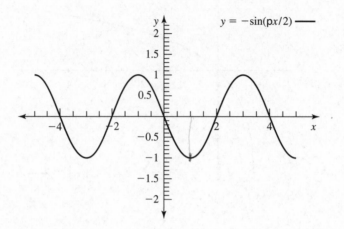

Prob. 23. (a) Shift two units down.

(b) Shift $y = x^2$ one unit to the right and then one unit up.

(c) Shift $y = x^2$ two units to the left, stretch by a factor of 2, and reflect about the x-axis.

Prob. 25. (a) First reflect $\frac{1}{x}$ about the x-axis, and then shift up one unit.

(b) First shift $\frac{1}{x}$ one unit to the right, and then reflect about the x-axis.

(c) We re-write the function as follows: $y = \frac{x}{x+1} = 1 - \frac{1}{x+1}$. First shift $y = \frac{1}{x}$ one unit to the left, then reflect about x-axis and finally shift up one unit.

Prob. 27. (a) Stretch $y = e^x$ by a factor of 2, then shift one unit down.

(b) Reflect $y = e^x$ about the y-axis, the reflect about the x-axis.

(c) Shift $y = e^x$ two units to the right, then shift one unit up.

Prob. 29. (a) Shift $y = \ln x$ one unit to the right.

(b) Reflect $y = \ln x$ about the x-axis, then shift up one unit.

(c) Shift $y = \ln x$ three units to the left, then down one unit.

Prob. 31. (a) Reflect $y = \sin x$ about the x-axis, then one unit up.

(b) Shift $y = \sin x$ by $\pi/4$ units to the right.

(c) Shift $y = \sin x$ by $\pi/3$ units to the left, then reflect about the x-axis.

Prob. 33. To locate the point, we have to compute the log of each of the numbers, which are:

$$\log 0.0002 \;=\; \log\left(2 \times 10^{-4}\right) = \log 2 + \log 10^{-4} = -4 + \log 2 \approx -3.7$$

$$
\begin{aligned}
\log 0.02 &= \log\left(2 \times 10^{-2}\right) = -2 + \log 2 \approx -1.7 \\
\log 1 &= 0 \\
\log 5 &\approx 0.7 \\
\log 50 &= \log(5 \times 10) = \log 5 + \log 10 = 1 + \log 5 \approx 1.7 \\
\log 100 &= \log 10^2 = 2\log 10 = 2 \\
\log 1000 &= \log 10^3 = 3\log 10 = 3 \\
\log 8000 &= \log\left(8 \times 10^3\right) = \log 8 + \log 10^3 = 3 + \log 8 \approx 3.9 \\
\log 20000 &= \log\left(2 \times 10^4\right) = \log 2 + \log 10^4 = 4 + \log 2 \approx 4.3
\end{aligned}
$$

Prob. 35. (a) The numbers on a logarithmic scale are $2, -3, -4, -7$ and -10 respectively.

(b) No,

(c) No

Prob. 37. seven

Prob. 39. one, three

Prob. 41. six to seven

Prob. 43. The slope of the line is given by $m = \frac{y - y_0}{x - x_0} = \frac{5-1}{0-3} = -\frac{4}{3}$ so the equation of $\log y$ in terms of x is

$$
\log y - \log y_0 = m(x - x_0)
$$

$$
\log y - \log 5 = -\frac{4}{3}(x - 0)
$$

$$
\log \frac{y}{5} = -\frac{4}{3}x
$$

$$
\frac{y}{5} = 10^{-\frac{4}{3}x}
$$

so finally

$$
y = 5 \times \left(10^{-4/3}\right)^x \approx 5(0.046)^x
$$

Prob. 45. The slope of the line is given by $m = \frac{y - y_0}{x - x_0} = \frac{3-1}{-2-1} = -\frac{2}{3}$ so the equation of $\log y$ in terms of x is

$$
\log y - \log y_0 = m(x - x_0)
$$

$$
\log y - \log 1 = -\frac{2}{3}(x - 1)
$$

$$\log y = -\frac{2}{3}x + \frac{2}{3}$$

$$y = 10^{-\frac{2}{3}x + \frac{2}{3}}$$

so finally

$$y = 10^{2/3} \cdot \left(10^{-2/3}\right)^x \approx 4.64(0.22)^x$$

Prob. 47. $\log y = \log(3 \times 10^{-2x}) = \log 3 + \log 10^{-2x} = \log 3 - 2x \log 10 = \log 3 - 2x$

Prob. 49. $\log y = \log(2e^{-1.2x}) = \log 2 + \log(e^{-1.2x}) = \log 2 - 1.2x \log e = \log 2 - (1.2 \log e)x$

Prob. 51. $\log y = \log(5 \times 2^{4x}) = \log 5 + 4x \log 2 = \log 5 + (4 \log 2)x$

Prob. 53. $\log y = \log(4 \times 3^{2x}) = \log 4 + 2x \log 3 = \log 4 + (2 \log 3)x$

Prob. 55. The slope of the line is given by $m = \frac{y - y_0}{x - x_0} = \frac{2-1}{1-5} = -\frac{1}{4}$ so the equation of $\log y$ in terms of $\log x$ is

$$\log y - \log y_0 = m(\log x - \log x_0)$$

$$\log y - \log 2 = -\frac{1}{4}(\log x - \log 1)$$

$$\log \frac{y}{2} = -\frac{1}{4} \log x$$

$$\log \frac{y}{2} = \log x^{-\frac{1}{4}}$$

$$\frac{y}{2} = x^{-\frac{1}{4}}$$

so finally

$$y = 2x^{-1/4}$$

Prob. 57. The slope of the line is given by $m = \frac{y - y_0}{x - x_0} = \frac{8-2}{8-4} = \frac{3}{2}$ so the equation of $\log y$ in terms of $\log x$ is

$$\log y - \log y_0 = m(\log x - \log x_0)$$

$$\log y - \log 2 = \frac{3}{2}(\log x - \log 4)$$

$$\log \frac{y}{2} = \frac{3}{2} \log \frac{x}{4}$$

$$\log \frac{y}{2} = \log \left(\frac{x^{\frac{3}{2}}}{4}\right)$$

$$\frac{y}{2} = \left(\frac{x^{\frac{3}{2}}}{4}\right)$$

so finally

$$y = \frac{2}{4^{3/2}} x^{3/2}$$

or

$$y = \frac{1}{4} x^{3/2}$$

Prob. 59. Starting with $y = 2x^5$ we take logarithms on both sides, to get $\log y = \log 2x^5 = \log 2 + 5 \log x$. Thus the linear relationship is $Y = 5X + \log 2$.

Prob. 61. Starting with $y = x^6$ we take logarithms on both sides, to get $\log y = \log x^6 = 6 \log x$. Thus the linear relationship is $Y = 6X$.

Prob. 63. Starting with $y = x^{-2}$ we take logarithms on both sides, to get $\log y = \log x^{-2} = -2 \log x$. Thus the linear relationship is $Y = -2X$.

Prob. 65. Starting with $y = 4x^{-3}$ we take logarithms on both sides, to get $\log y = \log 4x^{-3} = \log 4 - 3 \log x$. Thus the linear relationship is $Y = \log 4 - 3X$.

Prob. 67. Starting with $f(x) = 3x^{1.7}$ we take logarithms on both sides, to get $\log f = \log 3x^{1.7} = \log 3 + 1.7 \log x$. Thus the linear relationship is $Y = \log 3 + 1.7X$, and we should use a log-log transformation.

Prob. 69. Starting with $N(t) = 130 \cdot 2^{1.2t}$ we take logarithms on both sides, to get $\log N(t) = \log \left(130 \cdot 2^{1.2t}\right) = \log 130 + 1.2t \log 2$. Thus the linear relationship is $Y = \log 130 + (1.2 \log 2)t$, and we should use a log-linear plot.

Prob. 71. Starting with $R(t) = 3.6t^{1.2}$ we take logarithms on both sides, to get $\log R(t) = \log 3.6t^{1.2} = \log 3.6 + 1.2 \log t$. Thus the linear relationship is $Y = \log 3.6 + 1.2X$, and we should use a log-log transformation.

Prob. 73. Taking logarithms of both columns of data,

$\log x$	$\log y$
0	0.26
0.30	0.32
0.60	0.38
1	0.45
1.30	0.52

we can see a linear relation among the two columns, and the equation relating them can be checked to be $Y \approx 0.2X + 0.26$, that is:

$$\log y = 0.2 \log x + 0.26$$

so solving for y in terms of x we get:

$$\log y = \log x^{0.2} + 0.26$$

$$\log y - \log x^{0.2} = 0.26$$

or

$$\log \frac{y}{x^{0.2}} = 0.26$$

and thus the functional relationship is: $y = \left(10^{0.26}\right) x^{0.2} \approx 1.82 x^{0.2}$.

Prob. 75. Since the first column has negative entries, we take logarithms of the second column of data only.

x	$\log y$
-1	-0.40
-0.5	0.10
0	0.60
0.5	1.10
1	1.60

We can see a linear relation among the two columns, and the equation relating them can be checked to be $Y \approx X + 0.60$, that is:

$$\log y = x + 0.60$$

so solving for y in terms of x we get:

$$y = 10^{x+0.60}$$

$$y = 10^{0.60} \cdot 10^x$$

and thus the functional relationship is: $y = \left(10^{0.60}\right) 10^x \approx 4 \cdot 10^x$.

Prob. 77. Taking logarithms of both columns of data,

$\log x$	$\log y$
-1.00	-2.35
-0.30	0.12
0.00	0.76
0.18	1.13
0.30	1.39

we can see a linear relation among the two columns, and the equation relating them can be checked to be $Y \approx 2.1X + 0.76$, that is:

$$\log y = 2.1 \log x + 0.76$$

so solving for y in terms of x we get:

$$\log y = \log x^{2.1} + 0.76$$

$$\log y - \log x^{2.1} = 0.76$$

or

$$\log \frac{y}{x^{2.1}} = 0.76$$

and thus the functional relationship is: $y = \left(10^{0.76}\right) x^{2.1} \approx 5.7 x^{2.1}$.

Prob. 79. Taking logarithms in base 2 we have $\log_2 y = x$, so the linear relationship is $Y = X$ on a $\log_2$-linear plot.

Prob. 81. Taking logarithms in base 2 we have $\log_2 y = -x$, so the linear relationship is $Y = -X$ on a $\log_2$-linear plot.

Prob. 83. (a) Taking logarithms, we have $\log N = \log\left(2e^{3t}\right) = \log 2 + 3t \log e = (3 \log e)t + \log 2$. Setting $Y = \log N$ and $X = t$ we have $Y = (3 \log e)X + \log 2$, which is a linear relationship.

(b) Graphing gives the following line with slope $3\log(e)$.

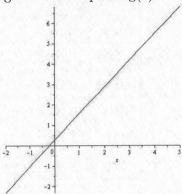

Prob. 85. (The relationship $S = CA^z$ leads to $\log S = \log(CA^z) = \log C + z\log A$. Plotting $Y = \log S$ and $X = \log A$ on a log-log scale would lead to a straight line given by $Y = zX + \log C$. Thus the model $S = CA^z$ is appropriate. The value of z is the slope of the straight line.

Prob. 87. Since $v_0 = \frac{v_{\max}s_0}{s_0+K_m}$, we can multiply both sides by $(s_0 + K_m)$ to obtain

$$v_0 s_0 + v_0 K_m = v_{\max} s_0$$

$$\frac{v_0 s_0 + v_0 K_m}{s_0 K_m} = \frac{v_{\max} s_0}{s_0 K_m}$$

$$\frac{v_0}{K_m} + \frac{v_0}{s_0} = \frac{v_{\max}}{K_m}$$

$$\frac{v_0}{s_0} = \frac{v_{\max}}{K_m} - \frac{1}{K_m} v_0$$

If we set $Y = \frac{v_0}{s_0}$ and $X = v_0$, this becomes $Y = \frac{v_{\max}}{K_m} - \frac{1}{K_m}X$, which is the equation of a straight line.

Here K_m is the negative reciprocal of the slope, and $v_{\max}$ is the Y-intercept multiplied by K_m.

Prob. 89. (a) Taking logarithms of the relation $S = 1.162B^{0.93}$, we have $\log S = \log 1.162 + 0.93\log B$. If we set $Y = \log S$ and $X = \log B$, we get the linear relationship $Y = \log 1.162 + 0.93X$. The slope is 0.93 and the y-intercept $\log 1.162$.

(b) When $B = 10$ cm, we have $S = 1.162(10)^{0.93} \approx 9.89$ and so the ratio of S to B is $S/B = 0.989$. When $B = 100$ cm, we have $S = 1.162(100)^{0.93} \approx 84.2$ and so the ratio of S to B is $S/B = 0.842$. When $B = 500$ cm, we have $S = 1.162(500)^{0.93} \approx 376$ and so the ratio of S to B is $S/B = 0.752$. Clearly, when B is smaller, the ratio is larger. As B increases (and the animal grows) the ratio decreases.

Prob. 91. (a) When $z = 1$, 90% of the intensity remains, and so we have $0.9 = I(1)/I(0) = e^{-\alpha z} = e^{-\alpha}$. This means $\alpha = -\ln 0.9$, measured in `meter`$^{-1}$.

(b) With $\alpha = -\ln 0.9$, we have $I(2)/I(1) = \frac{e^{-2\alpha}}{e^{-\alpha}} = e^{-\alpha} = e^{\ln 0.9} = 0.9$. Thus 90% of the intensity at the first meter remains at the second meter. That is, 10% is absorbed. A similar calculation shows that 10% of the intensity at the second meter is absorbed in the third meter.

(c) When $z = 1$m, we have $I(1)/I(0) = e^{-\alpha} = e^{\ln 0.9} = 0.9$, and so 90% remains. When $z = 2$m, we have $I(2)/I(0) = e^{-2\alpha} = e^{2\ln 0.9} = 0.9^2 = 0.81$, and so 81% remains. When $z = 3$m, we have $I(3)/I(0) = e^{-3\alpha} = e^{3\ln 0.9} = 0.9^3 = 0.729$, and so 72.9% remains.

(d) Please plot the graphs.

(e) Applying a logarithmic transformation to the equation $I(z) = I(0)e^{-\alpha z}$ we have

$$
\begin{aligned}
\log I(z) \quad &= \log(I(0)e^{-\alpha z}) \\
&= \log I(0) - \alpha z \log e \\
&= \log I(0) - (\alpha \log e)z
\end{aligned}
$$

Setting $Y = \log I(z)$, this gives the linear relationship $Y = \log I(0) - (\alpha \log e)z$. The slope of the line is given by $m = -\alpha \log e = \ln 0.9 \log e = \ln 0.9/\ln 10$.

(f) We want the ratio $I(z)/I(0)$ to equal 1%. That is, $0.01 = I(z)/I(0) = e^{-\alpha z} = e^{z \ln 0.9} = 0.9^z$. Thus $z = -\frac{\ln 0.01}{\alpha} = \frac{\ln(0.01)}{\ln(0.9)} \approx 43.7$m

(g) A clear lake would have small α, and a milky lake would have large α.

Prob. 93. $y = (100)(10^{1/3})^x$

Prob. 95. $y = (2^{1/3})(2^{2/3})^x$

Prob. 97. $Ay = \log x$

Prob. 99.

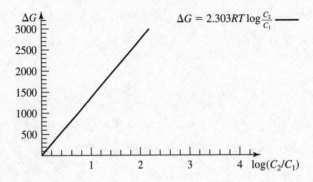

Prob. 101.

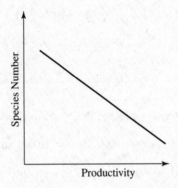

Prob. 103.

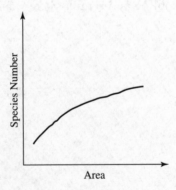

Prob. 105.

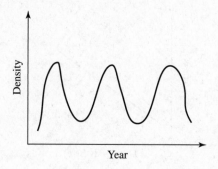

Prob. 107.

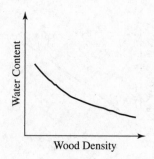

Prob. 109.

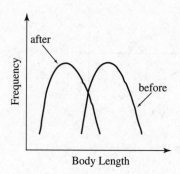

Prob. 111.

Prob. 113.

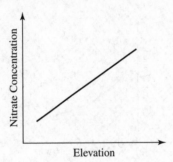

Prob. 115.

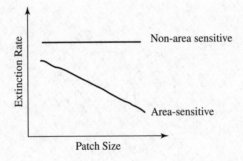

1.5 Review Problems

Prob. 1. (a) $10^4, 1.1 \times 10^4, 1.22 \times 10^4, 1.35 \times 10^4, 1.49 \times 10^4$

(b) $t = 10 \ln 10 \approx 23.0$

Prob. 3. (a) As the concentration of the reactants goes down as they are used up, the reaction rate slows down.

(b) $R(x) = -4kx^3 + 4k(a+b)x^2 - kb(4a-b)x + kab^2$, polynomial of degree 3

(c) $R(x) = (0.3)(5-x)(6-2x)^2$, $0 \le x \le 3$

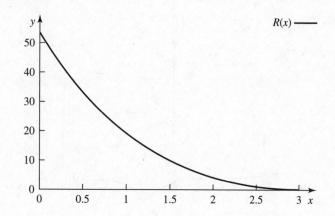

Prob. 5. (a) $L(t) = 0.69, 2.40, 4.62, 6.91$, $E(t) = 1.72, 2.20 \times 10^4, 2.69 \times 10^{43}, 10^{434}$

(b) 20.93 years, 3.13 ft

(c) $10^{536,000,000}$ years, 21.98 ft

(d) $L = 19.92$ ft, $E = 10^{195 \times 10^6}$ ft

Prob. 7. $T = \dfrac{\ln 2}{\ln(1+\frac{9}{100})}$, T goes to infinity as q gets closer to 0.

Prob. 9. (a)

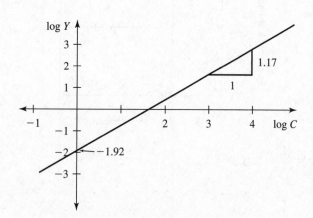

(b) $Y = C^{1.17}10^{-1.92}$

(c) $Y_p = 2.25Y_c$

(d) 8.5%

Prob. 11. (a) 400 days per year

(b) $y = 4.32 \times 10^9 - 1.8 \times 10^8 x$

(c) 376×10^6 to 563×10^6 years ago

Prob. 13. (a) males: $S(t) = \exp[-(0.019t)^{3.41}]$; females: $S(t) = \exp[-(0.022t)^{3.24}]$

(b) males: 47.27 days; females: 40.59 days

(c) males should live longer

Prob. 15. (a) $x = k$, $v = \frac{a}{2}$

(b) $x_{0.9} = 81x_{0.1}$

Prob. 17.

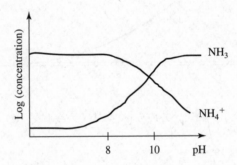

Prob. 19. $g(s) = \frac{v_{\max}}{S_k}S$

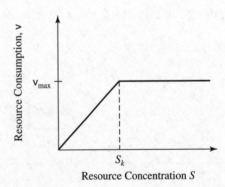

Prob. 21. (a) $\alpha = -\frac{\ln(0.01)}{18 \text{ m}} \approx 0.25\frac{1}{\text{m}}$

(b) 4.87 m

Chapter 2

Discrete Time Models, Sequences and Difference Equations

2.1 Exponential Growth and Decay

Prob. 1.

t	0	1	2	3	4	5
N_t	1	3	9	27	81	273

$N_t = 3^t$

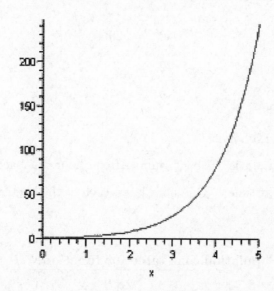

Prob. 3. $N_t = \frac{25}{4^t}$,

t	0	1	2	3	4	5
N_t	25	$\frac{25}{4}$	$\frac{25}{16}$	$\frac{25}{64}$	$\frac{25}{256}$	$\frac{25}{1024}$

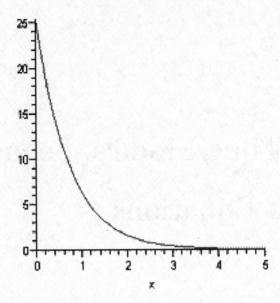

Prob. 5. We have $N(t) = 2 \cdot 2^t$

Prob. 7. We have $N(t) = 1 \cdot 2^{2t}$.

Prob. 9. We have $N(t) = 2 \cdot 4^{t/2}$

Prob. 11. Need to find t so that

$$
\begin{aligned}
40 &= 20.4^t \\
2 &= 4^t \\
2 &= 2^{2t}
\end{aligned}
$$

hence, $t = \frac{1}{2}$, meaning 1.5 hours.

Prob. 13. At time 0, there is one bacterium. After 1 hour, the bacterium splits into two, so there are two bacteria at time 1. One hour later, each of the bacteria splits again, resulting in 4 bacteria and so on.

Population Size table for function $N(t) = 2^t$

t	0	1	2	3	4	5
$N(t)$	1	2	4	8	16	32

Prob. 15. At time 0, there is one bacterium. After 23 min, the bacterium splits into two, so there are two bacteria at time 23 min. 23 minutes later, each of the bacteria splits again, resulting in 4 bacteria and so on.

Suppose one unit of time equals 23 minutes.

From the equation $N(t) = 2^t$, we get $128 = 2^t$, or $2^7 = 2^t$. Therefore, $t = 7$.

Seven units of time corresponds to $23 \cdot 7 = 161$ `minutes` $= 2$ hours and 41 minutes. Thus there are 128 bacteria in 2 hours and 41 minutes.

Prob. 17. At time 0, there are three bacteria. After 10 min, each bacteria splits into two, so there are 6 bacteria at time 10 min. Ten minutes later, each of the bacteria splits again, resulting in 12 bacteria and so on.

Suppose one unit of time equals 10 minutes.

From the equation $N(t) = N_0 2^t$, we get $96 = (3)2^t$, or $2^5 = 2^t$. Therefore, $t = 5$.

Five units of time corresponds to $10 \cdot 5 = 50$ minutes. Thus there are 96 bacteria in 50 minutes.

Prob. 19. Let $N(t)$ be the population size at time t, where t is measured in units of time, as in the previous exercises. If initially there is one bacterium and each bacteria splits into two in a unit time, then from the table shown below we will get:

Population Size table

t	0	1	2	3	4	5
N(t)	1	2	4	8	16	32

Here $N(t) = 2^t$, for $t = 0, 1, 2, \ldots$. The base of 2 reflects the fact that the population doubles in every unit of time. Now, if we had 40 bacteria in the beginning, then the new table will be:

Population Size table for the new function

t	0	1	2	3	4	5
N(t)	40	80	160	320	640	1280

We see that the initial population size appears as a multiplicative factor in front of the term 2^t. So we can now write the exponential growth equation as: $N(t) = 40 \cdot 2^t$, for $t = 0, 1, 2, \ldots$.

Prob. 21. As already discussed in the previous exercises, we know that the base 2 indicates that the population size doubles in every unit of time. For representing a population that triples in size in every unit of time, we initially modify our previous equation to be $N(t) = 3^t$, for $t = 0, 1, 2 \ldots$. Now, we have 20 bacteria in the beginning. This means the population table will be:

Population size table

t	0	1	2	3	4	5
$N(t)$	20	60	180	540	1620	4860

The initial population size appears as a multiplicative factor in front of the term 3^t. So we can write the exponential growth equation as $N(t) = 20 \cdot 3^t$, for $t = 0, 1, 2, \ldots$.

Prob. 23. As already discussed in the previous exercises, we know that the base 2 indicates that the population size doubles in every unit of time, and the base 3 indicates that the population size triples in every unit of time. For representing a population that quadruples in size in every unit of time, we initially modify our previous equation to be $N(t) = 4^t$, for $t = 0, 1, 2 \ldots$. Now, we have 5 bacteria in the beginning. This means the population table will be:

Population size table

t	0	1	2	3	4	5
$N(t)$	5	20	80	320	1280	5120

The initial population size appears as a multiplicative factor in front of the term 4^t. So we can write the exponential growth equation as $N(t) = 5 \cdot 4^t$, for $t = 0, 1, 2, \ldots$.

Prob. 25. While constructing population size tables for a population that doubles its size in every unit of time, we simply double the population size from one time step to the next. That is, we compute the population size at time $t + 1$ by doubling the population size at time t.

Let us denote N(t) by N_t. Here we begin with a population size of $N_0 = 20$ at $t = 0$, and so the population size at $t = 1$ is $N_1 = 40 = 2N_0$. The population at the end of the next time period is $N_2 = 80 = 2 \cdot 40 = 2N_1$, and so on. Thus the required recursion formula is $N_{t+1} = 2N_t$, with $N_0 = 20$.

Prob. 27. While constructing population size tables for a population that triples its size in every unit of time, we simply triple the population size from one time step to the next. That is, we compute the population size at time $t + 1$ by multiplying the population size at time t by 3.

Let us denote N(t) by N_t. Here we begin with a population size of $N_0 = 10$ at $t = 0$, and so the population size at $t = 1$ is $N_1 = 30 = 3 \cdot 10 = 3N_0$. The population at the end of the next time period is $N_2 = 90 = 3 \cdot 30 = 3N_1$, and so on. Thus the required recursion formula is $N_{t+1} = 3N_t$, with $N_0 = 10$.

Prob. 29. While constructing population size tables for a population that quadruples its size in every unit of time, we simply quadruple the population size from one time step to the next. That is, we compute the population size at time $t + 1$ by multiplying the population size at time t by 4.

Let us denote N(t) by N_t. Here we begin with a population size of $N_0 = 30$ at $t = 0$, and so the population size at $t = 1$ is $N_1 = 120 = 4 \cdot 30 = 4N_0$. The population at the end of the next time period is $N_2 = 480 = 4 \cdot 120 = 4N_1$, and so on. Thus the required recursion formula is $N_{t+1} = 4N_t$, with $N_0 = 30$.

Prob. 31.

Plot of the curve $f(x) = 2^x$ and the points $N_t = 2^t$.

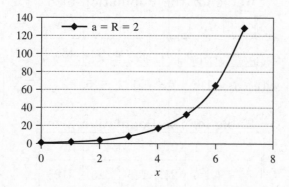

Prob. 33.

Plot of the curve $f(x) = 0.5^x$ and the points $N_t = 0.5^t$.

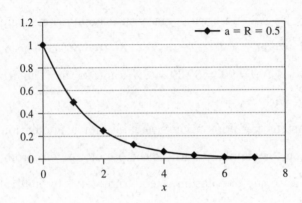

Prob. 35. $N_{t+1} = 2N_t$ with $N_0 = 3$, for $t = 0, 1, 2, \ldots 5$.

Table for the Population Size

t	0	1	2	3	4	5
N_t	3	6	12	24	48	96

Prob. 37. $N_{t+1} = 3N_t$ with $N_0 = 2$, for $t = 0, 1, 2, \ldots 5$.

Table for the Population Size

t	0	1	2	3	4	5
N_t	2	6	18	54	162	486

Prob. 39. $N_{t+1} = 5N_t$ with $N_0 = 1$, for $t = 0, 1, 2, \ldots 5$.

Table for the Population Size

t	0	1	2	3	4	5
N_t	1	5	25	125	625	3125

Prob. 41. $N_{t+1} = \frac{1}{2}N_t$ with $N_0 = 1024$, for $t = 0, 1, 2, \ldots 5$.

Table for the Population Size

t	0	1	2	3	4	5
N_t	1024	512	256	128	64	32

Prob. 43. $N_{t+1} = \frac{1}{3}N_t$ with $N_0 = 729$, for $t = 0, 1, 2, \ldots 5$.

Table for the Population Size

t	0	1	2	3	4	5
N_t	729	243	81	27	9	3

Prob. 45. $N_{t+1} = \frac{1}{5}N_t$ with $N_0 = 31250$, for $t = 0, 1, 2, \ldots 5$.

Table for the Population Size

t	0	1	2	3	4	5
N_t	31250	6250	1250	250	50	10

Prob. 47. Here $N_{t+1} = 2N_t$ with $N_0 = 15$. Thus N_t as a function of t would be $N_t = N(t) = 15 \cdot 2^t$.

Prob. 49. Here $N_{t+1} = 3N_t$ with $N_0 = 12$. Thus N_t as a function of t would be $N_t = N(t) = 12 \cdot 3^t$.

Prob. 51. Here $N_{t+1} = 4N_t$ with $N_0 = 24$. Thus N_t as a function of t would be $N_t = N(t) = 24 \cdot 4^t$.

Prob. 53. Here $N_{t+1} = \frac{1}{2}N_t$ with $N_0 = 5000$. Thus N_t as a function of t would be $N_t = N(t) = 5000(\frac{1}{2})^t = 5000 \cdot 2^{-t}$.

Prob. 55. Here $N_{t+1} = \frac{1}{3}N_t$ with $N_0 = 8000$. Thus N_t as a function of t would be $N_t = N(t) = 8000(\frac{1}{3})^t = 8000 \cdot 3^{-t}$.

Prob. 57. Here $N_{t+1} = \frac{1}{5}N_t$ with $N_0 = 1200$. Thus N_t as a function of t would be $N_t = N(t) = 1200(\frac{1}{5})^t = 1200 \cdot 5^{-t}$.

Prob. 59.

Plot of (N_t, N_{t+1}) for $R = 2, N_0 = 2$

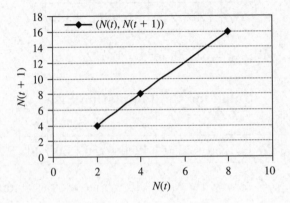

Prob. 61.

Plot of (N_t, N_{t+1}) for $R = 3, N_0 = 1$

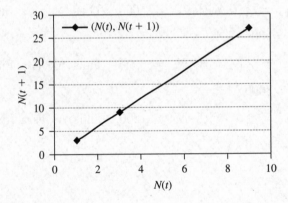

Prob. 63.

Plot of (N_t, N_{t+1}) for $R = \frac{1}{2}, N_0 = 16$

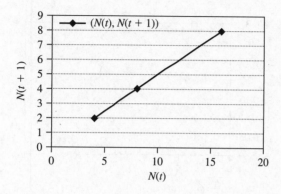

Prob. 65.

Plot of (N_t, N_{t+1}) for $R = \frac{1}{3}, N_0 = 81$

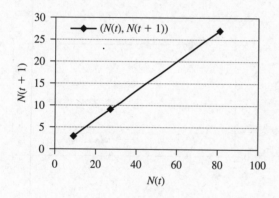

Prob. 67.

Plot of $\left(N_t, \frac{N_t}{N_{t+1}}\right)$ for $R = 2, N_0 = 2$

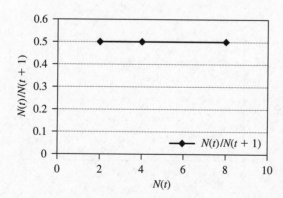

Prob. 69.

Plot of $(N_t, \frac{N_t}{N_{t+1}})$ for $R = 3, N_0 = 2$

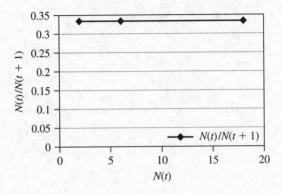

Prob. 71.

Plot of $(N_t, \frac{N_t}{N_{t+1}})$ for $R = \frac{1}{2}, N_0 = 16$

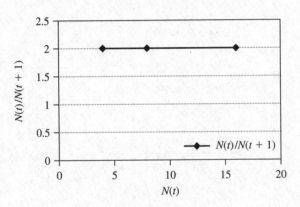

Prob. 73.

Plot of $(N_t, \frac{N_t}{N_{t+1}})$ for $R = \frac{1}{3}, N_0 = 27$

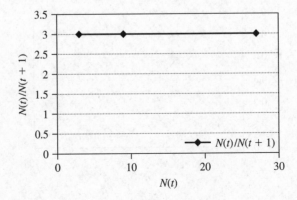

Prob. 75.

(a) As in the Section 2.1, we know that if the parent-offspring ratio is constant, then parents produce the same number of offsprings regardless of the current population density. But as the nesting sites is a limiting factor in population growth, if all the nesting sites are occupied then the parent-offspring ratio is probably not constant, and hence there would not be an exponential population growth.

(b) This is a suitable case as the number of nesting sites are much larger than the population. Assuming that all other conditions are favorable, population growth would be exponential.

(c) As a hurricane had killed a large number of birds, it is very probable that the nesting conditions and other important conditions necessary for survival are not suitable. There is a very high probability that the growth rate would not be exponential though such a situation cannot totally be ruled out.

Prob. 77. Population growth is density independent typically when conditions for the growth of the organism are constant and favorable throughout the time span. However, conditions like pollution, over-crowding, competition with other inhabitants or species, invasion of other species and natural disasters can be reasons for instability in the population growth.

2.2 Sequences

Prob. 1. Here the sequence is $a_n = n$. For $n = 0, 1, 2, \ldots, 5$ the sequence takes the values $0, 1, 2, 3, 4, 5$.

Prob. 3. Here the sequence is $a_n = \frac{1}{n+2}$. For $n = 0, 1, 2, \ldots, 5$ the sequence takes the values $\frac{1}{2}, \frac{1}{3}, \frac{1}{4}, \frac{1}{5}, \frac{1}{6}, \frac{1}{7}$.

Prob. 5. $f(n) = \frac{1}{(1+n)^2}$, we have

n	0	1	2	3	4	5
$f(n)$	1	$\frac{1}{4}$	$\frac{1}{9}$	$\frac{1}{16}$	$\frac{1}{25}$	$\frac{1}{36}$

Prob. 7. $f(n) = (n+1)^2$, we have

n	0	1	2	3	4	5
$f(n)$	1	4	9	16	25	36

Prob. 9. Here the sequence is $a_n = (-1)^n n$. For $n = 0, 1, 2, \ldots, 5$ the sequence takes the values $0, -1, 2, -3, 4, -5$.

Prob. 11. Here the sequence is $a_n = \frac{n^2}{n+1}$. For $n = 0, 1, 2, \ldots, 5$ the sequence takes the values $0, 1/2, 4/3, 9/4, 16/5, 25/6$. That is, $0, \frac{1}{2}, 1\frac{1}{3}, 2\frac{1}{4}, 3\frac{1}{5}, 4\frac{1}{6}$.

Prob. 13. $f(n) = e^{\sqrt{n}}$, we have

n	0	1	2	3	4	5
$f(n)$	1	e	$e^{\sqrt{2}}$	$e^{\sqrt{3}}$	e^2	$e^{\sqrt{5}}$

Prob. 15. $f(n) = (\frac{1}{3})^n$, we have

n	0	1	2	3	4	5
$f(n)$	1	$\frac{1}{3}$	$\frac{1}{9}$	$\frac{1}{27}$	$\frac{1}{81}$	$\frac{1}{243}$

Prob. 17. We can guess that the given sequence is $a_n = n + 1$, for $n = 0, 1, 2, 3, 4$. Thus for $n = 5, 6, 7, 8$ we have $a_n = 6, 7, 8, 9$.

Prob. 19. We can guess that the given sequence is $a_n = \frac{1}{(n+1)^2}$, for $n = 0, 1, 2, 3, 4$. Thus for $n = 5, 6, 7, 8$ we have $a_n = 1/36, 1/49, 1/64, 1/81$.

Prob. 21. We can guess that the given sequence is $a_n = \frac{n+1}{n+2}$, for $n = 0, 1, 2, 3, 4$. Thus for $n = 5, 6, 7, 8$ we have $a_n = 6/7, 7/8, 8/9, 9/10$.

Prob. 23. We can guess that the given sequence is $a_n = \sqrt{(n+1) + e^{n+1}}$, for $n = 0, 1, 2, 3, 4$. Thus for $n = 5, 6, 7, 8$ we have $a_n = \sqrt{6 + e^6}, \sqrt{7 + e^7}, \sqrt{8 + e^8}, \sqrt{9 + e^9}$.

Prob. 25. The given sequence is $a_0 = 0, a_1 = 1, a_2 = 2, a_3 = 3$ and $a_4 = 4$. Thus we can guess the expression to be $a_n = n$ for $n = 0, 1, 2, \ldots$.

Prob. 27. The given sequence is $a_0 = 1, a_1 = 2, a_2 = 4, a_3 = 8$ and $a_4 = 16$. Thus we can guess the expression to be $a_n = 2^n$ for $n = 0, 1, 2, \ldots$.

Prob. 29. The given sequence is $a_0 = 1, a_1 = \frac{1}{3}, a_2 = \frac{1}{9}, a_3 = \frac{1}{27}$ and $a_4 = \frac{1}{81}$. Thus we can guess the expression to be $a_n = \frac{1}{3^n}$ for $n = 0, 1, 2, \ldots$.

Prob. 31. The given sequence is $a_0 = -1, a_1 = 2, a_2 = -3, a_3 = 4$ and $a_4 = -5$. Thus we can guess the expression to be $a_n = (-1)^{n+1}(n + 1)$ for $n = 0, 1, 2, \ldots$.

Prob. 33. The given sequence is $a_0 = -\frac{1}{2}, a_1 = \frac{1}{3}, a_2 = -\frac{1}{4}, a_3 = \frac{1}{5}$ and $a_4 = -\frac{1}{6}$. Thus we can guess the expression to be $a_n = \frac{(-1)^{n+1}}{n+2}$ for $n = 0, 1, 2, \ldots$.

Prob. 35. The given sequence is $a_0 = \sin \pi, a_1 = \sin 2\pi, a_2 = \sin 3\pi, a_3 = \sin 4\pi$ and $a_4 = \sin 5\pi$. Thus we can guess the expression to be $a_n = \sin(n + 1)\pi$ for $n = 0, 1, 2, \ldots$.

Prob. 37. Plugging successive values of n into the given expression for a_n, we get the sequence

$$\frac{1}{2}, \frac{1}{3}, \frac{1}{4}, \frac{1}{5}, \frac{1}{6}, \ldots .$$

Also,

$$\lim_{n\to\infty} a_n = \lim_{n\to\infty} \frac{1}{n+2} = 0.$$

Prob. 39. Plugging successive values of n into the given expression for a_n, we get the sequence

$$0, \frac{1}{2}, \frac{2}{3}, \frac{3}{4}, \frac{4}{5}, \ldots .$$

Also,

$$\lim_{n\to\infty} a_n = \lim_{n\to\infty} \frac{n}{n+1} = 1.$$

Prob. 41. Plugging successive values of n into the given expression for a_n, we get the sequence

$$1, \frac{1}{2}, \frac{1}{5}, \frac{1}{10}, \frac{1}{17}, \ldots .$$

Also,

$$\lim_{n\to\infty} a_n = \lim_{n\to\infty} \frac{1}{n^2+1} = 0.$$

Prob. 43. Plugging successive values of n into the given expression for a_n, we get the sequence

$$1, -\frac{1}{2}, \frac{1}{3}, -\frac{1}{4}, \frac{1}{5}, \ldots .$$

Also,

$$\lim_{n\to\infty} a_n = \lim_{n\to\infty} \frac{(-1)^n}{n+1} = 0.$$

Prob. 45. Plugging successive values of n into the given expression for a_n we get the sequence

$$0, \frac{1}{2}, \frac{4}{3}, \frac{9}{4}, \frac{16}{5}, \ldots .$$

From this, we can guess that the terms will approach infinity as n tends to infinity. That is,

$$\lim_{n\to\infty} a_n = \lim_{n\to\infty} \frac{n^2}{n+1} = \infty.$$

Since the limiting value is not a number, we can say that the limit does not exist.

Prob. 47. Plugging successive values of n into the given expression for a_n we get the sequence

$$0, 1, \sqrt{2}, \sqrt{3}, 2, \ldots.$$

From this, we can guess that the terms will approach infinity as n tends to infinity. That is,

$$\lim_{n \to \infty} a_n = \lim_{n \to \infty} \sqrt{n} = \infty.$$

Since the limiting value is not a number, we can say that the limit does not exist.

Prob. 49. Plugging successive values of n into the given expression for a_n we get the sequence

$$1, 2, 4, 8, 16, \ldots.$$

From this, we can guess that the terms will approach infinity as n tends to infinity. That is,

$$\lim_{n \to \infty} a_n = \lim_{n \to \infty} 2^n = \infty.$$

Since the limiting value is not a number, we can say that the limit does not exist.

Prob. 51. Plugging successive values of n into the given expression for a_n we get the sequence

$$1, 3, 9, 27, 81, \ldots.$$

From this, we can guess that the terms will approach infinity as n tends to infinity. That is,

$$\lim_{n \to \infty} a_n = \lim_{n \to \infty} 3^n = \infty.$$

Since the limiting value is not a number, we can say that the limit does not exist.

Prob. 53. With $n = 1, 2, 3, 4, \ldots$ and $a_n = \frac{1}{n}$, we find that the sequence is $1, \frac{1}{2}, \frac{1}{3}, \frac{1}{4}, \ldots$, so we guess that the terms will approach the limit $a = 0$ as n tends to infinity. Now, we need to find an integer N such that $|\frac{1}{n} - 0| < \epsilon = 0.01$ whenever $n > N$. Solving the inequality $|\frac{1}{n} - 0| < 0.01$ for positive n, we find that $\frac{1}{n} < 0.01$ or $n > \frac{1}{0.01} = 100$. Thus the smallest value of N that we can choose seems to be $N = 100$. Successive values of $n > 100$ give us confidence that we are on the right track, but this doesn't prove that our choice is correct:

$$a_{101} = \frac{1}{101} \approx 0.00990, \quad a_{102} = \frac{1}{102} \approx 0.00980,$$

and so on. To prove that our choice for N works, we need to show that $n > N$ implies $|1/n - a| < \epsilon$. Indeed, we have that $n > 100$ implies $|\frac{1}{n} - 0| = \frac{1}{n} < \frac{1}{100} = 0.01$.

Prob. 55. With $n = 1, 2, 3, 4, \ldots$ and $a_n = \frac{1}{n^2}$, we find that the sequence is $1, \frac{1}{4}, \frac{1}{9}, \frac{1}{16}, \ldots$, so we guess that the terms will approach the limit $a = 0$ as n tends to infinity. Now, we need to find an integer N such that $|\frac{1}{n^2} - 0| < \epsilon = 0.01$ whenever $n > N$. Solving the inequality $|\frac{1}{n^2} - 0| < 0.01$ for positive n, we find that $\frac{1}{n^2} < 0.01$ or $n > \frac{1}{0.1} = 10$. Thus the smallest value of N that we can choose seems to be $N = 10$. Successive values of $n > 10$ give us confidence that we are on the right track, but this doesn't prove that our choice is correct:

$$a_{11} = \frac{1}{121} \approx 0.00826, \quad a_{12} = \frac{1}{144} \approx 0.00694,$$

and so on. To prove that our choice for N works, we need to show that $n > N$ implies $|1/n^2 - a| < \epsilon$. Indeed, we have that $n > 10$ implies $|\frac{1}{n^2} - 0| = \frac{1}{n^2} < \frac{1}{100} = 0.01$.

Prob. 57. With $n = 1, 2, 3, 4, \ldots$ and $a_n = \frac{1}{\sqrt{n}}$, we find that the sequence is $1, \frac{1}{\sqrt{2}}, \frac{1}{\sqrt{3}}, \frac{1}{2}, \ldots$, so we guess that the terms will approach the limit $a = 0$ as n tends to infinity. Now, we need to find an integer N such that $|\frac{1}{\sqrt{n}} - 0| < \epsilon = 0.1$ whenever $n > N$. Solving the inequality $|\frac{1}{\sqrt{n}} - 0| < 0.1$ for positive n, we find that $\frac{1}{\sqrt{n}} < 0.1$ or $n > \left(\frac{1}{0.1}\right)^2 = 100$. Thus the smallest value of N that we can choose seems to be $N = 100$. Successive values of $n > 100$ give us confidence that we are on the right track, but this doesn't prove that our choice is correct:

$$a_{101} = \frac{1}{\sqrt{101}} \approx 0.0995, \quad a_{102} = \frac{1}{\sqrt{102}} \approx 0.0990,$$

and so on. To prove that our choice for N works, we need to show that $n > N$ implies $|1/\sqrt{n} - a| < \epsilon$. Indeed, we have that $n > 100$ implies $|\frac{1}{\sqrt{n}} - 0| = \frac{1}{\sqrt{n}} < \frac{1}{\sqrt{100}} = 0.1$.

Prob. 59. With $n = 1, 2, 3, 4, \ldots$ and $a_n = \frac{(-1)^n}{n}$, we find that the sequence is $-1, \frac{1}{2}, -\frac{1}{3}, \frac{1}{4}, \ldots$, so we guess that the terms will approach the limit $a = 0$ as n tends to infinity. Now, we need to find an integer N such that $|\frac{(-1)^n}{n} - 0| < \epsilon = 0.01$ whenever $n > N$. Solving the inequality $|\frac{(-1)^n}{n} - 0| < 0.01$ for positive n, we find that $\frac{1}{n} < 0.01$ or $n > \frac{1}{0.01} = 100$. Thus the smallest value of N that we can choose seems to be $N = 100$. Successive values of $n > 100$ give us confidence that we are on the right track, but this doesn't prove that our choice is correct:

$$a_{101} = \frac{(-1)^{101}}{101} \approx -0.00990, \quad a_{102} = \frac{(-1)^{102}}{102} \approx 0.00980,$$

and so on. To prove that our choice for N works, we need to show that $n > N$ implies $|\frac{(-1)^n}{n} - a| < \epsilon$. Indeed, we have that $n > 100$ implies $|\frac{(-1)^n}{n} - 0| = \frac{1}{n} < \frac{1}{100} = 0.01$.

Prob. 61. With $n = 1, 2, 3, 4, \ldots$ and $a_n = \frac{n}{n+1}$, we find that the sequence is $\frac{1}{2}, \frac{2}{3}, \frac{3}{4}, \frac{4}{5}, \ldots$, so we guess that the terms will approach the limit $a = 1$ as n tends to infinity. Now, we need to find an integer N such that $|\frac{n}{n+1} - 1| < \epsilon = 0.01$ whenever $n > N$. Solving the inequality $|\frac{n}{n+1} - 1| < 0.01$ for positive n, we find that $|\frac{n-(n+1)}{n+1}| = \frac{1}{n+1} < 0.01$ or $n > \frac{1}{0.01} - 1 = 99$. Thus the smallest value of N that we can choose seems to be $N = 99$. Successive values of $n > 99$ give us confidence that we are on the right track, but this doesn't prove that our choice is correct:

$$a_{100} = \frac{100}{101} \approx 0.990099, \quad a_{101} = \frac{101}{102} \approx 0.990196,$$

and so on. To prove that our choice for N works, we need to show that $n > N$ implies $|\frac{n}{n+1} - a| < \epsilon$. Indeed, we have that $n > 99$ implies $|\frac{n}{n+1} - 1| = \frac{1}{n+1} < \frac{1}{100} = 0.01$.

Prob. 63. With $n = 1, 2, 3, 4, \ldots$ and $a_n = \frac{n^2}{n^2+1}$, we find that the sequence is $\frac{1}{2}, \frac{4}{5}, \frac{9}{10}, \frac{16}{17}, \ldots$, so we guess that the terms will approach the limit $a = 1$ as n tends to infinity. Now, we need to find an integer N such that $|\frac{n^2}{n^2+1} - 1| < \epsilon = 0.01$ whenever $n > N$. Solving the inequality $|\frac{n^2}{n^2+1} - 1| < 0.01$ for positive n, we find that $|\frac{n^2-(n^2+1)}{n^2+1}| = \frac{1}{n^2+1} < 0.01$ or $n > \sqrt{1/0.01 - 1} = \sqrt{99} \approx 9.95$. Thus the smallest value of N that we can choose seems to be $N = 9$. Successive values of $n > 9$ give us confidence that we are on the right track, but this doesn't prove that our choice is correct:

$$a_{10} = \frac{100}{101} \approx 0.990099, \quad a_{11} = \frac{121}{122} \approx 0.991803,$$

and so on. To prove that our choice for N works, we need to show that $n > N$ implies $|\frac{n^2}{n^2+1} - a| < \epsilon$. Indeed, we have that $n > 9$ implies $n \geq 10$ which implies that $|\frac{n^2}{n^2+1} - 1| = \frac{1}{n^2+1} \leq \frac{1}{100+1} < 0.01$.

Prob. 65. We have to show that for every $\epsilon > 0$ we can find N such that

$|1/n - 0| < \epsilon$ whenever $n > N$.

To find a candidate for N, we solve the inequality $|1/n| < \epsilon$. Since $1/n > 0$, we drop the absolute value and find that

$1/n < \epsilon$ or $n > 1/\epsilon$

Let's choose N to be the largest integer less than or equal to $1/\epsilon$. If $n > N$, then $n \geq N + 1$, which is equivalent to $1/n \leq 1/(N+1)$. Since N is the largest integer less than or equal to $1/\epsilon$, it follows that $1/\epsilon < N + 1$, and so $1/n \leq 1/(N+1) < \epsilon$ for $n > N$. This, together with $n > 0$, shows that if N is the largest integer less than or equal to $1/\epsilon$, then $|1/n - 0| < \epsilon$ whenever $n > N$.

Prob. 67. We have to show that for every $\epsilon > 0$ we can find N such that

$|1/n^2 - 0| < \epsilon$ whenever $n > N$.

To find a candidate for N, we solve the inequality $|1/n^2| < \epsilon$. Since $1/n^2 > 0$, we drop the absolute value and find that

$1/n^2 < \epsilon$ or $n > 1/\sqrt{\epsilon}$

Let's choose N to be the largest integer less than or equal to $1/\sqrt{\epsilon}$. If $n > N$, then $n \geq N + 1$, which is equivalent to $1/n \leq 1/(N+1)$. Since N is the largest integer less than or equal to $1/\sqrt{\epsilon}$, it follows that $1/\sqrt{\epsilon} < N + 1$, and so $1/n \leq 1/(N+1) < \sqrt{\epsilon}$ for $n > N$. Squaring, we get $1/n^2 \leq 1/(N+1)^2 < \epsilon$ for $n > N$. This shows that if N is the largest integer less than or equal to $1/\sqrt{\epsilon}$, then $|1/n^2 - 0| < \epsilon$ whenever $n > N$.

Prob. 69. We have to show that for every $\epsilon > 0$ we can find N such that

$|(n+1)/n - 1| < \epsilon$ whenever $n > N$.

To find a candidate for N, we solve the inequality $|(n+1)/n-1| < \epsilon$. Since $(n+1)/n-1 = 1/n > 0$, we drop the absolute value and find that

$1/n < \epsilon$ or $n > 1/\epsilon$.

Let's choose N to be the largest integer less than or equal to $1/\epsilon$. If $n > N$, then $n \geq N + 1$, which is equivalent to $1/n \leq 1/(N+1)$. Since N is the largest integer less than or equal to $1/\epsilon$, it follows that $1/\epsilon < N + 1$, and so $(n+1)/n - 1 = 1/n \leq 1/(N+1) < \epsilon$ for $n > N$. This, together with $n > 0$, shows that if N is the largest integer less than or equal to $1/\epsilon$, then $|(n+1)/n - 1| < \epsilon$ whenever $n > N$.

Prob. 71. We break $1/n^2$ into the product of two terms, namely $(1/n) \cdot (1/n)$. Since $\lim_{n \to \infty} \frac{1}{n}$ exists and is equal to 0, we find that

$$\lim_{n \to \infty} \left(\frac{1}{n} + \frac{1}{n^2} \right) = \lim_{n \to \infty} \frac{1}{n} + \left(\lim_{n \to \infty} \frac{1}{n} \right) \cdot \left(\lim_{n \to \infty} \frac{1}{n} \right) = 0 + 0 \cdot 0 = 0.$$

Prob. 73. We can write $\frac{n+1}{n}$ as $1 + 1/n$. Since both $\lim_{n\to\infty} 1$ and $\lim_{n\to\infty} (1/n)$ exist and are equal to 1 and 0 respectively, we have

$$\lim_{n\to\infty} \frac{n+1}{n} = \lim_{n\to\infty} (1 + 1/n) = \lim_{n\to\infty} 1 + \lim_{n\to\infty} (1/n) = 1 + 0 = 1.$$

Prob. 75. We can write $\frac{n^2+1}{n^2}$ as $1 + \frac{1}{n} \cdot \frac{1}{n}$. Since both $\lim_{n\to\infty} 1$ and $\lim_{n\to\infty} (1/n)$ exist and are equal to 1 and 0 respectively, we have

$$\lim_{n\to\infty} \frac{n^2+1}{n^2} = \lim_{n\to\infty} \left(1 + \frac{1}{n} \cdot \frac{1}{n}\right) = \lim_{n\to\infty} 1 + \lim_{n\to\infty} \left(\frac{1}{n} \cdot \frac{1}{n}\right)$$

$$= \lim_{n\to\infty} 1 + \lim_{n\to\infty} \frac{1}{n} \cdot \lim_{n\to\infty} \frac{1}{n} = 1 + 0 \cdot 0 = 1.$$

Prob. 77. We can factor the denominator of $\frac{n+1}{n^2-1}$ and write this as $\frac{n+1}{(n+1)(n-1)} = \frac{1}{n-1}$. Recalling that $\lim_{n\to\infty} \frac{1}{n}$ exists and is equal to 0, we find that

$$\lim_{n\to\infty} \frac{n+1}{n^2-1} = \lim_{n\to\infty} \frac{1}{n-1}$$

$$= \lim_{n\to\infty} \frac{1/n}{1 - 1/n}$$

$$= \frac{\lim_{n\to\infty} (1/n)}{\lim_{n\to\infty} (1 - 1/n)}$$

$$= \frac{\lim_{n\to\infty} (1/n)}{1 - \lim_{n\to\infty} (1/n)}$$

$$= \frac{0}{1-0} = 0.$$

Prob. 79. By Example 12, we know that $\lim_{n\to\infty} (1/3)^n$ and $\lim_{n\to\infty} (1/2)^n$ exist and are equal to 0. Thus we have

$$\lim_{n\to\infty} \left[\left(\frac{1}{3}\right)^n + \left(\frac{1}{2}\right)^n\right] = \lim_{n\to\infty} \left(\frac{1}{3}\right)^n + \lim_{n\to\infty} \left(\frac{1}{2}\right)^n = 0 + 0 = 0.$$

Prob. 81. First, notice that the expression $(n + 2^{-n})/n$ can also be written as $1 + \left(\frac{1}{2}\right)^n \cdot \frac{1}{n}$. Since the limits $\lim_{n\to\infty} 1$, $\lim_{n\to\infty} \left(\frac{1}{2}\right)^n$ and $\lim_{n\to\infty} \frac{1}{n}$ exist and are equal to 1, 0 and 0 respectively, we have that

$$\lim_{n\to\infty} \frac{n + 2^{-n}}{n} = \lim_{n\to\infty} \left[1 + \left(\frac{1}{2}\right)^n \cdot \frac{1}{n}\right]$$

$$= \lim_{n\to\infty} 1 + \lim_{n\to\infty} \left[\left(\frac{1}{2}\right)^n \cdot \frac{1}{n}\right]$$

$$= \lim_{n\to\infty} 1 + \lim_{n\to\infty} \left(\frac{1}{2}\right)^n \cdot \lim_{n\to\infty} \frac{1}{n}$$

$$= 1 + 0 \cdot 0 = 1.$$

Prob. 83. By repeatedly applying the recursion to the equation $a_{n+1} = 2a_n$ with $a_0 = 1$, we have

$$a_1 = 2a_0 = 2 \cdot 1 = 2,$$
$$a_2 = 2a_1 = 2 \cdot 2 = 4,$$
$$a_3 = 2a_2 = 2 \cdot 4 = 8,$$
$$a_4 = 2a_3 = 2 \cdot 8 = 16, \text{ and}$$
$$a_5 = 2a_4 = 2 \cdot 16 = 32.$$

Prob. 85. By repeatedly applying the recursion to the equation $a_{n+1} = 3a_n - 2$ with $a_0 = 1$, we have

$$a_1 = 3a_0 - 2 = 3 \cdot 1 - 2 = 1,$$
$$a_2 = 3a_1 - 2 = 3 \cdot 1 - 2 = 1,$$
$$a_3 = 3a_2 - 2 = 3 \cdot 1 - 2 = 1,$$
$$a_4 = 3a_3 - 2 = 3 \cdot 1 - 2 = 1, \text{ and}$$
$$a_5 = 3a_4 - 2 = 3 \cdot 1 - 2 = 1.$$

Prob. 87. By repeatedly applying the recursion to the equation $a_{n+1} = 4 - 2a_n$ with $a_0 = 5$, we have

$$a_1 = 4 - 2a_0 = 4 - 2 \cdot 5 = -6,$$
$$a_2 = 4 - 2a_1 = 4 - 2 \cdot (-6) = 16,$$
$$a_3 = 4 - 2a_2 = 4 - 2 \cdot 16 = -28,$$
$$a_4 = 4 - 2a_3 = 4 - 2 \cdot (-28) = 60, \text{ and}$$
$$a_5 = 4 - 2a_4 = 4 - 2 \cdot 60 = -116.$$

Prob. 89. By repeatedly applying the recursion to the equation $a_{n+1} = \frac{a_n}{1+a_n}$ with $a_0 = 1$, we have

$$a_1 = a_0/(1 + a_0) = 1/(1 + 1) = 1/2,$$
$$a_2 = a_1/(1 + a_1) = (1/2)/(3/2) = 1/3,$$
$$a_3 = a_2/(1 + a_2) = (1/3)/(4/3) = 1/4,$$
$$a_4 = a_3/(1 + a_3) = (1/4)/(5/4) = 1/5, \text{ and}$$
$$a_5 = a_4/(1 + a_4) = (1/5)/(6/5) = 1/6.$$

Prob. 91. By repeatedly applying the recursion to the equation $a_{n+1} = a_n + \frac{1}{a_n}$ with $a_0 = 1$, we have

$a_1 = a_0 + 1/a_0 = 1 + 1/1 = 2,$

$a_2 = a_1 + 1/a_1 = 2 + 1/2 = 5/2 = 2.5,$

$a_3 = a_2 + 1/a_2 = 5/2 + 2/5 = 29/10 = 2.9,$

$a_4 = a_3 + 1/a_3 = 29/10 + 10/29 = 941/290 \approx 3.2448,$ and

$a_5 = a_4 + 1/a_4 = 941/290 + 290/941 = 969581/272890 \approx 3.5530.$

Prob. 93. Following the method of Example 14, here we have $f(a) = \frac{1}{2}a + 2$. If a is a fixed point, then it must satisfy the equation $a = f(a)$. That is,

$$a = \frac{1}{2}a + 2 \quad \Rightarrow \quad \frac{1}{2}a = 2 \quad \Rightarrow \quad a = 4.$$

Thus the only fixed point is $a = 4$.

Prob. 95. Following the method of Example 14, here we have $f(a) = \frac{2}{5}a - \frac{9}{5}$. If a is a fixed point, then it must satisfy the equation $a = f(a)$. That is,

$$a = \frac{2}{5}a - \frac{9}{5} \quad \Rightarrow \quad \frac{3}{5}a = -\frac{9}{5} \quad \Rightarrow \quad a = -3.$$

Thus the only fixed point is $a = -3$.

Prob. 97. Following the method of Example 14, here we have $f(a) = 4/a$. If a is a fixed point, then it must satisfy the equation $a = f(a)$. That is,

$$a = \frac{4}{a} \quad \Rightarrow \quad a^2 = 4 \quad \Rightarrow \quad a = \pm 2.$$

Thus there are two fixed points, namely $a = 2$ and $a = -2$.

Prob. 99. Following the method of Example 14, here we have $f(a) = 2/(a + 2)$. If a is a fixed point, then it must satisfy the equation $a = f(a)$. That is,

$$a = \frac{2}{a + 2} \quad \Rightarrow \quad a(a + 2) = 2 \quad \Rightarrow \quad a^2 + 2a - 2 = 0 \quad \Rightarrow \quad a = -1 \pm \sqrt{3}.$$

Thus there are two fixed points, namely $a = -1 + \sqrt{3}$ and $a = -1 - \sqrt{3}$.

Prob. 101. Following the method of Example 14, here we have $f(a) = \sqrt{5a}$. If a is a fixed point, then it must satisfy the equation $a = f(a)$. That is,

$$a = \sqrt{5a} \quad \Rightarrow \quad a^2 = 5a \quad \Rightarrow \quad a(a - 5) = 0 \quad \Rightarrow \quad a = 0 \text{ or } 5.$$

Thus there are two fixed points, namely $a = 0$ and $a = 5$.

Prob. 103. Since the problem tells us that the limit exists, we shall not concern ourselves with the existence of the limit but instead try to identify the limiting value.

To do this, we first compute the fixed points. As in Example 14, we solve the equation $a = \frac{1}{2}(a + 5)$ to find that $\frac{a}{2} = \frac{5}{2}$, or that $a = 5$. Now when $a_0 = 1$, then $a_n > 1$ for all $n = 1, 2, 3, \cdots$ and so we conclude that

$$\lim_{n \to \infty} a_n = 5.$$

Using a calculator, we can find successive (approximated)values of a_n, which we collect in the following table. The tabulated values suggest that the limit is indeed 5.

n	a_n
0	1
1	3
2	4
3	4.5
4	4.75
5	4.875
6	4.9375

Prob. 105. Since the problem tells us that the limit exists, we shall not concern ourselves with the existence of the limit but instead try to identify the limiting value.

To do this, we first compute the fixed points. As in Example 14, we solve the equation $a = \sqrt{2a}$ to find that $a(a - 2) = 0$, which means that $a = 0$ or $a = 2$. Now when $a_0 = 1$, then $a_n > 1$ for all $n = 1, 2, 3, \cdots$ and so we could conclude that

$$\lim_{n \to \infty} a_n = 2.$$

Using a calculator, we can find successive values of a_n, which we collect in the following table. The tabulated values suggest that the fixed point $a = 2$ is the limit.

n	a_n
0	1
1	$\sqrt{2}$
2	$2^{3/4}$
3	$2^{7/8}$
4	$2^{15/16}$
5	$2^{31/32}$
6	$2^{63/64}$

Prob. 107. Since the problem tells us that the limit exists, we shall not concern ourselves with the existence of the limit but instead try to identify the limiting value.

To do this, we first compute the fixed points. As in Example 14, we solve the equation $a = 2a(1-a)$ to find that $a(2a-1) = 0$, which means that $a = 0$ or $a = 1/2$. Now when $a_0 = 0.1$, then $a_n > 0.1$ for all $n = 1, 2, 3, \cdots$ and so we could conclude that

$$\lim_{n \to \infty} a_n = 1/2.$$

Using a calculator, we can find successive (approximated) values of a_n, which we collect in the following table. The tabulated values suggest that the fixed point $a = 1/2$ is the limit.

n	a_n
0	0.1
1	0.18
2	0.2952
3	0.41611
4	0.48592
5	0.4996
6	0.49999

Prob. 109. Since the problem tells us that the limit exists, we shall not concern ourselves with the existence of the limit but instead try to identify the limiting value.

To do this, we first compute the fixed points. As in Example 14, we solve the equation $a = \frac{1}{2}(a + \frac{4}{a})$ to find that $a = \frac{4}{a}$, or that $a = \pm 2$. Now when $a_0 = 1$, then $a_n > 1$ for all

$n = 1, 2, 3, \cdots$ and so we could conclude that

$$\lim_{n \to \infty} a_n = 2.$$

Using a calculator, we can find successive (approximated) values of a_n, which we collect in the following table. The tabulated values suggest that the limit is indeed 2.

n	a_n
0	1
1	2.5
2	2.05
3	2.00061
4	2.00000
5	2.00000

2.3   More Population Models

Prob. 1. The recursion to the Beverton-Holt recruitment curve is

$$N_{t+1} = \frac{2N_t}{1 + \frac{1}{15}N_t}$$

Graph of $\frac{N_t}{N_{t+1}}$ as a function of N_t

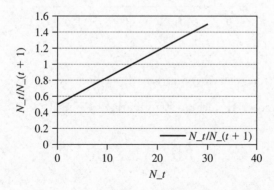

Prob. 3. The recursion to the Beverton-Holt recruitment curve is

$$N_{t+1} = \frac{1.5N_t}{1 + \frac{0.5}{40}N_t}$$

Graph of $\frac{N_t}{N_{t+1}}$ as a function of N_t

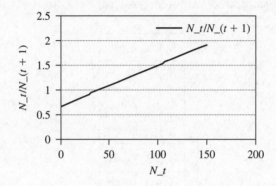

Prob. 5. The recursion to the Beverton-Holt recruitment curve is

$$N_{t+1} = \frac{2.5N_t}{1 + \frac{1.5}{90}N_t}$$

Graph of $\frac{N_t}{N_{t+1}}$ as a function of N_t

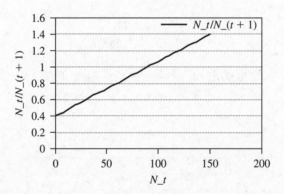

Prob. 7. The population growth equation described by the Beverton-Holt recruitment curve with growth parameter R and carrying capacity K is

$$N_{t+1} = \frac{RN_t}{1 + \frac{R-1}{K}N_t}$$

Now, the given equation

$$N_{t+1} = \frac{2N_t}{1 + \frac{1}{20}N_t}$$

can be written as

$$N_{t+1} = \frac{2N_t}{1 + \frac{2-1}{20}N_t}$$

and so we must have $R = 2$ and $K = 20$.

Prob. 9. The population growth equation described by the Beverton-Holt recruitment curve with growth parameter R and carrying capacity K is

$$N_{t+1} = \frac{RN_t}{1 + \frac{R-1}{K}N_t}$$

Now, the given equation

$$N_{t+1} = \frac{1.5N_t}{1 + \frac{0.5}{30}N_t}$$

can be written as

$$N_{t+1} = \frac{1.5N_t}{1 + \frac{1.5-1}{30}N_t}$$

and so we must have $R = 1.5$ and $K = 30$.

Prob. 11. The population growth equation described by the Beverton-Holt recruitment curve with growth parameter R and carrying capacity K is

$$N_{t+1} = \frac{RN_t}{1 + \frac{R-1}{K}N_t}$$

Now, the given equation

$$N_{t+1} = \frac{4N_t}{1 + \frac{1}{150}N_t}$$

can be written as

$$N_{t+1} = \frac{4N_t}{1 + \frac{4-1}{450}N_t}$$

and so we must have $R = 4$ and $K = 450$.

Prob. 13. From the given equation

$$N_{t+1} = \frac{4N_t}{1 + \frac{1}{30}N_t},$$

we can compute the fixed points of the equation by solving

$$N = \frac{4N}{1 + \frac{1}{30}N}$$

for N. We immediately find $N = 0$. If $N \neq 0$, we can divide both sides by N, to get

$$1 = \frac{4}{1 + \frac{1}{30}N}.$$

That is,

$$1 + \frac{1}{30}N = 4 \quad \Rightarrow \quad \frac{1}{30}N = 3 \quad \Rightarrow \quad N = 3 \cdot 30 = 90.$$

We thus have two fixed points; the fixed point $N = 0$, which we call trivial since it corresponds to the absence of the population, and the fixed point $N = 90$, which we call nontrivial since it corresponds to a positive population size.

Prob. 15. From the given equation

$$N_{t+1} = \frac{2N_t}{1 + \frac{1}{30}N_t},$$

we can compute the fixed points of the equation by solving

$$N = \frac{2N}{1 + \frac{1}{30}N}$$

for N. We immediately find $N = 0$. If $N \neq 0$, we can divide both sides by N, to get

$$1 = \frac{2}{1 + \frac{1}{30}N}.$$

That is,

$$1 + \frac{1}{30}N = 2 \quad \Rightarrow \quad \frac{1}{30}N = 1 \quad \Rightarrow \quad N = 1 \cdot 30 = 30.$$

We thus have two fixed points; the fixed point $N = 0$, which we call trivial since it corresponds to the absence of the population, and the fixed point $N = 30$, which we call nontrivial since it corresponds to a positive population size.

Prob. 17. From the given equation

$$N_{t+1} = \frac{3N_t}{1 + \frac{1}{30}N_t},$$

we can compute the fixed points of the equation by solving

$$N = \frac{3N}{1 + \frac{1}{30}N}$$

for N. We immediately find $N = 0$. If $N \neq 0$, we can divide both sides by N, to get

$$1 = \frac{3}{1 + \frac{1}{30}N}.$$

That is,

$$1 + \frac{1}{30}N = 3 \quad \Rightarrow \quad \frac{1}{30}N = 2 \quad \Rightarrow \quad N = 2 \cdot 30 = 60.$$

We thus have two fixed points; the fixed point $N = 0$, which we call trivial since it corresponds to the absence of the population, and the fixed point $N = 60$, which we call nontrivial since it corresponds to a positive population size.

Prob. 19. The population growth equation described by the Beverton-Holt recruitment curve with growth parameter R and carrying capacity K is given by

$$N_{t+1} = \frac{RN_t}{1 + \frac{R-1}{K}N_t}$$

If $R = 2$, $K = 10$ and $N_0 = 2$, then the population sizes for $t = 1, 2, 3, 4, 5$ are

$$N_1 = \frac{2N_0}{1 + \frac{2-1}{10}N_0} = \frac{2 \cdot 2}{1 + \frac{1}{10}2} = \frac{4}{6/5} = \frac{10}{3} \approx 3.33$$

$$N_2 = \frac{2N_1}{1 + \frac{2-1}{10}N_1} = \frac{2 \cdot (10/3)}{1 + \frac{1}{10}\frac{10}{3}} = \frac{20/3}{4/3} = 5$$

$$N_3 = \frac{2N_2}{1 + \frac{2-1}{10}N_2} = \frac{2 \cdot 5}{1 + \frac{1}{10}5} = \frac{10}{\frac{3}{2}} = \frac{20}{3} \approx 6.67$$

$$N_4 = \frac{2N_3}{1 + \frac{2-1}{10}N_3} = \frac{2 \cdot \frac{20}{3}}{1 + \frac{1}{10}\frac{20}{3}} = \frac{40/3}{5/3} = 8$$

$$N_5 = \frac{2N_4}{1 + \frac{2-1}{10}N_4} = \frac{2 \cdot 8}{1 + \frac{1}{10}8} = \frac{16}{9/5} = \frac{80}{9} \approx 8.89$$

Finally, since $K > 0$, $R > 1$ and $N_0 > 0$, we can say that $\lim_{t \to \infty} N_t = K = 10$.

Prob. 21. The population growth equation described by the Beverton-Holt recruitment curve with growth parameter R and carrying capacity K is given by

$$N_{t+1} = \frac{RN_t}{1 + \frac{R-1}{K}N_t}$$

If $R = 3$, $K = 15$ and $N_0 = 1$, then the population sizes for $t = 1, 2, 3, 4, 5$ are

$$N_1 = \frac{3N_0}{1 + \frac{3-1}{15}N_0} = \frac{3 \cdot 1}{1 + \frac{2}{15}1} = \frac{3}{17/15} = \frac{45}{17} \approx 2.647$$

$$N_2 = \frac{3N_1}{1 + \frac{3-1}{15}N_1} = \frac{3 \cdot (45/17)}{1 + \frac{2}{15}\frac{45}{17}} = \frac{135/17}{23/17} = \frac{135}{23} \approx 5.870$$

$$N_3 = \frac{3N_2}{1 + \frac{3-1}{15}N_2} = \frac{3 \cdot (135/23)}{1 + \frac{2}{15}\frac{135}{23}} = \frac{405/23}{41/23} = \frac{405}{41} \approx 9.878$$

$$N_4 = \frac{3N_3}{1 + \frac{3-1}{15}N_3} = \frac{3 \cdot (405/41)}{1 + \frac{2}{15}\frac{405}{41}} = \frac{1215/41}{95/41} = \frac{243}{19} \approx 12.790$$

$$N_5 = \frac{3N_4}{1 + \frac{3-1}{15}N_4} = \frac{3 \cdot (243/19)}{1 + \frac{2}{15}\frac{243}{19}} = \frac{729/19}{257/95} = \frac{3645}{257} \approx 14.183$$

Finally, since $K > 0$, $R > 1$ and $N_0 > 0$, we can say that $\lim_{t \to \infty} N_t = K = 15$.

Prob. 23. The population growth equation described by the Beverton-Holt recruitment curve with growth parameter R and carrying capacity K is given by

$$N_{t+1} = \frac{RN_t}{1 + \frac{R-1}{K}N_t}$$

If $R = 4$, $K = 40$ and $N_0 = 3$, then the population sizes for $t = 1, 2, 3, 4, 5$ are

$$N_1 = \frac{4N_0}{1 + \frac{4-1}{40}N_0} = \frac{4 \cdot 3}{1 + \frac{3}{40}3} = \frac{12}{49/40} = \frac{480}{49} \approx 9.796$$

$$N_2 = \frac{4N_1}{1 + \frac{4-1}{40}N_1} = \frac{4 \cdot (480/49)}{1 + \frac{3}{40}\frac{480}{49}} = \frac{1920/49}{85/49} = \frac{384}{17} \approx 22.588$$

$$N_3 = \frac{4N_2}{1 + \frac{4-1}{40}N_2} = \frac{4 \cdot (384/17)}{1 + \frac{3}{40}\frac{384}{17}} = \frac{1536/17}{229/85} = \frac{7680}{229} \approx 33.537$$

$$N_4 = \frac{4N_3}{1 + \frac{4-1}{40}N_3} = \frac{4 \cdot (7680/229)}{1 + \frac{3}{40}\frac{7680}{229}} = \frac{30720/229}{805/229} = \frac{6144}{161} \approx 38.161$$

$$N_5 = \frac{4N_4}{1 + \frac{4-1}{40}N_4} = \frac{4 \cdot (6144/161)}{1 + \frac{3}{40}\frac{6144}{161}} = \frac{24576/161}{3109/805} = \frac{122880}{3109} \approx 39.524$$

Finally, since $K > 0$, $R > 1$ and $N_0 > 0$, we can say that $\lim_{t \to \infty} N_t = K = 40$.

Prob. 25. The discrete logistic equation with parameter R and K is given by

$$N_{t+1} = N_t \left[1 + R \left(1 - \frac{N_t}{K} \right) \right].$$

Substituting the values $R = 1$ and $K = 10$, we can write the equation as

$$N_{t+1} = N_t \left[1 + \left(1 - \frac{N_t}{10} \right) \right].$$

In canonical form this can be written as

$$x_{t+1} = rx_t(1 - x_t) = (1 + R)x_t(1 - x_t)$$

where $r = R + 1$ and $x_t = \frac{R}{K(1+R)} N_t$. Substituting the given values of R and K we have that

$$r = 2 \quad \text{and} \quad x_t = \frac{1}{10(1+1)} N_t = \frac{1}{20} N_t$$

Prob. 27. The discrete logistic equation with parameter R and K is given by

$$N_{t+1} = N_t \left[1 + R \left(1 - \frac{N_t}{K} \right) \right].$$

Substituting the values $R = 2$ and $K = 15$, we can write the equation as

$$N_{t+1} = N_t \left[1 + 2 \left(1 - \frac{N_t}{15} \right) \right].$$

In canonical form this can be written as

$$x_{t+1} = rx_t(1 - x_t) = (1 + R)x_t(1 - x_t)$$

where $r = R + 1$ and $x_t = \frac{R}{K(1+R)} N_t$. Substituting the given values of R and K we have that

$$r = 3 \quad \text{and} \quad x_t = \frac{2}{15(1+2)} N_t = \frac{2}{45} N_t$$

Prob. 29. The discrete logistic equation with parameter R and K is given by

$$N_{t+1} = N_t \left[1 + R \left(1 - \frac{N_t}{K} \right) \right].$$

Substituting the values $R = 2.5$ and $K = 30$, we can write the equation as

$$N_{t+1} = N_t \left[1 + 2.5 \left(1 - \frac{N_t}{30} \right) \right].$$

In canonical form this can be written as

$$x_{t+1} = rx_t(1 - x_t) = (1 + R)x_t(1 - x_t)$$

where $r = R + 1$ and $x_t = \frac{R}{K(1+R)} N_t$. Substituting the given values of R and K we have that

$$r = 3.5 \quad \text{and} \quad x_t = \frac{2.5}{30(1 + 2.5)} N_t = \frac{1}{42} N_t$$

Prob. 31.

(a) The variable N_t and the parameter K both have units (or dimension) "number of individuals". Dividing N_t by K, the units cancel and we can say that $x_t = N_t/K$ is dimensionless.

(b) The variable M_t and the parameter L both have units (or dimension) "1000 individuals". Dividing M_t by L, the units cancel and we can say that $y_t = M_t/L$ is also dimensionless.

(c) The variable M_t has units (or dimension) "1000 individuals" while the variable N_t has units (or dimension) "number of individuals". Dividing M_t by N_t, the common units "number of individuals" cancel and we have

$$\frac{M_t}{N_t} = \frac{1}{1000}$$

Similarly, the variable L has units (or dimension) "1000 individuals" while the variable K has units (or dimension) "number of individuals". Dividing L by K, the common units "number of individuals" cancel and we have

$$\frac{L}{K} = \frac{1}{1000}$$

(d) From part (c), we know that $M_t/N_t = 1/1000$ and that $L/K = 1/1000$. We are given that $N_t = 20,000$ and $K = 5000$, and so $M_t = N_t/1000 = 20,000/1000 = 20$ and $L = K/1000 = 5000/1000 = 5$.

(e) We are given $M_t = 20$, $N_t = 20,000$, $L = 5$ and $K = 5000$. This means that

$$x_t = \frac{N_t}{K} = \frac{20,000}{5000} = 4 \quad \text{and} \quad y_t = \frac{M_t}{L} = \frac{20}{5} = 4.$$

Thus $x_t = y_t$.

Prob. 33. The unit or dimension of T is the characteristic time, and the unit or dimension of t is the time elapsed since the beginning of the experiment. We obtain z by dividing t by T, that is, the time elapsed by the characteristic time. As both t and T have the same units, upon division they will cancel and so $z = t/T$ is dimensionless.

If $t = 120$ minutes and $T = 20$ minutes, then $z = t/T = 120/20 = 6$. If instead t and T are measured in hours instead of minutes, then $t = 2$ and $T = 1/3$ hours. However, the value of z remains the same: $z = t/T = \frac{2}{1/3} = 6$.

Prob. 35. The discrete logistic equation is given by

$$x_{t+1} = rx_t(1 - x_t)$$

For $r = 2$ and $x_0 = 0.2$, we compute x_t for $t = 0, 1, 2, 3, 4, \ldots 20$.

$x_0 = 0.2$

$x_1 = rx_0(1 - x_0) = 2 \cdot 0.2(1 - 0.2) = 0.32$

$x_2 = rx_1(1 - x_1) = 2 \cdot 0.32(1 - 0.32) = 0.4352$

$x_3 = rx_2(1 - x_2) = 2 \cdot 0.4352(1 - 0.4352) = 0.49160192$

$x_4 = rx_3(1 - x_3) = 2 \cdot 0.49160192(1 - 0.49160192) = 0.4998589445$

$x_5 = rx_4(1 - x_4) = 2 \cdot 0.4998589445(1 - 0.4998589445) = 0.4999999602$

$x_6 = rx_5(1 - x_5) = 2 \cdot 0.4999999602(1 - 0.4999999602) = 0.5000000000$

$x_7 = rx_6(1 - x_6) = 2 \cdot 0.5(1 - 0.5) = 0.5$

$x_8 = rx_7(1 - x_7) = 2 \cdot 0.5(1 - 0.5) = 0.5$

$x_9 = rx_8(1 - x_8) = 2 \cdot 0.5(1 - 0.5) = 0.5$

$x_{10} = rx_9(1 - x_9) = 2 \cdot 0.5(1 - 0.5) = 0.5$

$x_{11} = rx_{10}(1 - x_{10}) = 2 \cdot 0.5(1 - 0.5) = 0.5$

$x_{12} = rx_{11}(1 - x_{11}) = 2 \cdot 0.5(1 - 0.5) = 0.5$

$x_{13} = rx_{12}(1 - x_{12}) = 2 \cdot 0.5(1 - 0.5) = 0.5$

$x_{14} = rx_{13}(1 - x_{13}) = 2 \cdot 0.5(1 - 0.5) = 0.5$

$x_{15} = rx_{14}(1 - x_{14}) = 2 \cdot 0.5(1 - 0.5) = 0.5$

$x_{16} = rx_{15}(1 - x_{15}) = 2 \cdot 0.5(1 - 0.5) = 0.5$

$x_{17} = rx_{16}(1 - x_{16}) = 2 \cdot 0.5(1 - 0.5) = 0.5$

$x_{18} = rx_{17}(1 - x_{17}) = 2 \cdot 0.5(1 - 0.5) = 0.5$

$x_{19} = rx_{18}(1 - x_{18}) = 2 \cdot 0.5(1 - 0.5) = 0.5$

$x_{20} = rx_{19}(1 - x_{19}) = 2 \cdot 0.5(1 - 0.5) = 0.5$

The graph of x_t as a function of t is given by:

Graph of x_t as a function of t with $r = 2, x_0 = 0.2$

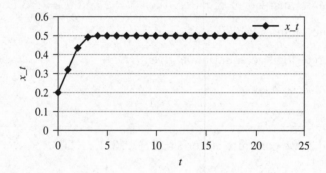

Prob. 37. The discrete logistic equation is given by

$$x_{t+1} = rx_t(1 - x_t)$$

For $r = 2$ and $x_0 = 0.1$, we compute x_t for $t = 0, 1, 2, 3, 4, \ldots 20$.

$x_0 = 0.9$

$x_1 = rx_0(1 - x_0) = 2 \cdot 0.9(1 - 0.9) = 0.18$

$x_2 = rx_1(1 - x_1) = 2 \cdot 0.18(1 - 0.18) = 0.2952$

$x_3 = rx_2(1 - x_2) = 2 \cdot 0.2952(1 - 0.2952) = 0.41611392$

$x_4 = rx_3(1 - x_3) = 2 \cdot 0.41611392(1 - 0.41611392) = 0.4859262512$

$x_5 = rx_4(1 - x_4) = 2 \cdot 0.4859262512(1 - 0.4859262512) = 0.4996038592$

$x_6 = rx_5(1 - x_5) = 2 \cdot 0.4996038592(1 - 0.4996038592) = 0.4999996861$

$x_7 = rx_6(1 - x_6) = 2 \cdot 0.4999996861(1 - 0.4999996861) = 0.5000000000$

$x_8 = rx_7(1 - x_7) = 2 \cdot 0.5(1 - 0.5) = 0.5$

$x_9 = rx_8(1 - x_8) = 2 \cdot 0.5(1 - 0.5) = 0.5$

$x_{10} = rx_9(1 - x_9) = 2 \cdot 0.5(1 - 0.5) = 0.5$

$x_{11} = rx_{10}(1 - x_{10}) = 2 \cdot 0.5(1 - 0.5) = 0.5$

$x_{12} = rx_{11}(1 - x_{11}) = 2 \cdot 0.5(1 - 0.5) = 0.5$

$x_{13} = rx_{12}(1 - x_{12}) = 2 \cdot 0.5(1 - 0.5) = 0.5$

$x_{14} = rx_{13}(1 - x_{13}) = 2 \cdot 0.5(1 - 0.5) = 0.5$

$x_{15} = rx_{14}(1 - x_{14}) = 2 \cdot 0.5(1 - 0.5) = 0.5$

$x_{16} = rx_{15}(1 - x_{15}) = 2 \cdot 0.5(1 - 0.5) = 0.5$

$x_{17} = rx_{16}(1 - x_{16}) = 2 \cdot 0.5(1 - 0.5) = 0.5$

$x_{18} = rx_{17}(1 - x_{17}) = 2 \cdot 0.5(1 - 0.5) = 0.5$

$x_{19} = rx_{18}(1 - x_{18}) = 2 \cdot 0.5(1 - 0.5) = 0.5$

$x_{20} = rx_{19}(1 - x_{19}) = 2 \cdot 0.5(1 - 0.5) = 0.5$

The graph of x_t as a function of t is given by:

Graph of x_t as a function of t with $r = 2, x_0 = 0.9$

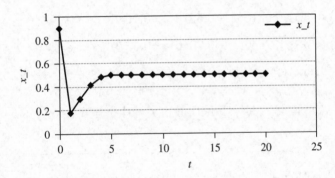

Prob. 39. The discrete logistic equation is given by

$$x_{t+1} = rx_t(1 - x_t)$$

For $r = 3.1$ and $x_0 = 0.5$, we compute x_t for $t = 0, 1, 2, 3, 4, \ldots 20$.

$x_0 = 0.5$

$x_1 = rx_0(1 - x_0) = 3.1 \cdot 0.5(1 - 0.5) = 0.775$

$x_2 = rx_1(1 - x_1) = 3.1 \cdot 0.775(1 - 0.775) \approx 0.5406$

$x_3 = rx_2(1 - x_2) \approx 3.1 \cdot 0.5406(1 - 0.5406) \approx 0.7699$

$$x_4 = rx_3(1 - x_3) \approx 3.1 \cdot 0.7699(1 - 0.7699) \approx 0.5491$$

$$x_5 = rx_4(1 - x_4) \approx 3.1 \cdot 0.5491(1 - 0.5491) \approx 0.7675$$

$$x_6 = rx_5(1 - x_5) \approx 3.1 \cdot 0.7675(1 - 0.7675) \approx 0.5531$$

$$x_7 = rx_6(1 - x_6) \approx 3.1 \cdot 0.5531(1 - 0.5531) \approx 0.7662$$

$$x_8 = rx_7(1 - x_7) \approx 3.1 \cdot 0.7662(1 - 0.7662) \approx 0.5552$$

$$x_9 = rx_8(1 - x_8) \approx 3.1 \cdot 0.5552(1 - 0.5552) \approx 0.7655$$

$$x_{10} = rx_9(1 - x_9) \approx 3.1 \cdot 0.7655(1 - 0.7655) \approx 0.5564$$

$$x_{11} = rx_{10}(1 - x_{10}) \approx 3.1 \cdot 0.5565(1 - 0.5565) \approx 0.7651$$

$$x_{12} = rx_{11}(1 - x_{11}) \approx 3.1 \cdot 0.7651(1 - 0.7651) \approx 0.5571$$

$$x_{13} = rx_{12}(1 - x_{12}) \approx 3.1 \cdot 0.5572(1 - 0.5572) \approx 0.7649$$

$$x_{14} = rx_{13}(1 - x_{13}) \approx 3.1 \cdot 0.7648(1 - 0.7648) \approx 0.5575$$

$$x_{15} = rx_{14}(1 - x_{14}) \approx 3.1 \cdot 0.5575(1 - 0.5575) \approx 0.7647$$

$$x_{16} = rx_{15}(1 - x_{15}) \approx 3.1 \cdot 0.7647(1 - 0.7647) \approx 0.5577$$

$$x_{17} = rx_{16}(1 - x_{16}) \approx 3.1 \cdot 0.5577(1 - 0.5577) \approx 0.7647$$

$$x_{18} = rx_{17}(1 - x_{17}) \approx 3.1 \cdot 0.7646(1 - 0.7646) \approx 0.5578$$

$$x_{19} = rx_{18}(1 - x_{18}) \approx 3.1 \cdot 0.5578(1 - 0.5578) \approx 0.7646$$

$$x_{20} = rx_{19}(1 - x_{19}) \approx 3.1 \cdot 0.7646(1 - 0.7646) \approx 0.5579$$

The graph of x_t as a function of t is given by:

Graph of x_t as a function of t with $r = 3.1$, $x_0 = 0.5$

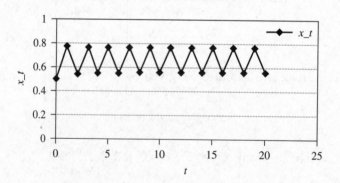

Prob. 41. The discrete logistic equation is given by

$$x_{t+1} = rx_t(1 - x_t)$$

For $r = 3.1$ and $x_0 = 0.9$, we compute x_t for $t = 0, 1, 2, 3, 4, \ldots 20$.

$x_0 = 0.9$

$x_1 = rx_0(1 - x_0) = 3.1 \cdot 0.9(1 - 0.9) = 0.279$

$x_2 = rx_1(1 - x_1) = 3.1 \cdot 0.279(1 - 0.279) \approx 0.6236$

$x_3 = rx_2(1 - x_2) \approx 3.1 \cdot 0.6236(1 - 0.6236) \approx 0.7276$

$x_4 = rx_3(1 - x_3) \approx 3.1 \cdot 0.7276(1 - 0.7276) \approx 0.6143$

$x_5 = rx_4(1 - x_4) \approx 3.1 \cdot 0.6143(1 - 0.6143) \approx 0.7345$

$x_6 = rx_5(1 - x_5) \approx 3.1 \cdot 0.7345(1 - 0.7345) \approx 0.6045$

$x_7 = rx_6(1 - x_6) \approx 3.1 \cdot 0.6045(1 - 0.6045) \approx 0.7412$

$x_8 = rx_7(1 - x_7) \approx 3.1 \cdot 0.7412(1 - 0.7412) \approx 0.5947$

$x_9 = rx_8(1 - x_8) \approx 3.1 \cdot 0.5947(1 - 0.5947) \approx 0.7472$

$x_{10} = rx_9(1 - x_9) \approx 3.1 \cdot 0.7472(1 - 0.7472) \approx 0.5856$

$x_{11} = rx_{10}(1 - x_{10}) \approx 3.1 \cdot 0.5856(1 - 0.5856) \approx 0.7522$

$x_{12} = rx_{11}(1 - x_{11}) \approx 3.1 \cdot 0.7522(1 - 0.7522) \approx 0.5778$

$x_{13} = rx_{12}(1 - x_{12}) \approx 3.1 \cdot 0.5778(1 - 0.5778) \approx 0.7562$

$x_{14} = rx_{13}(1 - x_{13}) \approx 3.1 \cdot 0.7562(1 - 0.7562) \approx 0.5714$

$x_{15} = rx_{14}(1 - x_{14}) \approx 3.1 \cdot 0.5714(1 - 0.5714) \approx 0.7592$

$x_{16} = rx_{15}(1 - x_{15}) \approx 3.1 \cdot 0.7592(1 - 0.7592) \approx 0.5668$

$x_{17} = rx_{16}(1 - x_{16}) \approx 3.1 \cdot 0.5668(1 - 0.5668) \approx 0.7612$

$x_{18} = rx_{17}(1 - x_{17}) \approx 3.1 \cdot 0.7612(1 - 0.7612) \approx 0.5634$

$x_{19} = rx_{18}(1 - x_{18}) \approx 3.1 \cdot 0.5634(1 - 0.5634) \approx 0.7625$

$x_{20} = rx_{19}(1 - x_{19}) \approx 3.1 \cdot 0.7625(1 - 0.7625) \approx 0.5614$

The graph of x_t as a function of t is given by:

Graph of x_t as a function of t with $r = 3.1$,$x_0 = 0.9$

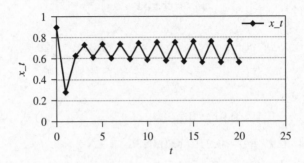

Prob. 43. The discrete logistic equation is given by

$$x_{t+1} = rx_t(1 - x_t)$$

For $r = 3.8$ and $x_0 = 0.5$, we compute x_t for $t = 0, 1, 2, 3, 4, \ldots 20$.

$x_0 = 0.5$

$x_1 = rx_0(1 - x_0) = 3.8 \cdot 0.5(1 - 0.5) = 0.95$

$x_2 = rx_1(1 - x_1) = 3.8 \cdot 0.95(1 - 0.95) = 0.1805$

$x_3 = rx_2(1 - x_2) = 3.8 \cdot 0.1805(1 - 0.1805) \approx 0.5621$

$x_4 = rx_3(1 - x_3) \approx 3.8 \cdot 0.5621(1 - 0.5621) \approx 0.9353$

$x_5 = rx_4(1 - x_4) \approx 3.8 \cdot 0.9353(1 - 0.9353) \approx 0.2298$

$x_6 = rx_5(1 - x_5) \approx 3.8 \cdot 0.2298(1 - 0.2298) \approx 0.6725$

$x_7 = rx_6(1 - x_6) \approx 3.8 \cdot 0.6725(1 - 0.6725) \approx 0.8369$

$x_8 = rx_7(1 - x_7) \approx 3.8 \cdot 0.8369(1 - 0.8369) \approx 0.5188$

$x_9 = rx_8(1 - x_8) \approx 3.8 \cdot 0.5188(1 - 0.5188) \approx 0.9486$

$x_{10} = rx_9(1 - x_9) \approx 3.8 \cdot 0.9486(1 - 0.9486) \approx 0.1851$

$x_{11} = rx_{10}(1 - x_{10}) \approx 3.8 \cdot 0.1851(1 - 0.1851) \approx 0.5732$

$x_{12} = rx_{11}(1 - x_{11}) \approx 3.8 \cdot 0.5732(1 - 0.5732) \approx 0.9297$

$x_{13} = rx_{12}(1 - x_{12}) \approx 3.8 \cdot 0.9297(1 - 0.9297) \approx 0.2485$

$x_{14} = rx_{13}(1 - x_{13}) \approx 3.8 \cdot 0.2485(1 - 0.2485) \approx 0.7097$

$x_{15} = rx_{14}(1 - x_{14}) \approx 3.8 \cdot 0.7097(1 - 0.7097) \approx 0.7829$

$x_{16} = rx_{15}(1 - x_{15}) \approx 3.8 \cdot 0.7829(1 - 0.7829) \approx 0.6458$

$x_{17} = rx_{16}(1 - x_{16}) \approx 3.8 \cdot 0.6458(1 - 0.6458) \approx 0.8693$

$x_{18} = rx_{17}(1 - x_{17}) \approx 3.8 \cdot 0.8693(1 - 0.8693) \approx 0.4319$

$x_{19} = rx_{18}(1 - x_{18}) \approx 3.8 \cdot 0.4319(1 - 0.4319) \approx 0.9324$

$x_{20} = rx_{19}(1 - x_{19}) \approx 3.8 \cdot 0.9324(1 - 0.9324) \approx 0.2396$

The Graph of x_t as a function of t is given by:

Graph of x_t as a function of t with $r = 3.8, x_0 = 0.5$

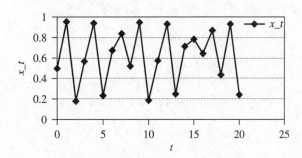

Prob. 45. The discrete logistic equation is given by

$$x_{t+1} = rx_t(1 - x_t)$$

For $r = 3.8$ and $x_0 = 0.9$, we compute x_t for $t = 0, 1, 2, 3, 4, \ldots 20$.

$x_0 = 0.9$

$x_1 = rx_0(1 - x_0) = 3.8 \cdot 0.9(1 - 0.9) = 0.342$

$x_2 = rx_1(1 - x_1) = 3.8 \cdot 0.342(1 - 0.342) \approx 0.8551$

$x_3 = rx_2(1 - x_2) \approx 3.8 \cdot 0.8551(1 - 0.8551) \approx 0.4707$

$x_4 = rx_3(1 - x_3) \approx 3.8 \cdot 0.4707(1 - 0.4707) \approx 0.9467$

$x_5 = rx_4(1 - x_4) \approx 3.8 \cdot 0.9467(1 - 0.9467) \approx 0.1916$

$x_6 = rx_5(1 - x_5) \approx 3.8 \cdot 0.1916(1 - 0.1916) \approx 0.5886$

$x_7 = rx_6(1 - x_6) \approx 3.8 \cdot 0.5886(1 - 0.5886) \approx 0.9202$

$x_8 = rx_7(1 - x_7) \approx 3.8 \cdot 0.9202(1 - 0.9202) \approx 0.2790$

$x_9 = rx_8(1 - x_8) \approx 3.8 \cdot 0.2790(1 - 0.2790) \approx 0.7645$

$x_{10} = rx_9(1 - x_9) \approx 3.8 \cdot 0.7645(1 - 0.7645) \approx 0.6842$

$x_{11} = rx_{10}(1 - x_{10}) \approx 3.8 \cdot 0.6842(1 - 0.6842) \approx 0.8211$

$$x_{12} = rx_{11}(1 - x_{11}) \approx 3.8 \cdot 0.8211(1 - 0.8211) \approx 0.5583$$

$$x_{13} = rx_{12}(1 - x_{12}) \approx 3.8 \cdot 0.5583(1 - 0.5583) \approx 0.9371$$

$$x_{14} = rx_{13}(1 - x_{13}) \approx 3.8 \cdot 0.9371(1 - 0.9371) \approx 0.2240$$

$$x_{15} = rx_{14}(1 - x_{14}) \approx 3.8 \cdot 0.2240(1 - 0.2240) \approx 0.6606$$

$$x_{16} = rx_{15}(1 - x_{15}) \approx 3.8 \cdot 0.6606(1 - 0.6606) \approx 0.8519$$

$$x_{17} = rx_{16}(1 - x_{16}) \approx 3.8 \cdot 0.8519(1 - 0.8519) \approx 0.4793$$

$$x_{18} = rx_{17}(1 - x_{17}) \approx 3.8 \cdot 0.4793(1 - 0.4793) \approx 0.9484$$

$$x_{19} = rx_{18}(1 - x_{18}) \approx 3.8 \cdot 0.9484(1 - 0.9484) \approx 0.1861$$

$$x_{20} = rx_{19}(1 - x_{19}) \approx 3.8 \cdot 0.1861(1 - 0.1861) \approx 0.5755$$

The graph of x_t as a function of t is given by:

Graph of x_t as a function of t with $r = 3.8$, $x_0 = 0.9$

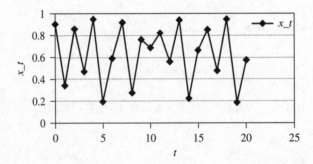

Prob. 47. The graph for the Ricker's curve in the N_t-N_{t+1} plane is given below.

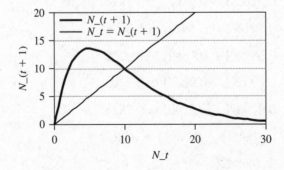

Graph of N_{t+1} as a function of N_t

The intersection of this graph with the line $N_{t+1} = N_t$ is at $N_t = 10$.

Prob. 49. The graph for the Ricker's curve in the N_t-N_{t+1} plane is given below.

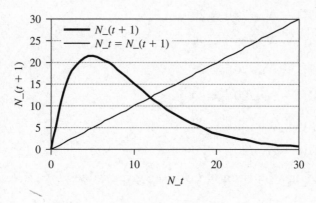

Graph of N_{t+1} as a function of N_t

The intersection of this graph with the line $N_{t+1} = N_t$ is at $N_t = 12$.

Prob. 51.

(a) We know that Ricker's curve is given by

$$N_{t+1} = N_t \exp\left[R\left(1 - \frac{N_t}{K}\right)\right]$$

When $R = 1$, $K = 20$ and $N_0 = 5$, we compute N_t for $t = 1, 2, 3, 4, \ldots 20$.

$N_1 = N_0 \exp[1(1 - N_0/20)] = 10.585$

$N_2 = N_1 \exp[1(1 - N_1/20)] = 16.94865$

$N_3 = N_2 \exp[1(1 - N_2/20)] = 19.74214$

$N_4 = N_3 \exp[1(1 - N_3/20)] = 19.99832$

$N_5 = N_4 \exp[1(1 - N_4/20)] = 20$

$N_6 = N_5 \exp[1(1 - N_5/20)] = 20$

$N_7 = N_6 \exp[1(1 - N_6/20)] = 20$

$N_8 = N_7 \exp[1(1 - N_7/20)] = 20$

$N_9 = N_8 \exp[1(1 - N_8/20)] = 20$

$N_{10} = N_9 \exp[1(1 - N_9/20)] = 20$

$$N_{11} = N_{10} \exp[1(1 - N_{10}/20)] = 20$$
$$N_{12} = N_{11} \exp[1(1 - N_{11}/20)] = 20$$
$$N_{13} = N_{12} \exp[1(1 - N_{12}/20)] = 20$$
$$N_{14} = N_{13} \exp[1(1 - N_{13}/20)] = 20$$
$$N_{15} = N_{14} \exp[1(1 - N_{14}/20)] = 20$$
$$N_{16} = N_{15} \exp[1(1 - N_{15}/20)] = 20$$
$$N_{17} = N_{16} \exp[1(1 - N_{16}/20)] = 20$$
$$N_{18} = N_{17} \exp[1(1 - N_{17}/20)] = 20$$
$$N_{19} = N_{18} \exp[1(1 - N_{18}/20)] = 20$$
$$N_{20} = N_{19} \exp[1(1 - N_{19}/20)] = 20$$

The graph of N_t as a function of t, for the given values $R = 1$, $K = 20$ and $N_0 = 5$ is given below.

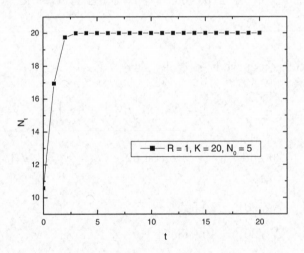

Graph of N_{t+1} as a function of t

(b) We know that Ricker's curve is given by

$$N_{t+1} = N_t \exp \left[R \left(1 - \frac{N_t}{K} \right) \right]$$

When $R = 1$, $K = 20$ and $N_0 = 10$, we compute N_t for $t = 1, 2, 3, 4, \ldots 20$.

$$N_1 = N_0 \exp[1(1 - N_0/20)] = 16.48721$$

$$N_2 = N_1 \exp[1(1 - N_1/20)] = 19.653$$

$$N_3 = N_2 \exp[1(1 - N_2/20)] = 19.99695$$

$$N_4 = N_3 \exp[1(1 - N_3/20)] = 20$$

$$N_5 = N_4 \exp[1(1 - N_4/20)] = 20$$

$$N_6 = N_5 \exp[1(1 - N_5/20)] = 20$$

$$N_7 = N_6 \exp[1(1 - N_6/20)] = 20$$

$$N_8 = N_7 \exp[1(1 - N_7/20)] = 20$$

$$N_9 = N_8 \exp[1(1 - N_8/20)] = 20$$

$$N_{10} = N_9 \exp[1(1 - N_9/20)] = 20$$

$$N_{11} = N_{10} \exp[1(1 - N_{10}/20)] = 20$$

$$N_{12} = N_{11} \exp[1(1 - N_{11}/20)] = 20$$

$$N_{13} = N_{12} \exp[1(1 - N_{12}/20)] = 20$$

$$N_{14} = N_{13} \exp[1(1 - N_{13}/20)] = 20$$

$$N_{15} = N_{14} \exp[1(1 - N_{14}/20)] = 20$$

$$N_{16} = N_{15} \exp[1(1 - N_{15}/20)] = 20$$

$$N_{17} = N_{16} \exp[1(1 - N_{16}/20)] = 20$$

$$N_{18} = N_{17} \exp[1(1 - N_{17}/20)] = 20$$

$$N_{19} = N_{18} \exp[1(1 - N_{18}/20)] = 20$$

$$N_{20} = N_{19} \exp[1(1 - N_{19}/20)] = 20$$

The graph of N_t as a function of t, for the given values $R = 1$, $K = 20$ and $N_0 = 10$ is given below.

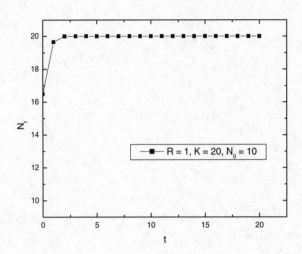

Graph of N_{t+1} as a function of t

(c) We know that Ricker's curve is given by

$$N_{t+1} = N_t \exp\left[R\left(1 - \frac{N_t}{K}\right)\right]$$

When $R = 1$, $K = 20$ and $N_0 = 20$, we compute N_t for $t = 1, 2, 3, 4, \ldots 20$.

$$N_1 = N_0 \exp[1(1 - N_0/20)] = 20$$
$$N_2 = N_1 \exp[1(1 - N_1/20)] = 20$$
$$N_3 = N_2 \exp[1(1 - N_2/20)] = 20$$
$$N_4 = N_3 \exp[1(1 - N_3/20)] = 20$$
$$N_5 = N_0 \exp[1(1 - N_4/20)] = 20$$
$$N_6 = N_5 \exp[1(1 - N_5/20)] = 20$$
$$N_7 = N_6 \exp[1(1 - N_6/20)] = 20$$
$$N_8 = N_7 \exp[1(1 - N_7/20)] = 20$$
$$N_9 = N_8 \exp[1(1 - N_8/20)] = 20$$
$$N_{10} = N_9 \exp[1(1 - N_9/20)] = 20$$
$$N_{11} = N_{10} \exp[1(1 - N_{10}/20)] = 20$$
$$N_{12} = N_{11} \exp[1(1 - N_{11}/20)] = 20$$

$N_{13} = N_{12} \exp[1(1 - N_{12}/20)] = 20$

$N_{14} = N_{13} \exp[1(1 - N_{13}/20)] = 20$

$N_{15} = N_{14} \exp[1(1 - N_{14}/20)] = 20$

$N_{16} = N_{15} \exp[1(1 - N_{15}/20)] = 20$

$N_{17} = N_{16} \exp[1(1 - N_{16}/20)] = 20$

$N_{18} = N_{17} \exp[1(1 - N_{17}/20)] = 20$

$N_{19} = N_{18} \exp[1(1 - N_{18}/20)] = 20$

$N_{20} = N_{19} \exp[1(1 - N_{19}/20)] = 20$

The graph of N_t as a function of t, for the given values $R = 1$, $K = 20$ and $N_0 = 20$ is given below.

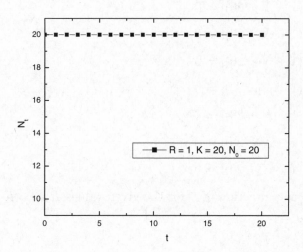

Graph of N_{t+1} as a function of t

(d) We know that Ricker's curve is given by

$$N_{t+1} = N_t \exp\left[R\left(1 - \frac{N_t}{K}\right)\right]$$

When $R = 1$, $K = 20$ and $N_0 = 0$, we compute N_t for $t = 1, 2, 3, 4, \ldots 20$.

$N_1 = N_0 \exp[1(1 - N_0/20)] = 0$

$N_2 = N_1 \exp[1(1 - N_1/20)] = 0$

$$N_3 = N_2 \exp[1(1 - N_2/20)] = 0$$

$$N_4 = N_3 \exp[1(1 - N_3/20)] = 0$$

$$N_5 = N_0 \exp[1(1 - N_4/20)] = 0$$

$$N_6 = N_5 \exp[1(1 - N_5/20)] = 0$$

$$N_7 = N_6 \exp[1(1 - N_6/20)] = 0$$

$$N_8 = N_7 \exp[1(1 - N_7/20)] = 0$$

$$N_9 = N_8 \exp[1(1 - N_8/20)] = 0$$

$$N_{10} = N_9 \exp[1(1 - N_9/20)] = 0$$

$$N_{11} = N_{10} \exp[1(1 - N_{10}/20)] = 0$$

$$N_{12} = N_{11} \exp[1(1 - N_{11}/20)] = 0$$

$$N_{13} = N_{12} \exp[1(1 - N_{12}/20)] = 0$$

$$N_{14} = N_{13} \exp[1(1 - N_{13}/20)] = 0$$

$$N_{15} = N_{14} \exp[1(1 - N_{14}/20)] = 0$$

$$N_{16} = N_{15} \exp[1(1 - N_{15}/20)] = 0$$

$$N_{17} = N_{16} \exp[1(1 - N_{16}/20)] = 0$$

$$N_{18} = N_{17} \exp[1(1 - N_{17}/20)] = 0$$

$$N_{19} = N_{18} \exp[1(1 - N_{18}/20)] = 0$$

$$N_{20} = N_{19} \exp[1(1 - N_{19}/20)] = 0$$

The graph of N_t as a function of t, for the given values $R = 1$, $K = 20$ and $N_0 = 0$ is given below.

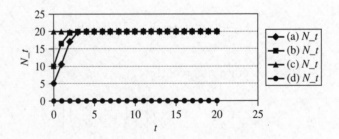

Graph of N_{t+1} as a function of t

Prob. 53.

(a) We know that Ricker's curve is given by

$$N_{t+1} = N_t \exp\left[R\left(1 - \frac{N_t}{K}\right)\right]$$

When $R = 2.1$, $K = 20$ and $N_0 = 5$, we compute N_t for $t = 1, 2, 3, 4, \ldots 20$.

$N_1 = N_0 \exp[2.1(1 - N_0/20)] = 24.15371$

$N_2 = N_1 \exp[2.1(1 - N_1/20)] = 15.61604$

$N_3 = N_2 \exp[2.1(1 - N_2/20)] = 24.74478$

$N_4 = N_3 \exp[2.1(1 - N_3/20)] = 15.03548$

$N_5 = N_4 \exp[2.1(1 - N_4/20)] = 25.32235$

$N_6 = N_5 \exp[2.1(1 - N_5/20)] = 14.48105$

$N_7 = N_6 \exp[2.1(1 - N_6/20)] = 25.85052$

$N_8 = N_7 \exp[2.1(1 - N_7/20)] = 13.98557$

$N_9 = N_8 \exp[2.1(1 - N_8/20)] = 26.29927$

$N_{10} = N_9 \exp[2.1(1 - N_9/20)] = 13.57348$

$N_{11} = N_{10} \exp[2.1(1 - N_{10}/20)] = 26.65302$

$N_{12} = N_{11} \exp[2.1(1 - N_{11}/20)] = 13.25448$

$N_{13} = N_{12} \exp[2.1(1 - N_{12}/20)] = 26.91316$

$N_{14} = N_{13} \exp[2.1(1 - N_{13}/20)] = 13.02322$

$N_{15} = N_{14} \exp[2.1(1 - N_{14}/20)] = 27.09396$

$N_{16} = N_{15} \exp[2.1(1 - N_{15}/20)] = 12.86451$

$N_{17} = N_{16} \exp[2.1(1 - N_{16}/20)] = 27.21311$

$N_{18} = N_{17} \exp[2.1(1 - N_{17}/20)] = 12.76009$

$N_{19} = N_{18} \exp[2.1(1 - N_{18}/20)] = 27.28980$

$N_{20} = N_{19} \exp[2.1(1 - N_{19}/20)] = 12.69343$

The graph of N_t as a function of t, for the given values $R = 2.1$, $K = 20$ and $N_0 = 5$ is given below.

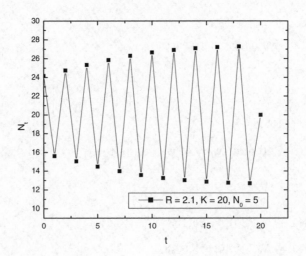

Graph of N_{t+1} as a function of t

(b) We know that Ricker's curve is given by

$$N_{t+1} = N_t \exp\left[R\left(1 - \frac{N_t}{K}\right)\right]$$

When $R = 2.1$, $K = 20$ and $N_0 = 10$, we compute N_t for $t = 1, 2, 3, 4, \ldots 20$.

$N_1 = N_0 \exp[2.1(1 - N_0/20)] = 28.57651$

$N_2 = N_1 \exp[2.1(1 - N_1/20)] = 11.61214$

$N_3 = N_2 \exp[2.1(1 - N_2/20)] = 28.01602$

$N_4 = N_3 \exp[2.1(1 - N_3/20)] = 12.07449$

$N_5 = N_4 \exp[2.1(1 - N_4/20)] = 27.75105$

$N_6 = N_5 \exp[2.1(1 - N_5/20)] = 12.29772$

$N_7 = N_6 \exp[2.1(1 - N_6/20)] = 27.60932$

$N_8 = N_7 \exp[2.1(1 - N_7/20)] = 12.41835$

$N_9 = N_8 \exp[2.1(1 - N_8/20)] = 27.52924$

$N_{10} = N_9 \exp[2.1(1 - N_9/20)] = 12.48688$

$N_{11} = N_{10} \exp[2.1(1 - N_{10}/20)] = 27.48268$

$N_{12} = N_{11} \exp[2.1(1 - N_{11}/20)] = 12.52685$

$N_{13} = N_{12} \exp[2.1(1 - N_{12}/20)] = 27.45518$

$N_{14} = N_{13} \exp[2.1(1 - N_{13}/20)] = 12.5505$

$N_{15} = N_{14} \exp[2.1(1 - N_{14}/20)] = 27.4388$

$N_{16} = N_{15} \exp[2.1(1 - N_{15}/20)] = 12.56461$

$N_{17} = N_{16} \exp[2.1(1 - N_{16}/20)] = 27.42898$

$N_{18} = N_{17} \exp[2.1(1 - N_{17}/20)] = 12.57307$

$N_{19} = N_{18} \exp[2.1(1 - N_{18}/20)] = 27.42308$

$N_{20} = N_{19} \exp[2.1(1 - N_{19}/20)] = 12.57816$

The graph of N_t as a function of t, for the given values $R = 2.1$, $K = 20$ and $N_0 = 10$ is given below.

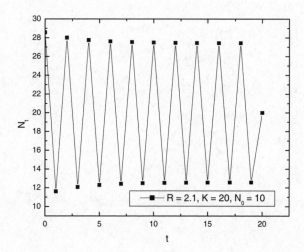

Graph of N_{t+1} as a function of t

(c) We know that Ricker's curve is given by

$$N_{t+1} = N_t \exp\left[R\left(1 - \frac{N_t}{K}\right)\right]$$

When $R = 2.1$, $K = 20$ and $N_0 = 20$, we compute N_t for $t = 1, 2, 3, 4, \ldots 20$.

$N_1 = N_0 \exp[2.1(1 - N_0/20)] = 20$

$N_2 = N_1 \exp[2.1(1 - N_1/20)] = 20$

$N_3 = N_2 \exp[2.1(1 - N_2/20)] = 20$

$$N_4 = N_3 \exp[2.1(1 - N_3/20)] = 20$$

$$N_5 = N_4 \exp[2.1(1 - N_4/20)] = 20$$

$$N_6 = N_5 \exp[2.1(1 - N_5/20)] = 20$$

$$N_7 = N_6 \exp[2.1(1 - N_6/20)] = 20$$

$$N_8 = N_7 \exp[2.1(1 - N_7/20)] = 20$$

$$N_9 = N_8 \exp[2.1(1 - N_8/20)] = 20$$

$$N_{10} = N_9 \exp[2.1(1 - N_9/20)] = 20$$

$$N_{11} = N_{10} \exp[2.1(1 - N_{10}/20)] = 20$$

$$N_{12} = N_{11} \exp[2.1(1 - N_{11}/20)] = 20$$

$$N_{13} = N_{12} \exp[2.1(1 - N_{12}/20)] = 20$$

$$N_{14} = N_{13} \exp[2.1(1 - N_{13}/20)] = 20$$

$$N_{15} = N_{14} \exp[2.1(1 - N_{14}/20)] = 20$$

$$N_{16} = N_{15} \exp[2.1(1 - N_{15}/20)] = 20$$

$$N_{17} = N_{16} \exp[2.1(1 - N_{16}/20)] = 20$$

$$N_{18} = N_{17} \exp[2.1(1 - N_{17}/20)] = 20$$

$$N_{19} = N_{18} \exp[2.1(1 - N_{18}/20)] = 20$$

$$N_{20} = N_{19} \exp[2.1(1 - N_{19}/20)] = 20$$

The graph of N_t as a function of t, for the given values $R = 2.1$, $K = 20$ and $N_0 = 20$ is given below.

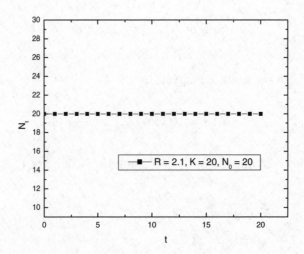

Graph of N_{t+1} as a function of t

(d) We know that Ricker's curve is given by

$$N_{t+1} = N_t \exp\left[R\left(1 - \frac{N_t}{K}\right)\right]$$

When $R = 2.1$, $K = 20$ and $N_0 = 0$, we compute N_t for $t = 1, 2, 3, 4, \ldots 20$.

$N_1 = N_0 \exp[2.1(1 - N_0/20)] = 0$

$N_2 = N_1 \exp[2.1(1 - N_1/20)] = 0$

$N_3 = N_2 \exp[2.1(1 - N_2/20)] = 0$

$N_4 = N_3 \exp[2.1(1 - N_3/20)] = 0$

$N_5 = N_4 \exp[2.1(1 - N_4/20)] = 0$

$N_6 = N_5 \exp[2.1(1 - N_5/20)] = 0$

$N_7 = N_6 \exp[2.1(1 - N_6/20)] = 0$

$N_8 = N_7 \exp[2.1(1 - N_7/20)] = 0$

$N_9 = N_8 \exp[2.1(1 - N_8/20)] = 0$

$N_{10} = N_9 \exp[2.1(1 - N_9/20)] = 0$

$N_{11} = N_{10} \exp[2.1(1 - N_{10}/20)] = 0$

$N_{12} = N_{11} \exp[2.1(1 - N_{11}/20)] = 0$

$N_{13} = N_{12} \exp[2.1(1 - N_{12}/20)] = 0$

$N_{14} = N_{13} \exp[2.1(1 - N_{13}/20)] = 0$

$N_{15} = N_{14} \exp[2.1(1 - N_{14}/20)] = 0$

$N_{16} = N_{15} \exp[2.1(1 - N_{15}/20)] = 0$

$N_{17} = N_{16} \exp[2.1(1 - N_{16}/20)] = 0$

$N_{18} = N_{17} \exp[2.1(1 - N_{17}/20)] = 0$

$N_{19} = N_{18} \exp[2.1(1 - N_{18}/20)] = 0$

$N_{20} = N_{19} \exp[2.1(1 - N_{19}/20)] = 0$

The graph of N_t as a function of t, for the given values $R = 2.1$, $K = 20$ and $N_0 = 0$ is given below.

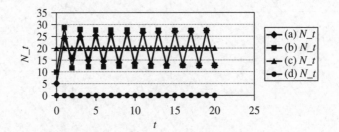

Graph of N_{t+1} as a function of t

Prob. 55. When $N_0 = 1$ and $N_1 = 1$, we have

$$t = 1 \quad \Rightarrow \quad N_2 = N_1 + N_0 = 1 + 1 = 2, \frac{N_t}{N_{t-1}} = \frac{N_1}{N_0} = 1$$

$$t = 2 \quad \Rightarrow \quad N_3 = N_2 + N_1 = 2 + 1 = 3, \frac{N_t}{N_{t-1}} = \frac{N_2}{N_1} = 2$$

$$t = 3 \quad \Rightarrow \quad N_4 = N_3 + N_2 = 3 + 2 = 5, \frac{N_t}{N_{t-1}} = \frac{N_3}{N_2} = \frac{3}{2} = 1.5$$

$$t = 4 \quad \Rightarrow \quad N_5 = N_4 + N_3 = 5 + 3 = 8, \frac{N_t}{N_{t-1}} = \frac{N_5}{N_3} = \frac{5}{3} \approx 1.66667$$

$$t = 5 \quad \Rightarrow \quad N_6 = N_5 + N_4 = 8 + 5 = 13, \frac{N_t}{N_{t-1}} = \frac{N_5}{N_4} = \frac{8}{5} \approx 1.60000$$

$$t = 6 \quad \Rightarrow \quad N_7 = N_6 + N_5 = 13 + 8 = 21, \frac{N_t}{N_{t-1}} = \frac{N_6}{N_5} = \frac{13}{8} \approx 1.62500$$

$$t = 7 \quad \Rightarrow \quad N_8 = N_7 + N_6 = 21 + 13 = 34, \frac{N_t}{N_{t-1}} = \frac{N_7}{N_6} = \frac{21}{13} \approx 1.61538$$

$$t = 8 \quad \Rightarrow \quad N_9 = N_8 + N_7 = 34 + 21 = 55, \frac{N_t}{N_{t-1}} = \frac{N_8}{N_7} = \frac{34}{21} \approx 1.61905$$

$$t = 9 \quad \Rightarrow \quad N_{10} = N_9 + N_8 = 55 + 34 = 89, \frac{N_t}{N_{t-1}} = \frac{N_9}{N_8} = \frac{55}{34} \approx 1.61765$$

$$t = 10 \quad \Rightarrow \quad N_{11} = N_{10} + N_9 = 89 + 55 = 144, \frac{N_t}{N_{t-1}} = \frac{N_{10}}{N_9} = \frac{89}{55} \approx 1.61818$$

$$t = 11 \quad \Rightarrow \quad N_{12} = N_{11} + N_{10} = 144 + 89 = 233, \frac{N_t}{N_{t-1}} = \frac{N_{11}}{N_{10}} = \frac{144}{89} \approx 1.61798$$

$$t = 12 \quad \Rightarrow \quad N_{13} = N_{12} + N_{11} = 233 + 144 = 377, \frac{N_t}{N_{t-1}} = \frac{N_{12}}{N_{11}} = \frac{233}{144} \approx 1.61806$$

$$t = 13 \quad \Rightarrow \quad N_{14} = N_{13} + N_{12} = 377 + 233 = 610, \frac{N_t}{N_{t-1}} = \frac{N_{13}}{N_{12}} = \frac{377}{233} \approx 1.61803$$

$$t = 14 \quad \Rightarrow \quad N_{15} = N_{14} + N_{13} = 610 + 377 = 987, \frac{N_t}{N_{t-1}} = \frac{N_{14}}{N_{13}} = \frac{610}{377} \approx 1.61804$$

$$t = 15 \quad \Rightarrow \quad N_{16} = N_{15} + N_{14} = 987 + 610 = 1597, \frac{N_t}{N_{t-1}} = \frac{N_{15}}{N_{14}} = \frac{987}{610} \approx 1.61803$$

$$t = 16 \quad \Rightarrow \quad N_{17} = N_{16} + N_{15} = 1597 + 987 = 2584, \frac{N_t}{N_{t-1}} = \frac{N_{16}}{N_{15}} = \frac{1597}{987} \approx 1.61803$$

$$t = 17 \quad \Rightarrow \quad N_{18} = N_{17} + N_{16} = 2584 + 1597 = 4181, \frac{N_t}{N_{t-1}} = \frac{N_{17}}{N_{16}} = \frac{2584}{1597} \approx 1.61803$$

$$t = 18 \quad \Rightarrow \quad N_{19} = N_{18} + N_{17} = 4181 + 2584 = 6765, \frac{N_t}{N_{t-1}} = \frac{N_{18}}{N_{17}} = \frac{4181}{2584} \approx 1.61803$$

$$t = 19 \quad \Rightarrow \quad N_{20} = N_{19} + N_{18} = 6765 + 4181 = 10946, \frac{N_t}{N_{t-1}} = \frac{N_{19}}{N_{18}} = \frac{6765}{4181} \approx 1.61803$$

And finally, we have $N_{20}/N_{19} = 10946/6765 \approx 1.61803$.

Prob. 57. This recursion describes the number of pairs of newborn rabbits at t months, if we assume that each pair produces one pair of rabbits at age 1 month and two pairs of rabbits at age 2 months, and none thereafter. We initially have one pair of newborn rabbits.

If N_t denotes the number of newborn rabbit pairs at time t (measured in months), then at time 0, there is one pair of rabbits ($N_0 = 1$). When $t = 1$, the pair of rabbits we started with is one month old and produces a pair of newborn rabbits, so $N_1 = 1$. When $t = 2$, there is one pair of two-month-old rabbits and one pair of one-month-old rabbits. The pair of two-month-old rabbits will produce two pairs of rabbits, and the pair of one-month-old rabbits will produce 1 pair of rabbits; thus $N_2 = 3$. When $t = 3$, our original pair of rabbits is now three months old and will stop reproducing; there is then one pair of two-month-old rabbits and three pairs of one-month-old rabbits. Since each pair of one-month-old rabbits produces a pair of newborn rabbits and the two-month-old rabbits produce two pairs of newborn rabbits, there will be $3 + 2 = 5$ newborn rabbits at time $t = 3$.

More generally, to find the number of pairs of newborn rabbits, we need to add up the number of pairs of one-month-old rabbits and 2 times the number of pairs of two-month-old rabbits. The one-month-old rabbits at time $t + 1$ were newborn rabbits at time t; the two-month-old rabbits were newborns at time $t - 1$. So the number of pairs of newborn rabbits at time $t + 1$ is

$$N_{t+1} = N_t + 2N_{t-1}, \quad \text{for } t = 1, 2, 3, 4, \ldots, \text{ and } N_0 = 1, N_1 = 1.$$

2.5 Review Problems

Prob. 1. The expression 2^{-n} can also be written as $(1/2)^n$. Thus, from Example 12 in Section 2.2.2, we conclude that

$$\lim_{n \to \infty} 2^{-n} = \lim_{n \to \infty} \left(\frac{1}{2}\right)^n = 0$$

Prob. 3. The expression $40(1 - 4^{-n})$ can be written as $40 - 40(1/4)^n$. From Example 12 in Section 2.2.2, we know that $\lim_{n \to \infty} (1/4)^n = 0$. Also, it is obvious that $\lim_{n \to \infty} 40 = 40$. Thus

$$\lim_{n \to \infty} 40(1 - 4^{-n}) = \lim_{n \to \infty} 40 - 40 \lim_{n \to \infty} (1/4)^n = 40 - 40 \cdot 0 = 40.$$

Prob. 5. Since $a > 1$, we can see that $a^2 > a > 1$, and the terms $a^3, a^4, a^5, a^6, \ldots$ are successively larger. This indicates that the terms continue to grow with n. Thus a^n goes to infinity as $n \to \infty$, and we can write $\lim_{n \to \infty} a^n = \infty$. Since infinity is not a real number, we say that the limit does not exist.

Prob. 7. We first note that the expression $\frac{n(n+1)}{n^2-1}$ can be written as $\frac{n(n+1)}{(n+1)(n-1)}$. Recalling that $\lim_{n\to\infty}(1/n)$ exists and is equal to 0, we have

$$\lim_{n\to\infty}\frac{n(n+1)}{n^2-1} = \lim_{n\to\infty}\frac{n(n+1)}{(n-1)(n+1)} = \lim_{n\to\infty}\frac{n}{n-1}$$

$$= \lim_{n\to\infty}\frac{1}{1-1/n} = \frac{\lim_{n\to\infty}1}{\lim_{n\to\infty}1-\lim_{n\to\infty}(1/n)} = \frac{1}{1-0} = 1.$$

Prob. 9. We first note that by dividing numerator and denominator by n, the expression $\frac{\sqrt{n}}{n+1}$ can be written as $\frac{1/\sqrt{n}}{1+1/n}$. Recalling that both $\lim_{n\to\infty}(1/n)$ and $\lim_{n\to\infty}(1/\sqrt{n})$ exist and are equal to 0, we have

$$\lim_{n\to\infty}\frac{\sqrt{n}}{n+1} = \lim_{n\to\infty}\frac{1/\sqrt{n}}{1+1/n}$$

$$= \frac{\lim_{n\to\infty}(1/\sqrt{n})}{\lim_{n\to\infty}(1+1/n)} = \frac{0}{1+0} = 0.$$

Prob. 11. Looking at the sequence, we can guess the next terms, namely, $\frac{11}{12}, \frac{13}{14}, \frac{15}{16}, \frac{17}{18}, \frac{19}{20}$ and so on. We thus find

$$a_n = \frac{2n+1}{2n+2} \quad \text{for } n = 0, 1, 2, 3, \ldots$$

Prob. 13. Looking at the sequence, we can guess the next terms, namely, $\frac{6}{37}, \frac{7}{50}, \frac{8}{65}, \frac{9}{82}, \frac{10}{101}$ and so on. The denominator of a_n is the sum of 2 and the first n odd numbers, beginning with the odd number 3. This sum is $2 + (n+1)^2 - 1 = (n+1)^2 + 2 = n^2 + 2n + 2$, and we thus have

$$a_n = \frac{n+1}{n^2+2n+2} \quad \text{for } n = 0, 1, 2, 3, \ldots$$

Prob. 15.

(a) The population growth equation described by the Beverton-Holt recruitment curve with growth parameter R and carrying capacity K is given by:

$$N_{t+1} = \frac{RN_t}{1 + \frac{R-1}{K}N_t}$$

The population sizes for $t = 1, 2, 3, 4 \ldots, 10$ and $\lim -t \to \infty N_t$ for the given values $R = 2$, $K = 100$ and $N_0 = 20$ are:

$$N_1 = \frac{2N_0}{1 + \frac{2-1}{100}N_0} = \frac{2\cdot 20}{1 + \frac{1}{100}20} = \frac{40}{6/5} \approx 33.3333$$

$$N_2 = \frac{2N_1}{1 + \frac{2-1}{100}N_1} = \frac{2 \cdot \frac{100}{3}}{1 + \frac{1}{100}\frac{100}{3}} = \frac{200/3}{4/3} = 50$$

$$N_3 = \frac{2N_2}{1 + \frac{2-1}{100}N_2} = \frac{2 \cdot 50}{1 + \frac{1}{100}50} = \frac{100}{3/2} \approx 66.6667$$

$$N_4 = \frac{2N_3}{1 + \frac{2-1}{100}N_3} = \frac{2 \cdot \frac{200}{3}}{1 + \frac{1}{100}\frac{200}{3}} = \frac{400/3}{5/3} = 80$$

$$N_5 = \frac{2N_4}{1 + \frac{2-1}{100}N_4} = \frac{2 \cdot 80}{1 + \frac{1}{100}80} = \frac{160}{9/5} \approx 88.8889$$

$$N_6 = \frac{2N_5}{1 + \frac{2-1}{100}N_5} = \frac{2 \cdot \frac{800}{9}}{1 + \frac{1}{100}\frac{800}{9}} = \frac{1600/9}{17/9} \approx 94.1176$$

$$N_7 = \frac{2N_6}{1 + \frac{2-1}{100}N_6} = \frac{2 \cdot \frac{1600}{17}}{1 + \frac{1}{100}\frac{1600}{17}} = \frac{3200/17}{33/17} \approx 96.9697$$

$$N_8 = \frac{2N_7}{1 + \frac{2-1}{100}N_7} = \frac{2 \cdot \frac{3200}{33}}{1 + \frac{1}{100}\frac{3200}{33}} = \frac{6400/33}{65/33} \approx 98.4615$$

$$N_9 = \frac{2N_8}{1 + \frac{2-1}{100}N_8} = \frac{2 \cdot \frac{1280}{13}}{1 + \frac{1}{100}\frac{1280}{13}} = \frac{2560/13}{258/130} \approx 99.2248$$

$$N_{10} = \frac{2N_9}{1 + \frac{2-1}{100}N_9} = \frac{2 \cdot \frac{12800}{129}}{1 + \frac{1}{100}\frac{12800}{129}} = \frac{25600/129}{257/129} \approx 99.6109$$

The graph for this function is:

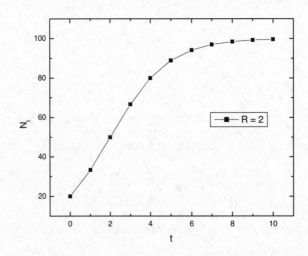

Graph of N_t as a function of t

(b) The population growth equation described by the Beverton-Holt recruitment curve with growth parameter R and carrying capacity K is given by:

$$N_{t+1} = \frac{RN_t}{1 + \frac{R-1}{K}N_t}$$

The population sizes for $t = 1, 2, 3, 4 \ldots, 10$ and $\lim -t \to \infty N_t$ for the given values $R = 5$, $K = 100$ and $N_0 = 20$ are:

$$N_1 = \frac{5N_0}{1 + \frac{5-1}{100}N_0} = \frac{5 \cdot 20}{1 + \frac{4}{100}20} = \frac{100}{9/5} \approx 55.5556$$

$$N_2 = \frac{5N_1}{1 + \frac{5-1}{100}N_1} = \frac{5 \cdot \frac{500}{9}}{1 + \frac{4}{100}\frac{500}{9}} = \frac{2500/9}{29/9} \approx 86.2069$$

$$N_3 = \frac{5N_2}{1 + \frac{5-1}{100}N_2} = \frac{5 \cdot \frac{2500}{29}}{1 + \frac{4}{100}\frac{2500}{29}} = \frac{12500/29}{129/29} \approx 96.8992$$

$$N_4 = \frac{5N_3}{1 + \frac{5-1}{100}N_3} = \frac{5 \cdot \frac{12500}{129}}{1 + \frac{4}{100}\frac{12500}{129}} = \frac{62500/129}{629/129} \approx 99.3641$$

$$N_5 = \frac{5N_4}{1 + \frac{5-1}{100}N_4} = \frac{5 \cdot \frac{62500}{629}}{1 + \frac{4}{100}\frac{62500}{629}} = \frac{312500/629}{3129/629} \approx 99.8722$$

$$N_6 = \frac{5N_5}{1 + \frac{5-1}{100}N_5} = \frac{5 \cdot \frac{312500}{3129}}{1 + \frac{4}{100}\frac{312500}{3129}} = \frac{1562500/3129}{15629/3129} \approx 99.9744$$

$$N_7 = \frac{5N_6}{1 + \frac{5-1}{100}N_6} = \frac{5 \cdot \frac{1562500}{15629}}{1 + \frac{4}{100}\frac{1562500}{15629}} = \frac{7812500/15629}{78129/15629} \approx 99.9949$$

$$N_8 = \frac{5N_7}{1 + \frac{5-1}{100}N_7} = \frac{5 \cdot \frac{7812500}{78129}}{1 + \frac{4}{100}\frac{7812500}{78129}} = \frac{39062500/78129}{390629/78129} \approx 99.9990$$

$$N_9 = \frac{5N_8}{1 + \frac{5-1}{100}N_8} = \frac{5 \cdot \frac{39062500}{390629}}{1 + \frac{4}{100}\frac{39062500}{390629}} = \frac{195312500/390629}{1953129/390629} \approx 99.9998$$

$$N_{10} = \frac{5N_9}{1 + \frac{5-1}{100}N_9} = \frac{5 \cdot \frac{195312500}{1953129}}{1 + \frac{4}{100}\frac{195312500}{1953129}} = \frac{976562500/1953129}{9765629/1953129} \approx 100.000$$

The graph for this function is:

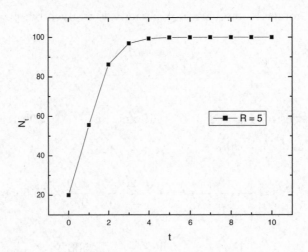

Graph of N_t as a function of t

(c) The population growth equation described by the Beverton-Holt recruitment curve with growth parameter R and carrying capacity K is given by:

$$N_{t+1} = \frac{RN_t}{1 + \frac{R-1}{K}N_t}$$

The population sizes for $t = 1, 2, 3, 4 \ldots, 10$ and $\lim -t \to \infty N_t$ for the given values $R = 10$, $K = 100$ and $N_0 = 20$ are:

$$N_1 = \frac{10N_0}{1 + \frac{10-1}{100}N_0} = \frac{10 \cdot 20}{1 + \frac{9}{100}20} = \frac{200}{14/5} \approx 71.4286$$

$$N_2 = \frac{10N_1}{1 + \frac{10-1}{100}N_1} = \frac{10 \cdot \frac{500}{7}}{1 + \frac{9}{100}\frac{500}{7}} = \frac{5000/7}{52/7} \approx 96.1538$$

$$N_3 = \frac{10N_2}{1 + \frac{10-1}{100}N_2} = \frac{10 \cdot \frac{5000}{52}}{1 + \frac{9}{100}\frac{5000}{52}} = \frac{50000/52}{502/52} \approx 99.6016$$

$$N_4 = \frac{10N_3}{1 + \frac{10-1}{100}N_3} = \frac{10 \cdot \frac{50000}{502}}{1 + \frac{9}{100}\frac{50000}{502}} = \frac{500000/502}{5002/502} \approx 99.9600$$

$$N_5 = \frac{10N_4}{1 + \frac{10-1}{100}N_4} = \frac{10 \cdot \frac{500000}{5002}}{1 + \frac{9}{100}\frac{500000}{5002}} = \frac{5000000/5002}{50002/5002} \approx 99.9960$$

$$N_6 = \frac{10N_5}{1 + \frac{10-1}{100}N_5} = \frac{10 \cdot \frac{5000000}{50002}}{1 + \frac{9}{100}\frac{5000000}{50002}} = \frac{50000000/50002}{500002/50002} \approx 99.9996$$

$$N_7 = \frac{10N_6}{1 + \frac{10-1}{100}N_6} = \frac{10 \cdot \frac{50000000}{500002}}{1 + \frac{9}{100}\frac{50000000}{500002}} = \frac{500000000/500002}{5000002/500002} \approx 100.000$$

$$N_8 = \frac{10N_7}{1 + \frac{10-1}{100}N_7} = \frac{10 \cdot \frac{500000000}{5000002}}{1 + \frac{9}{100}\frac{500000000}{5000002}} = \frac{5000000000/5000002}{50000002/5000002} \approx 100.000$$

$$N_9 = \frac{10N_8}{1 + \frac{10-1}{100}N_8} = \frac{10 \cdot \frac{5000000000}{50000002}}{1 + \frac{9}{100}\frac{5000000000}{50000002}} = \frac{50000000000/50000002}{500000002/50000002} \approx 100.000$$

$$N_{10} = \frac{10N_9}{1 + \frac{10-1}{100}N_9} = \frac{10 \cdot \frac{50000000000}{500000002}}{1 + \frac{9}{100}\frac{50000000000}{500000002}} = \frac{500000000000/500000002}{5000000002/500000002} \approx 100.000$$

The graph for this function is:

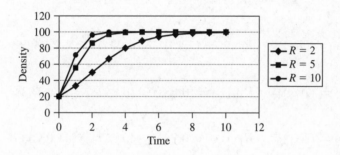

Graph of N_t as a function of t

Prob. 17. he long term behavior of the geometric mean of the growth parameter, $\hat{R}_t$, is defined as:

$$\hat{R}_t = (R_0 R_1 \cdots R_{t-1})^{1/t}$$

We now calculate $\hat{R}_t$ for $t = 1, 2, 3, \ldots 20$, using the fact that

$$\hat{R}_{t+1} = (R_0 R_1 \cdots R_t)^{1/(t+1)} = (R_t)^{\frac{1}{t+1}} \cdot (\hat{R}_t)^{\frac{t}{t+1}} \quad \textbf{for } t \geq 1.$$

$$\hat{R}_1 = R_0 = 2.78$$

$$\hat{R}_2 = (R_1)^{\frac{1}{2}} \cdot (\hat{R}_1)^{\frac{1}{2}} = (0.29)^{\frac{1}{2}} \cdot (2.78)^{\frac{1}{2}} \approx 0.8979$$

$$\hat{R}_3 = (R_2)^{\frac{1}{3}} \cdot (\hat{R}_2)^{\frac{2}{3}} = (0.43)^{\frac{1}{3}} \cdot (0.8979)^{\frac{2}{3}} \approx 0.7025$$

$$\hat{R}_4 = (R_3)^{\frac{1}{4}} \cdot (\hat{R}_3)^{\frac{3}{4}} = (0.25)^{\frac{1}{4}} \cdot (0.7025)^{\frac{3}{4}} \approx 0.5426$$

$$\hat{R}_5 = (R_4)^{\frac{1}{5}} \cdot (\hat{R}_4)^{\frac{4}{5}} = (2.90)^{\frac{1}{5}} \cdot (0.5426)^{\frac{4}{5}} \approx 0.7587$$

$$\hat{R}_6 = (R_5)^{\frac{1}{6}} \cdot (\hat{R}_5)^{\frac{5}{6}} = (1.67)^{\frac{1}{6}} \cdot (0.7587)^{\frac{5}{6}} \approx 0.8653$$

$$\hat{R}_7 = (R_6)^{\frac{1}{7}} \cdot (\hat{R}_6)^{\frac{6}{7}} = (1.17)^{\frac{1}{7}} \cdot (0.8653)^{\frac{6}{7}} \approx 0.9034$$

$$\hat{R}_8 = (R_7)^{\frac{1}{8}} \cdot (\hat{R}_7)^{\frac{7}{8}} = (0.69)^{\frac{1}{8}} \cdot (0.9034)^{\frac{7}{8}} \approx 0.8735$$

$$\hat{R}_9 = (R_8)^{\frac{1}{9}} \cdot (\hat{R}_8)^{\frac{8}{9}} = (1.45)^{\frac{1}{9}} \cdot (0.8735)^{\frac{8}{9}} \approx 0.9241$$

$$\hat{R}_{10} = (R_9)^{\frac{1}{10}} \cdot (\hat{R}_9)^{\frac{9}{10}} = (1.13)^{\frac{1}{10}} \cdot (0.9241)^{\frac{9}{10}} \approx 0.9428$$

$$\hat{R}_{11} = (R_{10})^{\frac{1}{11}} \cdot (\hat{R}_{10})^{\frac{10}{11}} = (0.08)^{\frac{1}{11}} \cdot (0.9428)^{\frac{10}{11}} \approx 0.7534$$

$$\hat{R}_{12} = (R_{11})^{\frac{1}{12}} \cdot (\hat{R}_{11})^{\frac{11}{12}} = (0.88)^{\frac{1}{12}} \cdot (0.7534)^{\frac{11}{12}} \approx 0.7633$$

$$\hat{R}_{13} = (R_{12})^{\frac{1}{13}} \cdot (\hat{R}_{12})^{\frac{12}{13}} = (2.69)^{\frac{1}{13}} \cdot (0.7218)^{\frac{12}{13}} \approx 0.8409$$

$$\hat{R}_{14} = (R_{13})^{\frac{1}{14}} \cdot (\hat{R}_{13})^{\frac{13}{14}} = (0.36)^{\frac{1}{14}} \cdot (0.6897)^{\frac{13}{14}} \approx 0.7915$$

$$\hat{R}_{15} = (R_{14})^{\frac{1}{15}} \cdot (\hat{R}_{14})^{\frac{14}{15}} = (0.08)^{\frac{1}{15}} \cdot (0.7112)^{\frac{14}{15}} \approx 0.6793$$

$$\hat{R}_{16} = (R_{15})^{\frac{1}{16}} \cdot (\hat{R}_{15})^{\frac{15}{16}} = (2.34)^{\frac{1}{16}} \cdot (0.6821)^{\frac{15}{16}} \approx 0.7339$$

$$\hat{R}_{17} = (R_{16})^{\frac{1}{17}} \cdot (\hat{R}_{16})^{\frac{16}{17}} = (2.13)^{\frac{1}{17}} \cdot (0.5611)^{\frac{16}{17}} \approx 0.7814$$

$$\hat{R}_{18} = (R_{17})^{\frac{1}{18}} \cdot (\hat{R}_{17})^{\frac{17}{18}} = (2.20)^{\frac{1}{18}} \cdot (0.4965)^{\frac{17}{18}} \approx 0.8276$$

$$\hat{R}_{19} = (R_{18})^{\frac{1}{19}} \cdot (\hat{R}_{18})^{\frac{18}{19}} = (2.80)^{\frac{1}{19}} \cdot (0.4645)^{\frac{18}{19}} \approx 0.8825$$

$$\hat{R}_{20} = (R_{19})^{\frac{1}{20}} \cdot (\hat{R}_{19})^{\frac{19}{20}} = (0.29)^{\frac{1}{20}} \cdot (0.4569)^{\frac{19}{20}} \approx 0.8347$$

Prob. 19.

(a) We have the equation:

$$N_{t+1} = (1-c)N_t \exp\left[R\left(1 - \frac{(1-c)N_t}{K}\right)\right]$$

When $R = 1$, $K = 100$, $N_0 = 50$ and $c = 0.1$, we compute N_t for $t = 1, 2, \ldots 20$ using

$$N_{t+1} = (1-0.1)N_t \exp\left[1\left(1 - \frac{(1-0.1)N_t}{100}\right)\right] = 0.9N_t \exp[1 - 0.009N_t]$$

$N_1 = 0.9N_0 \exp[1 - 0.009N_0] \approx 77.99639$

$N_2 = 0.9N_1 \exp[1 - 0.009N_1] \approx 94.56945$

$N_3 = 0.9N_2 \exp[1 - 0.009N_2] \approx 98.77543$

$N_4 = 0.9N_3 \exp[1 - 0.009N_3] \approx 99.33615$

$N_5 = 0.9N_4 \exp[1 - 0.009N_4] \approx 99.39717$

$N_6 = 0.9N_5 \exp[1 - 0.009N_5] \approx 99.40363$

$N_7 = 0.9N_6 \exp[1 - 0.009N_6] \approx 99.40431$

$N_8 = 0.9N_7 \exp[1 - 0.009N_7] \approx 99.40438$

$N_9 = 0.9N_8 \exp[1 - 0.009N_8] \approx 99.40439$

$N_{10} = 0.9N_9 \exp[1 - 0.009N_9] \approx 99.40439$

$N_{11} = 0.9N_{10} \exp[1 - 0.009N_{10}] \approx 99.40439$

$N_{12} = 0.9N_{11} \exp[1 - 0.009N_{11}] \approx 99.40439$

$N_{13} = 0.9N_{12} \exp[1 - 0.009N_{12}] \approx 99.40439$

$N_{14} = 0.9N_{13} \exp[1 - 0.009N_{13}] \approx 99.40439$

$N_{15} = 0.9N_{14} \exp[1 - 0.009N_{14}] \approx 99.40439$

$N_{16} = 0.9N_{15} \exp[1 - 0.009N_{15}] \approx 99.40439$

$N_{17} = 0.9N_{16} \exp[1 - 0.009N_{16}] \approx 99.40439$

$N_{18} = 0.9N_{17} \exp[1 - 0.009N_{17}] \approx 99.40439$

$N_{19} = 0.9N_{18} \exp[1 - 0.009N_{18}] \approx 99.40439$

$N_{20} = 0.9N_{19} \exp[1 - 0.009N_{19}] \approx 99.40439$

(b) We have the equation:

$$N_{t+1} = (1-c)N_t \exp\left[R\left(1 - \frac{(1-c)N_t}{K}\right)\right]$$

When $R = 1$, $K = 100$, $N_0 = 50$ and $c = 0.5$, we compute N_t for $t = 1, 2, \ldots 20$ using

$$N_{t+1} = (1-0.5)N_t \exp\left[1\left(1 - \frac{(1-0.5)N_t}{100}\right)\right] = 0.5N_t \exp[1 - 0.005N_t]$$

$N_1 = 0.5N_0 \exp[1 - 0.005N_0] \approx 52.92500$

$N_2 = 0.5N_1 \cdot \exp[1 - 0.005N_1] \approx 55.20776$

$N_3 = 0.5N_2 \exp[1 - 0.005N_2] \approx 56.93542$

$N_4 = 0.5N_3 \exp[1 - 0.005N_3] \approx 58.21211$

$N_5 = 0.5N_4 \exp[1 - 0.005N_4] \approx 59.13871$

$N_6 = 0.5N_5 \exp[1 - 0.005N_5] \approx 59.80235$

$N_7 = 0.5N_6 \exp[1 - 0.005N_6] \approx 60.27311$

$N_8 = 0.5N_7 \exp[1 - 0.005N_7] \approx 60.60475$

$N_9 = 0.5N_8 \exp[1 - 0.005N_8] \approx 60.83726$

$N_{10} = 0.5N_9 \exp[1 - 0.005N_9] \approx 60.99970$

$N_{11} = 0.5N_{10} \exp[1 - 0.005N_{10}] \approx 61.11292$

$N_{12} = 0.5N_{11} \exp[1 - 0.005N_{11}] \approx 61.19170$

$N_{13} = 0.5N_{12} \exp[1 - 0.005N_{12}] \approx 61.24645$

$N_{14} = 0.5N_{13} \exp[1 - 0.005N_{13}] \approx 61.284470$

$N_{15} = 0.5N_{14} \exp[1 - 0.005N_{14}] \approx 61.310854$

$N_{16} = 0.5N_{15} \exp[1 - 0.005N_{15}] \approx 61.329161$

$N_{17} = 0.5N_{16} \exp[1 - 0.005N_{16}] \approx 61.341858$

$N_{18} = 0.5N_{17} \exp[1 - 0.005N_{17}] \approx 61.350663$

$N_{19} = 0.5N_{18} \exp[1 - 0.005N_{18}] \approx 61.356768$

$N_{20} = 0.5N_{19} \exp[1 - 0.005N_{19}] \approx 61.361000$

(c) We have the equation:

$$N_{t+1} = (1-c)N_t \exp\left[R\left(1 - \frac{(1-c)N_t}{K}\right)\right]$$

When $R = 1$, $K = 100$, $N_0 = 50$ and $c = 0.9$, we compute N_t for $t = 1, 2, \ldots 20$ using

$$N_{t+1} = (1-0.9)N_t \exp\left[1\left(1 - \frac{(1-0.9)N_t}{100}\right)\right] = 0.1N_t \exp[1 - 0.001N_t]$$

$N_1 = 0.1N_0 \exp[1 - 0.001N_0] \approx 12.92855$

$N_2 = 0.1N_1 \exp[1 - 0.001N_1] \approx 3.46920$

$N_3 = 0.1N_2 \exp[1 - 0.001N_2] \approx 0.93976$

$N_4 = 0.1N_3 \exp[1 - 0.001N_3] \approx 0.25521$

$N_5 = 0.1N_4 \exp[1 - 0.001N_4] \approx 0.06936$

$N_6 = 0.1N_5 \exp[1 - 0.001N_5] \approx 0.01885$

$N_7 = 0.1N_6 \exp[1 - 0.001N_6] \approx 0.00512$

$N_8 = 0.1N_7 \exp[1 - 0.001N_7] \approx 0.00139$

$N_9 = 0.1N_8 \exp[1 - 0.001N_8] \approx 0.00038$

$N_{10} = 0.1N_9 \exp[1 - 0.001N_9] \approx 0.00010$

$N_{11} = 0.1N_{10} \exp[1 - 0.001N_{10}] \approx 2.7977 \cdot 10^{-5}$

$N_{12} = 0.1N_{11} \exp[1 - 0.001N_{11}] \approx 7.6051 \cdot 10^{-6}$

$N_{13} = 0.1N_{12} \exp[1 - 0.001N_{12}] \approx 2.0673 \cdot 10^{-6}$

$N_{14} = 0.1N_{13} \exp[1 - 0.001N_{13}] \approx 5.6195 \cdot 10^{-7}$

$N_{15} = 0.1N_{14} \exp[1 - 0.001N_{14}] \approx 1.5275 \cdot 10^{-7}$

$N_{16} = 0.1N_{15} \exp[1 - 0.001N_{15}] \approx 4.1523 \cdot 10^{-8}$

$N_{17} = 0.1N_{16} \exp[1 - 0.001N_{16}] \approx 1.1287 \cdot 10^{-8}$

$N_{18} = 0.1N_{17} \exp[1 - 0.001N_{17}] \approx 3.0681 \cdot 10^{-9}$

$N_{19} = 0.1N_{18} \exp[1 - 0.001N_{18}] \approx 8.3400 \cdot 10^{-10}$

$N_{20} = 0.1N_{19} \exp[1 - 0.001N_{19}] \approx 2.2671 \cdot 10^{-10}$

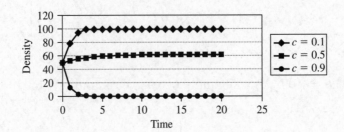

Chapter 3

Limits and Continuity

3.1 Limits

Prob. 1. We will approach $x = 2$ from the left and right:

x	$x^2 - 4x + 1$	x	$x^2 - 4x + 1$
1.9	-2.99	2.1	-2.99
1.99	-2.999	2.01	-2.999
1.999	-2.999	2.001	-2.999

Hence the limit is -3.

Prob. 3. We will approach $x = -1$ from the left and right

X	$\frac{2x}{1+x^2}$	x	$\frac{2x}{1+2^x}$
-1.1	0.999	-0.9	-0.99
-1.01	-0.99	-0.99	-0.99
-1.001	-0.99	-0.99	-0.99

Hence the limit is -1.

Prob. 5. In the following table, we compute the values of $3\cos\frac{x}{4}$ for x close to π (but not equal to π). In the left half of the table, we approach $x = \pi$ from the left $(x \to \pi^-)$; in the right half of the table, we approach $x = \pi$ from the right $(x \to \pi^+)$. We would guess that the value of this limit is 2.12 as x approaches π, from either side.

117

x	$3\cos\frac{x}{4}$	x	$3\cos\frac{x}{4}$
3.13985	2.122	3.14334	2.120
3.14142	2.121	3.14177	2.12
3.14157	2.121	3.14161	2.12
3.141591	2.121	3.141594	2.12

Since this limit is a finite number, we say that the limit exists and that the function converges to 2.12 as x tends to π.

Prob. 7. In the following table, we compute the values of $2\sec(x/3)$ for x close to $\pi/2$ (but not equal to $\pi/2$). In the left half of the table, we approach $x = \pi/2$ from the left $(x \to (\pi/2)^-)$; in the right half of the table, we approach $x = \pi/2$ from the right $(x \to (\pi/2)^+)$. We would guess that the value of this limit is 2.309 as x approaches $\pi/2$, from either side.

x	$2\sec(x/3)$	x	$2\sec(x/3)$
1.56905	2.308	1.57254	2.3102
1.57062	2.3093	1.57097	2.30947
1.57077	2.3093	1.57081	2.3094
1.57079	2.3093	1.57080	2.3094

Since this limit is a finite number, we say that the limit exists and that the function converges to 2.309 as x tends to $\pi/2$.

Prob. 9. In the following table, we compute the values of $e^{-x^2/2}$ for x close to -1 (but not equal to -1). In the left half of the table, we approach $x = -1$ from the left $(x \to -1^-)$; in the right half of the table, we approach $x = -1$ from the right $(x \to -1^+)$. We would guess that the value of this limit is 0.606 as x approaches -1, from either side.

x	$e^{-x^2/2}$	x	$e^{-x^2/2}$
-1.1	0.5461	-0.9	0.667
-1.01	0.6005	-0.99	0.6125
-1.001	0.6059	-0.999	0.6071
-1.0001	0.6064	-0.9999	0.6066

Since this limit is a finite number, we say that the limit exists and that the function converges to 0.606 as x tends to -1.

Prob. 11. In the following table, we compute the values of $\ln(x+1)$ for x close to 0 (but not equal to 0). In the left half of the table, we approach $x = 0$ from the left ($x \to 0^-$); in the right half of the table, we approach $x = 0$ from the right ($x \to 0^+$). We would guess that the value of this limit is 0 as x approaches 0, from either side.

x	$\ln(x+1)$	x	$\ln(x+1)$
-0.1	-0.10536	0.1	0.0953
-0.01	-0.01005	0.01	0.00995
-0.001	-0.0010005	0.001	0.000999
-0.0001	-0.000100005	0.0001	0.0000999

Since this limit is a finite number, we say that the limit exists and that the function converges to 0 as x tends to 0.

Prob. 13. In the following table, we compute the values of $\frac{x^2-16}{x-4}$ for x close to 3 (but not equal to 3). In the left half of the table, we approach $x = 3$ from the left ($x \to 3^-$); in the right half of the table, we approach $x = 3$ from the right ($x \to 3^+$). We would guess that the value of this limit is 7 as x approaches 3, from either side.

x	$\frac{x^2-16}{x-4}$	x	$\frac{x^2-16}{x-4}$
2.9	6.9	3.1	7.1
2.99	6.99	3.01	7.01
2.999	6.999	3.001	7.001
2.9999	6.9999	3.0001	7.0001

Since this limit is a finite number, we say that the limit exists and that the function converges to 7 as x tends to 3.

Prob. 15. We will approach $x = \frac{\pi}{2}$ from the left and right:

x	$\sin(2x)$	x	$\sin(2x)$
89.9	0.003	90.1	-0.003
89.99	0.0003	90.01	-0.0003
89.999	0.00003	90.001	-0.00003

Hence the limit is 0.

Prob. 17. We will approach $x0$ from the left and right:

x	$\frac{1}{1+x^2}$	x	$\frac{1}{1+x^2}$
-0.1	0.99	0.1	0.99
-0.01	0.99	0.01	0.99
-0.001	0.99	0.001	0.99

Hence the limit is 1.

Prob. 19. We will approach $x0$ from right only:

x	$1 - e^{-x}$
0.1	0.095
0.01	0.0099
0.001	0.00099

Hence the limit is 0.

Prob. 21. As we approach the vertical asymptote $x = 4$ from the left, the graph of $f(x) = \frac{2}{x-4}$ falls sharply off. This reveals that

$$\lim_{x \to 4^-} \frac{2}{x - 4} = -\infty$$

We arrive at the same conclusion when we compute values of $f(x)$ for x close to, and slightly smaller than, 4. When x is slightly smaller than 4, the term $x - 4$ is negative and decreases to zero. From this we can conclude that $f(x)$ grows negatively without bound as x approaches 4 from the left.

Prob. 23. As we approach the vertical asymptote $x = 1$ from the left, the graph of $f(x) = \frac{2}{1-x}$ grows steeply. This reveals that

$$\lim_{x \to 1^-} \frac{2}{1 - x} = \infty$$

We arrive at the same conclusion when we compute values of $f(x)$ for x close to, and slightly smaller than, 1. When x is slightly smaller than 1, the term $1 - x$ is positive and decreases to zero. From this we can conclude that $f(x)$ grows positively without bound as x approaches 1 from the left.

Prob. 25. As we approach the vertical asymptote $x = 1$ from the left, the graph of $f(x) = \frac{1}{1-x^2}$ grows steeply. This reveals that

$$\lim_{x \to 1^-} \frac{1}{1-x^2} = \infty$$

We arrive at the same conclusion when we compute values of $f(x)$ for x close to, and slightly smaller than, 1. When x is slightly smaller than 1, the term $1 - x^2$ is positive and decreases to zero. From this we can conclude that $f(x)$ grows positively without bound as x approaches 1 from the left.

Prob. 27. We will approach $x = 3$ from the left and right:

x	$\frac{1}{(x-3)^2}$	x	$\frac{1}{(x-3)^2}$
2.9	100	3.1	100
2.99	10000	3.01	10000
2.999	10^6	3.001	10^6

Hence, the limit does not exist, diverging to infinity.

Prob. 29. The graph of the function $f(x) = \frac{\sqrt{x^2+9}-3}{x^2}$ ($x \neq 0$), indicates that the limit exists and, based on the graph, we conjecture that it is equal to 0.16667. If, instead, we use a calculator to produce a table for values of $f(x)$ close to 0, something strange seems to happen.

x	$f(x)$	x	$f(x)$
0.1	0.1666	0.00001	0.15
0.01	0.1666	0.000001	0.1
0.001	0.1666	0.0000001	0
0.0001	0.1666	0.00000001	0

As we move closer to 0, we first find that $f(x)$ gets close to 0.16667, but when x gets very close to 0, $f(x)$ seems to drop to 0. What is going on? First, before you worry too much, it is true that $\lim_{x \to 0} f(x) = \frac{1}{6}$. In the next section, we will learn how to compute this limit without resorting to the (somewhat dubious) help of a calculator. The strange behaviour of the calculated values happens because when x is very small, the difference in the numerator is so close to 0 that the calculator can no longer accurately determine its value, and simply returns 0. The calculator can only accurately compute a certain number of digits, which is good enough for most cases, but here we need greater accuracy. The same strange thing happens when you try to graph this function on a graphing calculator. When the x-axis interval of the viewing window is too small, the graph is no longer accurate.

Prob. 31. The graph of the function $f(x) = \frac{1-\sqrt{1-x^2}}{x^2}$ ($x \neq 0$), indicates that the limit exists and, based on the graph, we conjecture that it is equal to 0.5. If, instead, we use a calculator to produce a table for values of $f(x)$ close to 0, something strange seems to happen.

x	$f(x)$	x	$f(x)$
0.1	0.50126	0.00001	1.00000
0.01	0.50001	0.000001	0
0.001	0.50000	0.0000001	0
0.0001	0.50000	0.00000001	0

As we move closer to 0, we first find that $f(x)$ gets close to 0.5, but when x gets very close to 0, $f(x)$ seems to drop to 0. What is going on? First, before you worry too much, it is true that $\lim_{x \to 0} f(x) = 0.5$. In the next section, we will learn how to compute this limit without resorting to the (somewhat dubious) help of a calculator. The strange behaviour of the calculated values happens because when x is very small, the difference in the numerator is so close to 0 that the calculator can no longer accurately determine its value, and simply returns 0. The calculator can only accurately compute a certain number of digits, which is good enough for most cases, but here we need greater accuracy. The same strange thing happens when you try to graph this function on a graphing calculator. When the x-axis interval of the viewing window is too small, the graph is no longer accurate.

Prob. 33.

(a) From both the graph and the table for $2/x^2$, we conjecture that the limit of this function is 0 as x approaches ∞. We compute the values of $2/x^2$ for larger and larger values of x, since this is what $x \to \infty$ means.

x	$2/x^2$
100	0.0002
1000	0.000002
10000	$2 \cdot 10^{-8}$
100000	$2 \cdot 10^{-10}$

Since this limit is a finite number, we say that the limit exists and that the function converges to 0 as x tends to ∞.

(b) From both the graph and the table for $2/x^2$, we conjecture that the limit of this function is 0 as x approaches $-\infty$. We compute the values of $2/x^2$ for larger and larger negative values of x, since this is what $x \to -\infty$ means.

x	$2/x^2$
-100	0.0002
-1000	0.000002
-10000	$2 \cdot 10^{-8}$
-100000	$2 \cdot 10^{-10}$

Since this limit is a finite number, we say that the limit exists and that the function converges to 0 as x tends to $-\infty$.

(c) The graph of $f(x) = 2/x^2$ $(x \neq 0)$, reveals that $f(x)$ increases positively without bound as $x \to 0$. We can confirm this when we select values for x that are close to 0. By choosing values sufficiently close to 0, we can get arbitrarily large values of $2/x^2$.

x	$2/x^2$	x	$2/x^2$
-0.1	200	0.1	200
-0.01	20,000	0.01	20,000
-0.001	$2 \cdot 10^6$	0.001	$2 \cdot 10^6$
-0.0001	$2 \cdot 10^8$	0.0001	$2 \cdot 10^8$

This indicates that $\lim_{x \to 0} \frac{1}{x^2}$ does not exist.

Prob. 35. Simply using a calculator and plugging in values to find limits can yield unclear answers if we do not exercise some caution. If we produced a table of values of $f(x) = \sin \frac{1}{x-1}$ for $x = 1.1, 1.01, 1.001, \ldots$, we would find that $\sin \frac{1}{1.1-1} = -0.5440$, $\sin \frac{1}{1.01-1} = -0.5064$, $\sin \frac{1}{1.001-1} = 0.8269$, and so on. (Note that we measure angles in radians.) These values seem to be not approaching any fixed value. Looking at the graph of $f(x)$, we find that the values of $f(x)$ oscillate infinitely often between -1 and $+1$ as $x \to 1$. This is because as $x \to 1^+$, the argument in the sine function goes to infinity (likewise, as $x \to 1^-$, the argument goes to negative infinity), namely,

$$\lim_{x \to 1^+} \frac{1}{x-1} = \infty \quad \text{and} \quad \lim_{x \to 1^-} \frac{1}{x-1} = -\infty,$$

and so the function values oscillate between -1 and $+1$. Therefore, $\sin \frac{1}{x-1}$ continues to oscillate between -1 and $+1$ as $x \to 1$.

Prob. 37. Using Limit Laws 1 and 2, we can write

$$\lim_{x \to -1} \left(x^3 + 7x - 1\right) = \lim_{x \to -1} x^3 + 7 \lim_{x \to -1} x - \lim_{x \to -1} 1$$

provided the individual limits exist. For the first term, we use Limit Law 3 twice, to get

$$\lim_{x \to -1} x^3 = \lim_{x \to -1} x \cdot \lim_{x \to -1} x^2 = \lim_{x \to -1} x \cdot \lim_{x \to -1} x \cdot \lim_{x \to -1} x$$

provided that $\lim_{x \to -1} x$ exists. Using Formula (3.3), it follows that $\lim_{x \to -1} x = -1$ and we find that

$$\lim_{x \to -1} x \cdot \lim_{x \to -1} x \cdot \lim_{x \to -1} x = (-1)(-1)(-1) = -1.$$

To compute the second term, we use Formula (3.3) again to get $\lim_{x \to -1} x = -1$. For the last term, it is clear that $\lim_{x \to -1} 1 = 1$. Thus each of the individual limits exist, and we have

$$\lim_{x \to -1} \left(x^3 + 7x - 1\right) = \lim_{x \to -1} x^3 + 7 \lim_{x \to -1} x - \lim_{x \to -1} 1 = -1 + (7)(-1) - 1 = -9$$

Prob. 39. Using Limit Laws 1 and 2, we can write

$$\lim_{x \to -5} \left(4 + 2x^2\right) = \lim_{x \to -5} 4 + 2 \lim_{x \to -5} x^2$$

provided the individual limits exist. For the second term, we use Limit Law 3, to get

$$\lim_{x \to -5} x^2 = \lim_{x \to -5} x \cdot \lim_{x \to -5} x$$

provided $\lim_{x \to -5} x$ exists. Using Formula (3.3), it follows that $\lim_{x \to -5} x = -5$ and we find that

$$\lim_{x \to -5} x \cdot \lim_{x \to -5} x = (-5)(-5) = 25$$

To compute the first term, we find that $\lim_{x \to -5} 4 = 4$. Thus each of the individual limits exist, and we have

$$\lim_{x \to -5} \left(4 + 2x^2\right) = \lim_{x \to -5} 4 + 2 \lim_{x \to -5} x^2 = 4 + 2(25) = 54 \qquad (3.1.1)$$

Prob. 41. Using Limit Laws 1 and 4, we can write

$$\lim_{x \to 3} \left(2x^2 - \frac{1}{x}\right) = 2 \lim_{x \to 3} x^2 - \frac{\lim_{x \to 3} 1}{\lim_{x \to 3} x}$$

provided the individual limits exist. For the first term, we use Limit Law 3, to get

$$\lim_{x \to 3} x^2 = \lim_{x \to 3} x \cdot \lim_{x \to 1} x$$

provided $\lim_{x \to 3} x$ exists. Using Formula (3.3), it follows that $\lim_{x \to 3} x = 3$ and we find that

$$\lim_{x \to 3} x \cdot \lim_{x \to 3} x = (3)(3) = 9$$

To compute the second term, we use Formula (3.3) again to get $\lim_{x \to 3} x = 3$, and we also have $\lim_{x \to 3} 1 = 1$. Thus each of the individual limits exist, and we have

$$\lim_{x \to 3} \left(2x^2 - \frac{1}{x}\right) = 2 \lim_{x \to 3} x^2 - \frac{\lim_{x \to 3} 1}{\lim_{x \to 3} x} = 2(9) - \frac{1}{3} = \frac{53}{3}$$

Prob. 43. Using Limit Law 4, we find that

$$\lim_{x \to -3} \frac{x^3 - 20}{x + 1} = \frac{\lim_{x \to -3}(x^3 - 20)}{\lim_{x \to -3}(x + 1)}$$

provided the limits in the numerator and denominator exist and the limit in the denominator is not equal to 0. Using Limit Laws 2 and 3, and Formula (3.3) in the numerator, we find

$$\lim_{x \to -3}(x^3 - 20) = \lim_{x \to -3} x^3 - \lim_{x \to -3} 20 = (-3)(-3)(-3) - 20 = -47$$

Note that breaking up the limit of the sum in the numerator into a sum of limits is only justified once we show that the individual limits exist. Using Limit Law 2 and Formula (3.3) in the denominator, we find

$$\lim_{x \to -3} (x + 1) = \lim_{x \to -3} x + \lim_{x \to -3} 1 = -3 + 1 = -2$$

Again using the limit laws is only justified once we demonstrate that the individual limits exist. Since the limits in both the denominator and the numerator exist and the limit in the denominator is not equal to 0, we have

$$\lim_{x \to -3} \frac{x^3 - 20}{x + 1} = \frac{\lim_{x \to -3}(x^3 - 20)}{\lim_{x \to -3}(x + 1)} = \frac{-47}{-2} = 23.5$$

Prob. 45. Using Limit Law 4, we find

$$\lim_{x \to 3} \frac{3x^2 + 1}{2x - 3} = \frac{\lim_{x \to 3}(3x^2 + 1)}{\lim_{x \to 3}(2x - 3)}$$

provided the limits in the numerator and denominator exist and the limit in the denominator is not equal to 0. Using Limit Laws 1, 2 and 3, and Formula (3.3) in the numerator, we find

$$\lim_{x \to 3} (3x^2 + 1) = 3 \lim_{x \to 3} x^2 + \lim_{x \to 3} 1 = 3(3)(3) + 1 = 28$$

Note that breaking up the limit of the sum in the numerator into a sum of limits is only justified once we show that the individual limits exist. Using Limit Laws 1 and 2, and Formula (3.3) in the denominator, we find

$$\lim_{x \to 3} (2x - 3) = 2 \lim_{x \to 3} x - \lim_{x \to 3} 3 = 2(3) - 3 = 3$$

Again using the limit laws is only justified once we demonstrate that the individual limits exist. Since the limits in both the denominator and the numerator exist and the limit in the denominator is not equal to 0, we obtain

$$\lim_{x \to 3} \frac{3x^2 + 1}{2x - 3} = \frac{\lim_{x \to 3}(3x^2 + 1)}{\lim_{x \to 3}(2x - 3)} = \frac{28}{3}$$

Prob. 47. The function $f(x) = \frac{1 - x^2}{1 - x}$ is a rational function, but since $\lim_{x \to 1}(1 - x) = 0$, we cannot use Limit Law 4. Instead, we need to try to simplify $f(x)$ first. Notice that

$$\lim_{x \to 1} \frac{1 - x^2}{1 - x} = \lim_{x \to 1} \frac{(1 - x)(1 + x)}{1 - x}$$

Since $x \neq 1$, we can cancel the non-zero factor $1 - x$ in the numerator and denominator, which yields

$$\lim_{x \to 1} \frac{(1-x)(1+x)}{1-x} = \lim_{x \to 1}(1+x) = \lim_{x \to 1} 1 + \lim_{x \to 1} x = 1 + 1 = 2$$

where we have used Limit Law 2 and Formula (3.3) when computing the limit.

Prob. 49. The function $f(x) = \frac{x^2 - 2x - 3}{x-3}$ is a rational function, but since $\lim_{x \to 3}(x - 3) = 0$, we cannot use Limit Law 4. Instead, we need to try to simplify $f(x)$ first. Notice that

$$\lim_{x \to 3} \frac{x^2 - 2x - 3}{x - 3} = \lim_{x \to 3} \frac{(x-3)(x+1)}{x-3}$$

Since $x \neq 3$, we can cancel the non-zero factor $x - 3$ in the numerator and denominator, which yields

$$\lim_{x \to 3} \frac{(x-3)(x+1)}{x-3} = \lim_{x \to 3}(x+1) = \lim_{x \to 3} x + \lim_{x \to 3} 1 = 3 + 1 = 4$$

where we used Limit Law 2 and Formula (3.3) when computing the limit.

Prob. 51.

$$\lim_{x \to 2} \frac{2-x}{x^2 - 4} = \lim_{x \to 2} \frac{2-x}{(x-2)(x+2)} = \lim_{x \to 2} \frac{-(x-2)}{(x-2)(x+2)}.$$

Since $x \neq 2$, we have:

$$\lim_{x \to 2} \frac{-1}{x+2} = \frac{\lim_{x \to 2}(-1)}{\lim_{x \to 2}(x+2)} = \frac{-1}{\lim_{x \to 2} x + \lim_{x \to 2} 2} = \frac{-1}{2+2} = -\frac{1}{4}.$$

Prob. 53. The function $f(x) = \frac{2x^2 + 3x - 2}{x+2}$ is a rational function, but since $\lim_{x \to -2}(x + 2) = 0$, we cannot use Limit Law 4. Instead, we need to try to simplify $f(x)$ first. Notice that

$$\lim_{x \to -2} \frac{2x^2 + 3x - 2}{x + 2} = \lim_{x \to -2} \frac{(2x-1)(x+2)}{x+2}$$

Since $x \neq -2$, we can cancel the non-zero factor $x + 2$ in the numerator and denominator, which yields

$$\lim_{x \to -2} \frac{(2x-1)(x+2)}{x+2} = \lim_{x \to -2}(2x - 1) = 2 \lim_{x \to -2} x - \lim_{x \to -2} 1 = 2(-2) - 1 = -5$$

where we used Limit Laws 1 and 2, and Formula (3.3) when computing the limit.

3.2   Continuity

Prob. 1. Given $f(x) = 2x$, $c = \frac{1}{2}$, to show that it is continuous at c we must show:

(1) $f(c)$ exists: $f(c) = f(\frac{1}{2}) = 2 \cdot \frac{1}{2} = 1$

(2) $\lim_{x \to c} f(x)$ exits: $\lim_{x \to \frac{1}{2}} 2x = 2 \cdot \frac{1}{2} = 1$

(3) $f(c) = \lim_{x \to c} f(x):$ $1 = 1.$

Hence, this function is continuous at $c = \frac{1}{2}$.

Prob. 3. We must check all three conditions on page 129.

1. $f(x)$ is defined at $x = 2$ since $f(2) = 2^3 - 2 \cdot 2 + 1 = 5$.

2. We can repeatedly use the fact that $\lim_{x \to 2} x = 2$ to conclude that $\lim_{x \to 2} f(x)$ exists.

3. Using the limit laws, we find that $\lim_{x \to 2} f(x) = 5$. This is same as $f(2)$.

Since all three conditions are satisfied, $f(x) = x^3 - 2x + 1$ is continuous at $x = 2$.

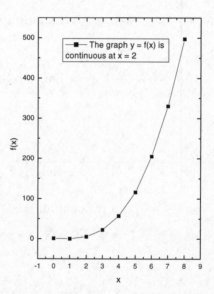

Prob. 5. The graph of $f(x)$ is shown below. We have

$$\lim_{x \to 2} \frac{x^2 - x - 2}{x - 2} = \lim_{x \to 2} \frac{(x + 1)(x - 2)}{x - 2} = \lim_{x \to 2} (x + 1) = 2 + 1 = 3$$

which is equal to $f(2)$. Thus the required conditions are satisfied and we can conclude that the function is continuous at $x = 2$.

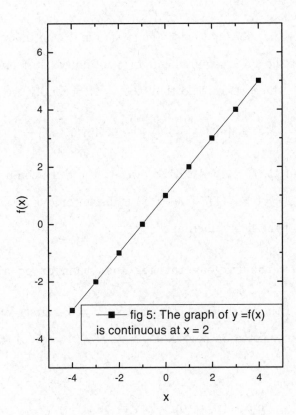

fig 5: The graph of y =f(x) is continuous at x = 2

Prob. 7. First, we compute the limit of $f(x)$ as $x \to 3$. We have

$$\lim_{x \to 3} \frac{x^2 - 9}{x - 3} = \lim_{x \to 3} \frac{(x - 3)(x + 3)}{x - 3} = \lim_{x \to 3} (x + 3) = 3 + 3 = 6.$$

To ensure that $f(x)$ is continuous at $x = 3$, we require that

$$\lim_{x \to 3} f(x) = f(3) = a.$$

We therefore need to choose $a = 6$. This is the only choice for a that would make $f(x)$ continuous. Any other value of a would result in $f(x)$ being discontinuous.

Prob. 9. We consider the three requirements for continuity listed on page 129. First, $f(x)$ is defined at $x = 3$, since we are told that $f(3) = 0$. The second condition, however, fails to be satisfied, since we know that

$$\lim_{x \to 3^-} f(x) = \lim_{x \to 3^-} \frac{1}{x - 3} = -\infty$$

and that

$$\lim_{x \to 3^+} f(x) = \lim_{x \to 3^+} \frac{1}{x - 3} = \infty.$$

This means that $\lim_{x \to 3} f(x)$ does not exist. Thus $f(x)$ is discontinuous at $x = 3$.

Prob. 11. We consider the three requirements for continuity listed on page 129. First, $f(x)$ is defined at $x = 1$, since we are told that $f(1) = 1$. Secondly, we have

$$\lim_{x \to 1} \frac{x^2 - 3x + 2}{x - 2} = \lim_{x \to 1} \frac{(x - 1)(x - 2)}{x - 2} = \lim_{x \to 1} (x - 1) = 1 - 1 = 0$$

and so we know that $\lim_{x \to 1} f(x)$ does exist. However, the third condition for continuity fails, because $\lim_{x \to 1} f(x) = 0 \neq f(1)$. Thus $f(x)$ is discontinuous at $x = 1$.

Prob. 13. The graph of the floor function

$$f(x) = \lfloor x \rfloor = \texttt{the largest integer less than or equal to } x$$

is given in Figure 3.12 of the textbook. Let us look at $x = 5/2$ first. We consider the three requirements for continuity listed on page 129. First, $f(5/2) = \lfloor 5/2 \rfloor = 2$ and so the function exists at $x = 5/2$. Next,

$$\lim_{x \to 5/2} f(x) = \lim_{x \to 5/2} \lfloor x \rfloor = \lim_{x \to 5/2} 2 = 2$$

since the function takes the value 2 a little to the left and right of $x = 5/2$. Since $\lim_{x \to 5/2} f(x) = 2 = f(5/2)$, we conclude that $f(x)$ is indeed continuous at $x = 5/2$. Next, what happens at $x = 3$? Notice from the graph that the function jumps whenever x is an integer. Now $f(3) = 3$, but

$$\lim_{x \to 3^+} f(x) = 3 \quad \text{and} \quad \lim_{x \to 3^-} = 2$$

which means that $\lim_{x \to 3} f(x)$ cannot exist. The function is therefore discontinuous at $x = 3$.

Prob. 15. Since $f(x) = 3x^4 - x^2 + 4$ is a polynomial, it is continuous for all $x \in \mathbb{R}$.

Prob. 17. Since $f(x) = \frac{x^2+1}{x-1}$ is a rational function, it is defined for all $x \neq 1$. Thus it is continuous for all $x \neq 1$.

Prob. 19. Setting $g(x) = -x$ and $h(x) = e^x$, we have $f(x) = (h \circ g)(x)$. Since both g and h are continuous for all $x \in \mathbb{R}$, the given function f is continuous for all $x \in \mathbb{R}$.

Prob. 21. The given function $f(x) = \ln \frac{x}{x+1}$ is logarithmic and is defined as long as $\frac{x}{x+1} > 0$ and $x + 1 \neq 0$. This means that $x > 0$ or $x < -1$. It is therefore continuous over the intervals $(-\infty, -1)$ and $(0, \infty)$.

Prob. 23. Since $f(x)$ is a trigonometric function, it is continuous wherever it is defined. The tangent function $\tan u$ is defined for all $u \neq \frac{\pi}{2} + k\pi$, where k is an integer. Setting $2\pi x = \frac{\pi}{2} + k\pi$, we find that $x = \frac{1}{4} + \frac{k}{2}$. Thus the given function is defined (and therefore continuous) for all $x \neq \frac{1}{4} + \frac{k}{2}$, where k is an integer.

Prob. 25.

(a) The graph of $f(x)$ when $c = 1$ is shown in the figure below. The function is not continuous at $x = 0$, because the left- and right-hand limits are not equal to each other.

$$\lim_{x \to 0^-} f(x) = \lim_{x \to 0^-} (x^2 + 2) = 0^2 + 2 = 2 \neq 1 = c = \lim_{x \to 0^+} (x + c) = \lim_{x \to 0^+} f(x)$$

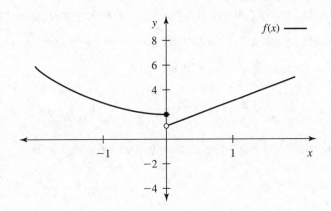

(b) We know that $f(0) = 2$. If we choose $c = 2$, then the function will be continuous for all $x \in \mathbb{R}$ because in this case

$$\lim_{x \to 0^-} f(x) = \lim_{x \to 0^-} (x^2 + 2) = 2 = f(0) = \lim_{x \to 0^+} (x + c) = \lim_{x \to 0^+} f(x)$$

Prob. 27.

(a) The function $f(x) = \sqrt{x - 1}$ is the composition of a polynomial $(x - 1)$ with the square root function. Since both functions are continuous from the right over their domains, the given function f must also be continuous from the right at $x = 1$.

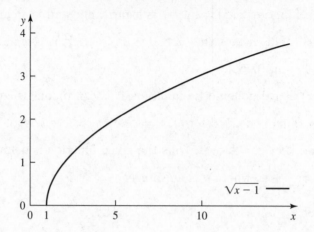

(b)

(c) Since $f(x)$ is not defined for $x < 1$, it does not make sense to look at continuity from the left at $x = 1$.

Prob. 29. The function $\sin(x/2)$ is a composition of a polynomial and a trigonometric function, and is defined for all $x \in \mathbb{R}$, and therefore it is continuous for all $x \in \mathbb{R}$. In particular, it is continuous at $x = \pi/3$, and we have

$$\lim_{x \to \pi/3} \sin\left(\frac{x}{2}\right) = \sin\left(\frac{\pi/3}{2}\right) = \sin\left(\frac{\pi}{6}\right) = 0.5$$

Prob. 31. Here, the limits of both the numerator and the denominator are equal to zero, since

$$\lim_{x \to \pi/2} \cos^2 x = \cos^2(\pi/2) = 0 \quad \text{and} \quad \lim_{x \to \pi/2} (1 - \sin^2 x) = 1 - \sin^2(\pi/2) = 1 - 1 = 0.$$

However, this does not mean that the limit does not exist. We use a trick that will allow us to find the limit; since $\cos^2 x + \sin^2 x = 1$, we have

$$\lim_{x \to \pi/2} \frac{\cos^2 x}{1 - \sin^2 x} = \lim_{x \to \pi/2} \frac{\cos^2 x}{\cos^2 x} = \lim_{x \to \pi/2} 1 = 1.$$

Prob. 33.

$$\lim_{x \to -1} \sqrt{4 + 5x^4} = \sqrt{4 + 5(-1)^4} = \sqrt{4 + 5} = 3.$$

Prob. 35. The function $\sqrt{x^2 + 2x + 2} = \sqrt{(x+1)^2 + 1}$ is a composition of a polynomial and a power function, and is defined for all $x \in \mathbb{R}$, and therefore it is continuous for all

$x \in \mathbb{R}$. In particular, it is continuous at $x = -1$, and we have

$$\lim_{x \to -1} \sqrt{x^2 + 2x + 2} = \sqrt{1 - 2 + 2} = 1.$$

Prob. 37. $\lim_{x \to 0} e^{-x^2/3}$, continuous at $x = 0$, hence

$$\lim_{x \to 0} e^{-x^2/3} = e^{\frac{-0^2}{3}} = e^0 = 1$$

Prob. 39. $f(x) = e^{x^2-9}$ is continuous at $x = 3$, hence

$$\lim_{x \to 3} e^{x^2-9} = e^{3^2-9} = e^0 = 1.$$

Prob. 41. Here, the limits of both the numerator and the denominator are zero, since

$$\lim_{x \to 0} (e^{2x} - 1) = e^0 - 1 = 0 \quad \text{and} \quad \lim_{x \to 0} (e^x - 1) = e^0 - 1 = 0.$$

However, this does not mean that the limit does not exist. We use a trick that will allow us to find the limit; namely, we factor the numerator.

$$\lim_{x \to 0} \frac{e^{2x} - 1}{e^x - 1} == \lim_{x \to 0} \frac{(e^x - 1)(e^x + 1)}{e^x - 1} = \lim_{x \to 0} (e^x + 1) = e^0 + 1 = 2.$$

Prob. 43. The function $1/\sqrt{5x^2 - 4}$ is the composition of a power function and a polynomial, both of which are continuous at every point where they are defined. Since this function is defined at $x = -2$, it is also continuous at $x = -2$. Hence,

$$\lim_{x \to -2} \frac{1}{\sqrt{5x^2 - 4}} = \frac{1}{\sqrt{5(-2)^2 - 4}} = \frac{1}{\sqrt{16}} = \frac{1}{4}.$$

Prob. 45. Here, the limits of both the numerator and the denominator are zero, since

$$\lim_{x \to 0} \left(\sqrt{x^2 + 9} - 3 \right) = \sqrt{0 + 9} - 3 = 0 \quad \text{and} \quad \lim_{x \to 0} x^2 = 0^2 = 0.$$

However, this does not mean that the limit does not exist. We use a trick that will allow us to find the limit; namely, we rationalize the numerator.

$$\frac{\sqrt{x^2 + 9} - 3}{x^2} = \frac{\sqrt{x^2 + 9} - 3}{x^2} \cdot \frac{\sqrt{x^2 + 9} + 3}{\sqrt{x^2 + 9} + 3} = \frac{x^2 + 9 - 3^2}{x^2(\sqrt{x^2 + 9} + 3)} = \frac{1}{\sqrt{x^2 + 9} + 3}$$

Now we evaluate the limit:

$$\lim_{x \to 0} \frac{\sqrt{x^2 + 9} - 3}{x^2} = \lim_{x \to 0} \frac{1}{\sqrt{x^2 + 9} + 3} = \frac{1}{\sqrt{0 + 9} + 3} = \frac{1}{6}.$$

Prob. 47. The function $\ln(1-x)$ is a composition of a polynomial and a logarithmic function, and is defined for all $x < 1$, and therefore it is continuous for all $x < 1$. In particular, it is continuous at $x = 0$, and we have

$$\lim_{x \to 0} \ln(1-x) = \ln(1-0) = \ln 1 = 0.$$

3.3   Limits at Infinity

Prob. 1. The degree of the denominator is greater than the degree of the numerator, so the limit is 0.

Prob. 3. The degree of the numerator is greater than the degree of the denominator, so the limit does not exist. The rational function tends to ∞ since it behaves like the function $\frac{x^3}{x} = x^2$.

Prob. 5. $\lim_{x \to \infty} \frac{1 - x^3 + 2x^4}{2x^2 + x^4}$. Divide by the highest power of the denominator to get:

$$\lim_{x \to \infty} \frac{\frac{1}{x^4} - \frac{x^3}{x^4} + \frac{2x^4}{x^4}}{\frac{2x^2}{x^1} + \frac{x^4}{x^4}} = \lim_{x \to \infty} \frac{\frac{1}{x^4} - \frac{1}{x} + 2}{\frac{2}{x^2} + 1}.$$

Hence we have: $\frac{2}{1} = 2$.

Prob. 7. The degree of the numerator is greater than that of the denominator, so the limit does not exist. The rational function tends to ∞ since it behaves like the function $\frac{x^2}{2x} = \frac{x}{2}$.

Prob. 9. The degree of the numerator is greater than that of the denominator, so the limit does not exist. The rational function tends to ∞ since it behaves like the function $\frac{x^2}{-x} = -x$ as $x \to -\infty$.

Prob. 11. Divide by the highest power of the denominator

$$\lim_{x \to -\infty} \frac{\frac{2}{x^2} + \frac{x^2}{x^2}}{\frac{1}{x^2} - \frac{x^2}{x^2}} = \lim_{x \to -\infty} \frac{\frac{2}{x^2} + 1}{\frac{1}{X^2} - 1},$$

hence we have: $\frac{1}{-1} = -1$.

Prob. 13. Using the limit laws, we have

$$\lim_{x \to \infty} \frac{4}{1 + e^{-2x}} = \frac{\lim_{x \to \infty} 4}{\lim_{x \to \infty}(1 + e^{-2x})} = \frac{4}{1 + 0} = 4$$

since e^{-2x} tends to 0 as $x \to \infty$.

Prob. 15. Using the limit laws, we have

$$\lim_{x \to \infty} \frac{2e^x}{e^x + 3} = \lim_{x \to \infty} \frac{2}{1 + 3e^{-x}} = \frac{\lim_{x \to \infty} 2}{\lim_{x \to \infty}(1 + 3e^{-x})} = \frac{2}{1 + 3(0)} = 2$$

since e^{-x} tends to 0 as $x \to \infty$.

Prob. 17. Since $\exp[x]$ is the same function as e^x, the limit is 0.

Prob. 19. Divide by the highest power at the denominator:

$$\lim_{x \to \infty} \frac{\frac{3e^{2x}}{e^{2x}}}{\frac{2e^{2x}}{e^{2x}} - \frac{e^x}{e^{2x}}} = \lim_{x \to \infty} \frac{3}{2 - \frac{1}{e^x}}.$$

Hence, we have: $\frac{3}{2} = 1.5$.

Prob. 21. Using the limit laws, we have

$$\lim_{x \to \infty} \frac{3}{2 + e^{-x}} = \frac{\lim_{x \to \infty} 3}{\lim_{x \to \infty}(2 + e^{-x})} = \frac{3}{2 + 0} = \frac{3}{2}$$

since e^{-x} tends to 0 as $x \to \infty$.

Prob. 23. Using the limit laws, we have

$$\lim_{x \to -\infty} \frac{e^x}{1 + x} = \lim_{x \to -\infty} e^x \cdot \lim_{x \to -\infty} \frac{1}{1 + x} = 0 \cdot 0 = 0$$

since both e^x and $\frac{1}{1+x}$ tend to 0 as $x \to -\infty$.

Prob. 25. We have

$$\lim_{N \to \infty} r(N) = \lim_{N \to \infty} a \frac{N}{k + N} = a \lim_{N \to \infty} \frac{N}{k + N} = a \cdot 1 = a$$

since the degree of the denominator $(k + N)$ equals that of the numerator (N).

Prob. 27.

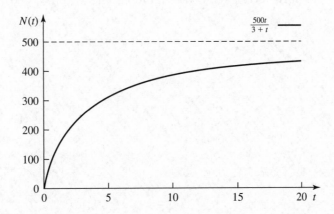

(a)

(b) Since both numerator and denominator have the same degree, we have

$$\lim_{t \to \infty} N(t) = \lim_{t \to \infty} \frac{500t}{3+t} = \frac{500}{1} = 500.$$

(c) When $t = 3$, we have $N(t) = N(3) = \frac{500 \cdot 3}{3+3} = \frac{500}{2} = 250$. This is half of 500, which is the limiting population size.

Prob. 29.

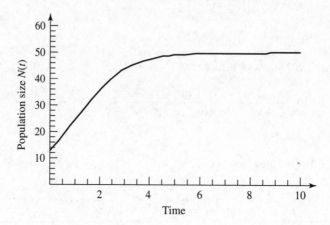

(a)

(b) Since $e^{-t} \to 0$ as t tends to ∞, we have

$$\lim_{t \to \infty} N(t) = \lim_{t \to \infty} \frac{50}{1 + 3e^{-t}} = \frac{\lim_{t \to \infty} 50}{\lim_{t \to \infty}(1 + 3e^{-t})} = \frac{50}{1 + 3(0)} = 50.$$

This matches the horizontal asymptote found on the graph.

3.4 The Sandwich Theorem and Some Trigonometric Limits

Prob. 1.

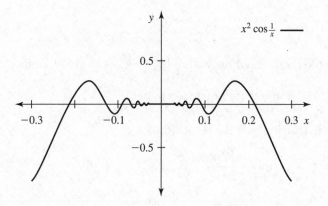

(a)

(b) Since $-1 \leq \cos \frac{1}{x} \leq 1$ for any x such that $\frac{1}{x}$ is defined (that is, for $x \neq 0$), we can multiply both inequalities by the positive number x^2, obtaining

$$-x^2 \leq x^2 \cos \frac{1}{x} \leq x^2.$$

(c) The role of $g(x)$ in the Sandwich Theorem is played by $x^2 \cos \frac{1}{x}$, and the sandwiching functions $f(x)$ and $h(x)$ are $-x^2$ and x^2, respectively. Since $\lim_{x \to 0} -x^2 = 0 = \lim_{x \to 0} x^2$, we conclude that $\lim_{x \to 0} x^2 \cos \frac{1}{x} = 0$ as well.

Prob. 3.

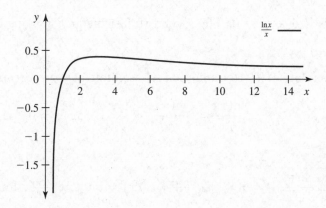

(a)

(b) The inequalities hold for $x \geq e$. This means that the inequalities can be applied as x tends to ∞.

(c) Since

$$\frac{1}{x} \leq \frac{\ln x}{x} \leq \frac{1}{\sqrt{x}}$$

we apply the Sandwich Theorem with $f(x) = \frac{1}{x}$, $g(x) = \frac{\ln x}{x}$ and $h(x) = \frac{1}{\sqrt{x}}$.

Since $\lim_{x \to \infty} \frac{1}{x} = 0 = \lim_{x \to \infty} \frac{1}{\sqrt{x}}$, we deduce that $\lim_{x \to \infty} \frac{\ln x}{x} = 0$.

Prob. 5. Use the substitution $y = 2x$ to obtain

$$\lim_{x \to 0} \frac{\sin(2x)}{2x} = \lim_{(y/2) \to 0} \frac{\sin y}{y}$$
$$= \lim_{y \to 0} \frac{\sin y}{y} = 1.$$

The last equality was proved in Section 3.4, and the second is justified because $y \to 0$ if and only if $(y/2) \to 0$.

Prob. 7. Use the substitution $y = 5x$ to obtain

$$\lim_{x \to 0} \frac{\sin(5x)}{x} = \lim_{x \to 0} \left(5 \cdot \frac{\sin(5x)}{5x} \right) = 5 \cdot \lim_{x \to 0} \frac{\sin(5x)}{5x}$$
$$= 5 \cdot \lim_{(y/5) \to 0} \frac{\sin y}{y} = 5 \cdot \lim_{y \to 0} \frac{\sin y}{y} = 5$$

since $\frac{\sin y}{y}$ tends to 1 as $y \to 0$. The second to last inequality is justified because $y \to 0$ if and only if $(y/5) \to 0$.

Prob. 9. Use the substitution $y = \pi x$ to obtain

$$\lim_{x \to 0} \frac{\sin(\pi x)}{x} = \lim_{x \to 0} \left(\pi \cdot \frac{\sin(\pi x)}{\pi x} \right) = \pi \cdot \lim_{x \to 0} \frac{\sin(\pi x)}{\pi x}$$
$$= \pi \cdot \lim_{(y/\pi) \to 0} \frac{\sin y}{y} = \pi \cdot \lim_{y \to 0} \frac{\sin y}{y} = \pi$$

since $\frac{\sin y}{y}$ tends to 1 as $y \to 0$. The second to last inequality is justified because $y \to 0$ if and only if $(y/\pi) \to 0$.

Prob. 11. We relate $\frac{\sin \pi x}{\sqrt{x}}$ to a function whose limit at 0 we already know, as follows:

$$\frac{\sin \pi x}{\sqrt{x}} = \frac{\sin \pi x}{\pi x} \cdot \pi \sqrt{x} \quad \text{for } x \neq 0.$$

Hence,

$$\lim_{x \to 0} \frac{\sin \pi x}{\sqrt{x}} = \lim_{x \to 0} \left(\frac{\sin \pi x}{\pi x} \cdot \pi \sqrt{x} \right) = \lim_{x \to 0} \frac{\sin \pi x}{\pi x} \cdot \lim_{x \to 0} \pi \sqrt{x}$$
$$= \lim_{\pi x \to 0} \frac{\sin \pi x}{\pi x} \cdot \pi \lim_{x \to 0} \sqrt{x} = 1 \cdot 0 = 0$$

Here we have used the fact that $\frac{\sin y}{y}$ tends to 1 as $y \to 0$, where $y = \pi x$, and that $(y/\pi) \to 0$ if and only if $y \to 0$.

Prob. 13. We relate $\frac{\sin x \cos x}{x(1-x)}$ to functions whose limit at 0 we already know, as follows:

$$\frac{\sin x \cos x}{x(1-x)} = \frac{\sin x}{x} \cdot \cos x \cdot \frac{1}{1-x}$$

Hence,

$$\lim_{x \to 0} \frac{\sin x \cos x}{x(1-x)} = \lim_{x \to 0} \left(\frac{\sin x}{x} \cdot \cos x \cdot \frac{1}{1-x} \right)$$

$$= \lim_{x \to 0} \frac{\sin x}{x} \cdot \lim_{x \to 0} \cos x \cdot \lim_{x \to 0} \frac{1}{1-x} = 1 \cdot 1 \cdot 1 = 1$$

Here we have used the fact that $\frac{\sin x}{x}$ tends to 1 as $x \to 0$.

Prob. 15. We have

$$\lim_{x \to 0} \frac{1 - \cos x}{2x} = \lim_{x \to 0} \left(\frac{1}{2} \cdot \frac{1 - \cos x}{x} \right)$$

$$= \frac{1}{2} \cdot \lim_{x \to 0} \frac{1 - \cos x}{x} = \frac{1}{2} \cdot 0 = 0.$$

The fact that $\frac{1-\cos x}{x}$ tends to 0 as $x \to 0$ was proved in Section 3.4.

Prob. 17.

$$\lim_{x \to 0} \frac{1 - \cos(5x)}{2x} = \lim_{x \to 0} \frac{1 - \cos(5x)}{\frac{2}{5} \cdot 5x}$$

$$= \lim_{x \to 0} \frac{5}{2} \cdot \frac{1 - \cos(5x)}{5x} = \lim_{x \to 0} \frac{5}{2} \cdot \lim_{x \to 0} \frac{1 - \cos(5x)}{5x} = \frac{5}{2} \cdot 0 = 0.$$

Prob. 19. We relate $\frac{\sin x(1-\cos x)}{x^2}$ to functions whose limit at 0 we already know, as follows:

$$\frac{\sin x(1 - \cos x)}{x^2} = \frac{\sin x}{x} \cdot \frac{1 - \cos x}{x}$$

Hence,

$$\lim_{x \to 0} \frac{\sin x(1 - \cos x)}{x^2}$$

$$= \lim_{x \to 0} \left(\frac{\sin x}{x} \cdot \frac{(1 - \cos x)}{x} \right)$$

$$= \lim_{x \to 0} \frac{\sin x}{x} \cdot \lim_{x \to 0} \frac{1 - \cos x}{x}$$

$$= 1 \cdot 0 = 0.$$

Here we use the fact that as $x \to 0$, $\frac{1-\cos x}{x}$ tends to 0 and $\frac{\sin x}{x}$ tends to 1. This was proved in Section 3.4.

Prob. 21.

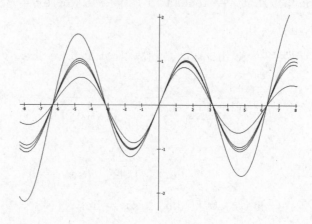

(a)

(b) $\sin(x)$ produces the oscillations

(c) a changes the amplitude of the oscillations.

(d) Since $-1 \leq \sin x \leq 1$ we have:

$$-e^{ax} \leq eax \sin x \leq e^{ax} \sin x$$

3.5   Properties of Continuous Functions

Prob. 1.

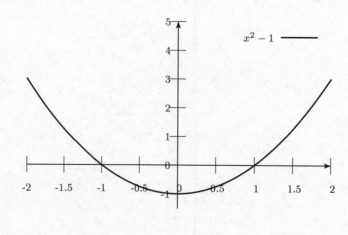

(a)

(b) We have $f(0) = -1$ and $f(2) = 3$, and so $f(0) = -1 < 0 < 3 = f(2)$. The function is continuous on the interval $[0, 2]$ since it a polynomial. Since $f(0) < 0 < f(2)$, by the

Intermediate Value Theorem there is some value c of the independent variable such that $0 < c < 2$ and $f(c) = 0$.

Prob. 3.

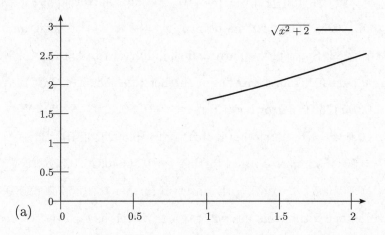

(a)

(b) We have $f(1) = \sqrt{1^2 + 2} = \sqrt{3}$ and $f(2) = \sqrt{2^2 + 2} = \sqrt{6}$. Now $\sqrt{3} < 2 < \sqrt{6}$ because $3 < 4 < 6$. Thus $f(1) < 2 < f(2)$, and since f is continuous on the interval $[1, 2]$ (being a square root of a polynomial that only takes positive values), we can apply the Intermediate Value Theorem to conclude that there is some value c of the independent variable such that $1 < c < 2$ and $f(c) = \sqrt{c^2 + 2} = 2$.

Prob. 5. Any solution to $e^{-x} = x$ is a root of the function $f(x) = e^{-x} - x$. Now, $f(1) = e^{-1} - 1$ is negative, since $e^{-1} = 1/e$ is less than 1, while $f(0) = e^0 - 0 = 1$ is positive. Since $f(x) = e^{-x} - x$ is the difference of two continuous functions, it is also continuous and this together with the fact that $f(0) > 0 > f(1)$ allows us to apply the Intermediate Value Theorem. We can conclude that there is a value c of the independent variable such that $0 < c < 1$ and $f(c) = 0$, that is, $e^{-c} = c$.

Prob. 7. We define the function $f(x) = e^{-x} - x$, which is continuous since it is the difference of two continuous functions. We are looking for a zero of this function. Now, $f(0) = e^0 - 0 = 1 > 0$ and $f(1) = e^{-1} - 1 < 0$, since $e^{-1} = 1/e < 1$. Therefore $f(x)$ must have a zero in the open interval $(0, 1)$, by the Intermediate Value Theorem. We next try the midpoint $x = \frac{0+1}{2} = 0.5$, for which $f(0.5) = e^{-0.5} - 0.5 \approx 0.106531$. Since this value is positive and $f(1)$ is negative, the function must have a zero between $x = 0.5$ and $x = 1$. We then try the midpoint $x = \frac{0.5+1}{2} = 0.75$, for which

$f(0.75) = e^{-0.75} - 0.75 \approx -0.277633$. Since this value is negative and $f(0.5)$ is positive, we have narrowed the location of a zero to the interval $(0.5, 0.75)$. We next test the midpoint between 0.5 and 0.75, namely $x = \frac{0.5+0.75}{2} = 0.625$. This gives $f(0.625) \approx -0.08973857$. Since this value is negative and $f(0.5)$ is positive, we have narrowed the location of a zero to the interval $(0.5, 0.625)$. The next midpoint is $x = \frac{0.5+0.625}{2} = 0.5625$, and $f(0.5625) \approx 0.00728282$. So, we have narrowed the interval to $(0.5625, 0.625)$. We next have $f(0.59375) \approx -0.04149755$, narrowing it further to $(0.5625, 0.59375)$. We next have $f(0.578125) \approx -0.01717584$, narrowing it further to $(0.5625, 0.578125)$. We next have $f(0.5703125) \approx -0.00496376$, narrowing it further to $(0.5625, 0.5703125)$. We next have $f(0.56640625) \approx 0.0011552$, narrowing it further to $(0.56640625, 0.5703125)$. We next have $f(0.5683594) \approx -0.0019054$, narrowing the interval further to $(0.56640625, 0.5683594)$.

Notice that any number within this interval can be written as 0.56 to two decimal places. Therefore our solution, accurate to two decimal places, is 0.56.

We note here that if we are not forced to choose the middle point at each step, then it might be possible to speed things up if we choose a point closer to an appropriate endpoint instead.

Prob. 9.

(a) We must start with some value of x, so we test $f(0)$, which gives 2. Next we test $f(1)$, which gives exactly 0. With this luck, we have found a solution $x = 1$ which is exact.

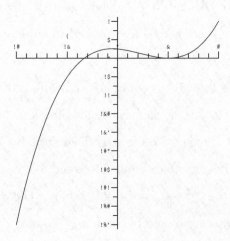

(b)

(c) The graph has a point of tangency with the x-axis at $x = 1$, and $f(x)$ is positive on either side of $x = 1$. So, if we had not found the solution $x = 1$ already, we would not be able to find it using the bisection method, since we would not suspect (by looking at the sign of $f(x)$ at $x = 0$ and $x = 2$, say) that there is a solution in between. On the other hand, having found the solution $x = 1$ and having plotted the graph, we can easily find the other root using the bisection method, starting, for example, with the interval $[-1, 0]$.

Prob. 11.

(a) The number should be reported as 23 individuals.

(b) We must decide to what accuracy formula (3.7) models the population. It is reasonable to think that the coefficients 54 and 13 that appear in the formula have 2 significant digits only (otherwise they would be written 54.0 and so on). Thus the result of formula (3.7) is also limited to two significant digits, and the correct way to express the result at $t = 10$ is 23 million individuals, or 2.3×10^7 individuals.

(c) A small population, such as that given in part (a), is not very well approximated by a continuous function. In particular, formula (3.7) purports to be valid for $t \geq 0$. At the lower end of this interval, $t = 0$, we have $N(t) = 4.15385$, and for some slightly bigger t we have $N(t) = 4.5$. Since the real population is measured by integers, this value of $N(t)$ involves a relative error of at least 0.5 in 4.5, or about 11%.

In part (b) on the other hand, the size of the population is large and the issue of the discreteness of the population does not play a significant role in the approximation. Other sources of error, including counting errors and the empirical nature of the modelling function in equation (3.7) to fit the data, would instead be the limiting factors in the precision of the approximation.

Prob. 13. A polynomial $f(x) = Ax^3 + Bx^2 + Cx + D$ of degree 3 takes on opposite signs as x approaches $+\infty$ and $-\infty$. This is because, for x very large in absolute value, the behavior of the polynomial is controlled by the term of highest degree (called the leading term); in other words, for x very large in absolute value, the polynomial has the same sign as Ax^3 (where A is either positive or negative). By starting with an interval of values of x such that $f(x)$ has opposite signs at the two endpoints, we can apply the Intermediate Value Theorem to deduce that somewhere in the interval, $f(x) = 0$.

Prob. 15. We present three different explanations.

First, we know the two roots; since $y = x^2 - 4 = (x+2)(x-2)$, they're obviously -2 and 2.

Second, since any second degree equation $y = ax^2 + bx + c$ of positive discriminant $D = b^2 - 4ac$ has two roots, the given equation must have two roots because it has $D = 0^2 - 4(1)(-4) = 16 > 0$.

Third, we note that for x large in absolute value (say $x = \pm 100$), we have $y = x^2 - 4$ is positive, while for $x = 0$, y is negative. Since y is a continuous function, by the Intermediate Value Theorem there is a root in the interval $(0, 100)$ and a root in the interval $(-100, 0)$.

3.6   Formal Definition of Limits

Prob. 1. The condition $\sqrt{2x - 1} < 0.01$ is equivalent to the following two inequalities:

$$2x - 1 < 0.01 \quad \text{and} \quad 2x - 1 > -0.01$$

These reduce to

$$x < 0.505 \quad \text{and} \quad x > 0.495$$

respectively. Therefore the desired values of x are those in the interval $(0.495, 0.505)$.

Prob. 3. From $|x^2 - 9| < 0.1$ we see that

$$-0.1 < x^2 - 9 < 0.1$$

$$-0.1 + 9 < x^2 < 0.1 + 9$$

$$8.9 < x^2 < 9.1$$

So, we have: $x^2 > 8.9$, meaning $X \in (-\infty, \sqrt{8.9}) \cup (\sqrt{8.9}, +\infty)$ and: $x^2 < 9.1$, meaning $x \in (-\sqrt{9.1}, \sqrt{9.1})$.

Putting it together gives: $X \in (-\sqrt{9.1}, -\sqrt{8.9}) \cup (\sqrt{8.9}, \sqrt{9.1})$.

Prob. 5.

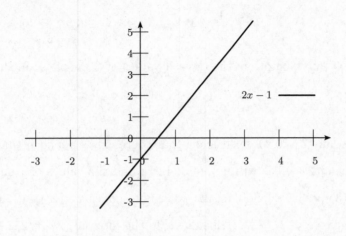

    (a)

(b) We must find all x such that $\sqrt{f(x)-3}$ is at most 0.1. Thus the inequalities to be satisfied are

$$(2x-1)-3 < 0.1 \quad \text{and} \quad (2x-1)-3 > -0.1$$

These become

$$x < 2.05 \quad \text{and} \quad x > 1.95$$

respectively. Thus the desired values of x are those in the interval $(1.95, 2.05)$.

    (c)

Prob. 7.

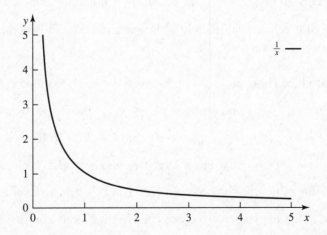

    (a)

(b) We must find all x such that $f(x) > 4$. This inequality becomes

$$\frac{1}{x} > 4 \quad \Rightarrow \quad x < \frac{1}{4} = 0.25$$

Since $x > 0$ by assumption, the desired values of x are those in the interval $(0, 0.25)$.

(c)

Prob. 9. Given any $\epsilon > 0$ we must show that there exists a δ such that if x is within δ of 2, then the value of $2x - 1$ is within ϵ of 3. Let's find the values of x such that $2x - 1$ lies in the interval $(3 - \epsilon, 3 + \epsilon)$. Solving $2x - 1 < 3 + \epsilon$ gives $x < 2 + \epsilon/2$ and solving $2x - 1 > 3 - \epsilon$ gives $x > 2 - \epsilon/2$. Thus any x in the interval $(2 - \epsilon/2, 2 + \epsilon/2)$ will give us a value of $2x - 1$ in the interval $(3 - \epsilon, 3 + \epsilon)$. This shows that we can choose $\delta = \epsilon/2$.

Prob. 11. Let $\epsilon > 0$, then we must have $|f(x) - f(c)| < \epsilon$ meaning $|x^5 - 0| < \epsilon$ $|x^5| < \epsilon$. $|x| < \sqrt[5]{\epsilon}$. Now, letting $\delta = \sqrt[5]{\epsilon}$ means that for every $\epsilon > 0$, there is $\delta = \sqrt[5]{\epsilon}$ so that $|x| < \sqrt[5]{\epsilon}$ implies that $|x^5| < \epsilon$. Therefore, the limit is indeed 0.

Prob. 13. Given any large $M > 0$ we must show that there exists a δ such that if x is within δ of 0, then the value of $4/x^2$ is larger than M. Let's find the values of x such that $4/x^2 > M$. Solving this gives $\sqrt{x} < 2/\sqrt{M}$. Thus any x in the interval $(-2/\sqrt{M}, 2/\sqrt{M})$ will give us a value of $4/x^2$ larger than M. This shows that we can choose $\delta = 2/\sqrt{M}$.

Prob. 15. Given any large $M > 0$ we must show that there exists a δ such that if x is within δ of 0, then the value of $1/x^4$ is larger than M. Let's find the values of x such that $1/x^4 > M$. Solving this gives $\sqrt{x} < 1/\sqrt[4]{M} = M^{-1/4}$. Thus any x in the interval $(-M^{-1/4}, M^{-1/4})$ will give us a value of $1/x^4$ larger than M. This shows that we can choose $\delta = M^{-1/4}$.

Prob. 17. We must show that for any x sufficiently large the value of $f(x)$ is close to 0. Let $\epsilon > 0$, we must have $|\frac{2}{x^2}| < \epsilon$, meaning $-\epsilon < \frac{2}{x^2} < \epsilon$, since $\frac{2}{x^2} > 0$, we have $\frac{2}{x^2} < \epsilon \Rightarrow x > \sqrt{\frac{2}{\epsilon}}$. Letting $L = \sqrt{\frac{2}{\epsilon}}$, any $x > L$ wraps ϵ-close to 0.

Prob. 19. Given any $\epsilon > 0$ we must show that there exists an $L > 0$ such that if x is larger than L, then the value of $\frac{x}{x+1}$ is within ϵ of 1. Let's find the values of x such that $\sqrt{x/(x+1) - 1} < \epsilon$. This inequality becomes the simultaneous conditions

$$\frac{x}{x+1} < 1 + \epsilon \quad \text{and} \quad \frac{x}{x+1} > 1 - \epsilon.$$

Since we are interested in large values of x, we can assume that $x + 1 > 0$. Then the inequalities above turn into

$$x < (x+1)(1+\epsilon) \quad \text{and} \quad x > (x+1)(1-\epsilon)$$

which have solutions

$$x > -\frac{1}{\epsilon} - 1 \quad \text{and} \quad x > \frac{1}{\epsilon} - 1$$

respectively. Evidently the second condition is the more stringent one and we conclude that the desired values of x are those satisfying $x > \frac{1}{\epsilon} - 1$. Thus any x larger than $\epsilon^{-1} - 1$ will give us a value of $\frac{x}{x+1}$ within a distance ϵ of 1. This shows that we can choose $L = \epsilon^{-1} - 1$.

Prob. 21. Let $\epsilon > 0$, then we must have

$$|mx - mc| < \epsilon$$

$$|m||x - c| < \epsilon$$

$$|x - c| < \frac{\epsilon}{|m|}.$$

Now, letting $\delta = \frac{\epsilon}{|m|}$, we have $|X - C| < \delta$ implies that $|mx - mc| < \epsilon$. Hence, the limit is indeed mc.

3.8 Review Problems

Prob. 1. The function $f(x) = e^{-\sqrt{x}}$ is continuous for all $x \in \mathbb{R}$. This is because it is a composition of three continuous functions, all with domain $\mathbb{R}$, namely $y = f_1(x) = |x|$, $z = f_2(y) = -y$, and $f_3(z) = e^z$.

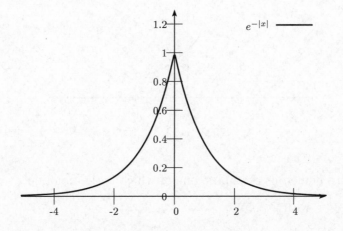

Prob. 3. The given function is a quotient of two functions, $g(x) = 2$ and $h(x) = e^x + e^{-x}$. Obviously $g(x)$ is a continuous function. Now $h(x)$ is continuous for every $x \in \mathbb{R}$, because it is the sum of two continuous functions, and furthermore it takes only positive values because both e^x and e^{-x} are always positive. Thus $h(x)$ is never equal to 0, and so $g(x)/h(x)$ is continuous for all $x \in \mathbb{R}$.

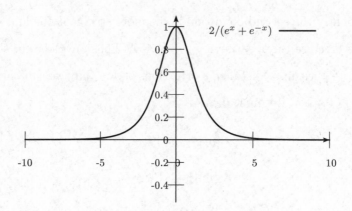

Prob. 5. An example of such a function is

$$f(x) = \begin{cases} 0 & \text{for } x < 1, \\ 1 & \text{for } x \geq 1. \end{cases}$$

and the graph of yet another one is:

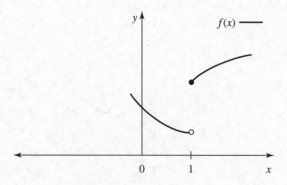

Prob. 7. There are numerous examples, one of which is the function $(e^x - e^{-x})/(e^x + e^{-x})$, with graph:

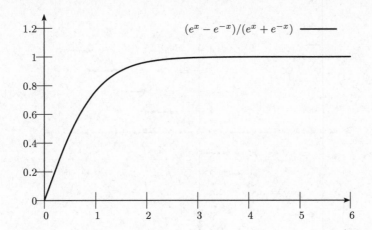

Another one is give by the graph:

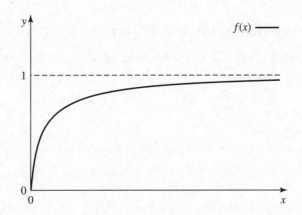

Prob. 9. First we show that the function $f(x) = \lfloor x \rfloor$ is continuous from the right at $x = -2$. We have $f(x) = -2$ for every $x \in [-2, -1)$. Therefore the function is constant and equal to -2 in an interval whose left endpoint is -2. This means that $\lim_{x \to -2+} f(x) = -2$. Since we also have $f(-2) = -2$, the function is continuous from the right at $x = -2$.

Next we show that the function is discontinuous from the left at $x = -2$. We have $f(x) = -3$ for every $x \in [-3, -2)$. Therefore the function is constant and equal to -3 in an interval whose right endpoint is -2. This means that $\lim_{x \to -2-} f(x) = -3$. But we have $f(-2) = -2$, and so the limit from the left does not agree with the value of the function at $x = -2$. Thus the function is discontinuous from the left at $x = -2$.

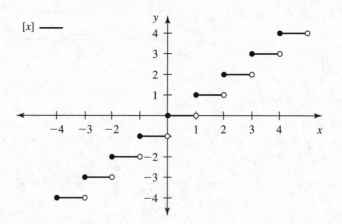

Prob. 11. The limit of $N(t)$ as $t \to \infty$ equals a, because we can rewrite $N(t)$ as

$$N(t) = \frac{at}{k+t} = \frac{a}{\frac{k}{t}+1}$$

and as $t \to \infty$ the denominator has the limit 1. Therefore $a = 1.24 \times 10^6$. We also know that at $t = 5$ the value of $N(t)$ is half of its limiting value, which is to say, $\frac{a}{2}$. Thus we have

$$\frac{a}{2} = N(5) = \frac{5a}{k+5} \quad \Rightarrow \quad k+5 = 10$$

and so $k = 5$.

Prob. 13.

(a) The function $g(t)$ should take the value 1 when $s(t) \geq 1/2$ and the value 0 when $s(t) < 1/2$. Now, $s(t) = \sin(\pi t)$ is a periodic function, with period 2. Thus we can consider only the interval $[0, 2)$ for the moment. Within the interval $[0, 2)$ the function $s(t)$ is greater than $1/2$ whenever t lies between $1/6$ and $5/6$ (corresponding to $\sin(\pi/6) = 0.5$ and $\sin(5\pi/6) = 0.5$).

Therefore, one way to construct the function g is the following: given a value of t, first reduce it to lie in the interval $[0, 2)$ by subtracting multiples of 2. This can be done by subtracting from t (which is positive) as great a multiple of 2 as required, namely $t - 2\lfloor t/2 \rfloor$. Then we check if this reduced value lies in the interval $[1/6, 5/6]$. If so, set $g(t) = 1$; if not, set $g(t) = 0$. Thus we have $g(t) = 1$ if $t - 2\lfloor t/2 \rfloor \in [1/6, 5/6]$, and $g(t) = 0$ otherwise.

Another way to describe the function is as follows:

$$g(t) = \begin{cases} 1 & \text{for } \frac{1}{6} + 2k \le x \le \frac{5}{6} + 2k, k = 0, 1, 2, \ldots \\ 0 & \text{otherwise} \end{cases}$$

(b) The function $s(t)$ is continuous, since the sine function is continuous. However, $g(t)$ is discontinuous, since it only takes two discrete values, 0 and 1. We can also see this by noticing that the floor function is discontinuous.

Prob. 15.

(a) The graph of $f(N) = aTN = 0.2N$ is linear.

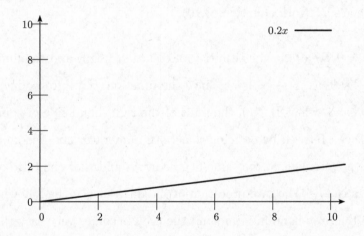

(b) In the modified model in which prey handling time is taken into account, the actual searching time is

$$T - T_h \frac{N_e}{P}$$

and this value is proportional to the number of prey encounters:

$$\frac{N_e}{P} = a\left(T - T_h \frac{N_e}{P}\right)N.$$

Solving for N_e/p we have

$$\frac{N_e}{P} = aTN - aT_h \frac{N_e}{P}N \;\Rightarrow\; \frac{N_e}{P}(1 + aT_hN) = aTN \;\Rightarrow\; \frac{N_e}{P} = \frac{aTN}{1 + aT_hN}.$$

We now graph $g(N) = 0.2N/(1 + 0.01N)$ for $N \ge 0$.

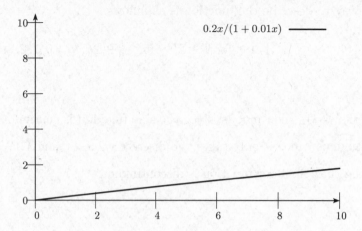

(c) When $T_h = 0$ the denominator of the right-hand side of equation (3.11) becomes 1, and so we are left with equation (3.10).

(d) When $T_h = 0$, we are in the situation of equation (3.10) and the limit of the right-hand side as $N \to \infty$ is ∞, since the function $f(N)$ grows linearly with N. On the other hand, when $T_h > 0$, the limit of the right-hand side of equation (3.11) is T/T_h; this can be seen by noticing that both numerator and denominator in equation (3.11) are polynomials of the same degree in N, and that the ratio of their leading coefficients is $\frac{aT}{aT_h}$. The interpretation of the difference is the following: if we disregard the prey-handling time and the prey becomes very easy to find, there will be larger and larger numbers of prey encounters. But if we take into account the prey-handling time, things change: even though another prey can be found immediately after one is eaten, the total number of prey that can be eaten is limited by the quotient of the available time T by the per-prey handling time T_h.

Prob. 17.

(a) The function sinh and cosh must be continuous because they are defined by taking sums, differences, and constant multiples of continuous functions. To see that tanh is continuous we must also observe that the denominator $e^x + e^{-x}$ never vanishes; the quotient of a continuous numerator by a continuous, nonvanishing denominator is again continuous.

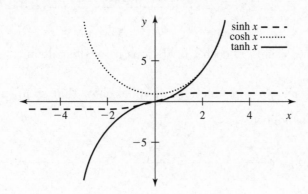

(b) As $x \to \infty$, the term e^{-x} in the definition of the hyperbolic sine becomes very small, and the term e^x very large, so the difference $e^x - e^{-x}$ is very large and the limit of $\sinh x$ is ∞. On the other hand, when $x \to -\infty$, the term e^{-x} becomes very large and the term e^x becomes very small; now the difference $e^x - e^{-x}$ is large and negative, and the limit of $\sinh x$ is $-\infty$.

With the hyperbolic cosine we have a sum $(e^x + e^{-x})$ instead of a difference, and either e^x or e^{-x} gets very large depending on whether $x \to \infty$ or $x \to -\infty$ respectively, while the other gets very small. Therefore the sum gets very large whether we take $x \to \infty$ or $x \to -\infty$. The limit is of $\cosh x$ ∞ in both cases.

With the hyperbolic tangent we cannot use this simple reasoning, because both numerator $(e^x - e^{-x})$ and denominator $(e^x + e^{-x})$ get very large. Instead, dividing numerator and denominator by e^x, we rewrite the quotient as follows:

$$\tanh x = \frac{e^x - e^{-x}}{e^x + e^{-x}} = \frac{1 - e^{-2x}}{1 + e^{-2x}}$$

As $x \to \infty$, e^{-2x} tends to zero, and so the limit of $\tanh x$ is 1. To find the limit of $\tanh x$ as $x \to -\infty$ we do yet another rewriting, by multiplying top and bottom by e^x:

$$\tanh x = \frac{e^x - e^{-x}}{e^x + e^{-x}} = \frac{e^{2x} - 1}{e^{2x} + 1}$$

As $x \to -\infty$, e^{2x} tends to zero, and so the limit of $\tanh x$ is -1.

(c) We have

$$\cosh^2 x - \sinh^2 x \qquad = \frac{(e^x + e^{-x})^2}{4} - \frac{(e^x - e^{-x})^2}{4}$$

$$= \frac{(e^{2x}+2+e^{-2x})-(e^{2x}-2+e^{-2x})}{4} = \frac{4}{4} = 1.$$

The equality $\tanh x = \sinh x / \cosh x$ is obvious from their definitions.

(d) To show that $\sinh x$ is an odd function, we must show that $\sinh(-x) = -\sinh x$. This can be seen as follows:

$$\sinh(-x) = \frac{e^{-x} - e^{-(-x)}}{2} = \frac{e^{-x} - e^{x}}{2} = -\frac{e^{x} - e^{-x}}{2} = -\sinh x.$$

Similarly, to show that $\cosh x$ is an even function, we must show that $\cosh(-x) = \cosh x$. This can be seen as follows:

$$\cosh(-x) = \frac{e^{-x} + e^{-(-x)}}{2} = \frac{e^{-x} + e^{x}}{2} = \frac{e^{x} + e^{-x}}{2} = \cosh x.$$

For tanh, we can use the fact that the quotient of an odd function and an even function is odd.

Chapter 4

Differentiation

4.1 Formal Definition of the Derivative

Prob. 1. $f'(x) = 0$, hence $f'(1) = 0$.

Prob. 3. $f'(x) = 4$, hence $f'(-1) = 4$.

Prob. 5. $f'(x) = 4x$, hence $f'(0) = 4 - 0 = 0$.

Prob. 7. The function $f(x) = \cos x$ is not constant, nor is it linear. Thus we must use the definition of the derivative here. We have

$$f'(0) = \lim_{h \to 0} \frac{f(0+h) - f(0)}{h} = \lim_{h \to 0} \frac{\cos h - \cos 0}{h} = \lim_{h \to 0} \frac{\cos h - 1}{h} = 0.$$

Here the last equality comes from Section 3.4.

Prob. 9. $f'(x) = -6x$, so we need to have

$$-6c = 0$$

$$c = 0.$$

Prob. 11. $f'(x) = 2(x - 2)$, we need to have:

$$2(c - 2) = 0$$

$$c = 2.$$

Prob. 13. $f'(x) = 2x - 6$, we need to have:

$$2c - 6 = 0$$

$$2c = 6$$

$$c = 3.$$

Prob. 15. $f'(x) = \cos(\frac{\pi}{2}x) \cdot \frac{\pi}{2}$, so we need to have

$$\frac{\pi}{2}\cos(\frac{\pi}{2}c) = 0$$

$$c = 2k + 1, \quad k \in Z.\texttt{(integers)}$$

Prob. 17. We have

$$
\begin{aligned}
f(c+h) - f(c) \quad &= f(2+h) - f(2) \\
&= -2(2+h) + 1 - (-2(2) + 1) \\
&= -4 - 2h + 1 + 3 \\
&= -2h
\end{aligned}
$$

Prob. 19. We have

$$
\begin{aligned}
f(c+h) - f(c) \quad &= f(4+h) - f(4) \\
&= \sqrt{4+h} - \sqrt{4} \\
&= \sqrt{4+h} - 2
\end{aligned}
$$

Prob. 21.

(a) We have to find the value of $f'(-1)$ when $f(x) = 5x^2$, and we have

$$
\begin{aligned}
f'(-1) \quad &= \lim_{h\to 0} \frac{f(-1+h) - f(-1)}{h} = \lim_{h\to 0} \frac{5(h-1)^2 - 5}{h} \\
&= \lim_{h\to 0} \frac{5h^2 - 10h}{h} = \lim_{h\to 0}(5h - 10) = -10.
\end{aligned}
$$

(b) When $x = -1$, we have $y = f(-1) = 5(-1)^2 = 5$. From part (a), we know that the slope of the tangent line is $m = -10$. Thus the equation of the tangent line is

$$y - 5 = -10(x - -1) \quad \textbf{or} \quad y = -10x - 5$$

(c) Graphing gives the following plot.

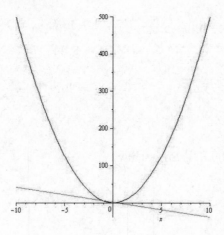

Prob. 23.

(a) We have to find the value of $f'(2)$ when $f(x) = 1 - x^3$, and we have

$$f'(2) \quad = \lim_{h \to 0} \frac{f(2+h) - f(2)}{h} = \lim_{h \to 0} \frac{1 - (h+2)^3 - (1 - 2^3)}{h}$$

$$= \lim_{h \to 0} \frac{-h^3 - 6h^2 - 12h}{h} = \lim_{h \to 0} (-h^2 - 6h - 12) = -12.$$

(b) When $x = 2$, we have $y = f(2) = 1 - (2)^3 = -7$. From part (a), we know that the slope of the tangent line is -12. Thus the slope of the normal line is $m = 1/12$, and the equation of the normal line is

$$y - (-7) = \frac{1}{12}(x - 2) \quad \text{or} \quad y = \frac{1}{12}x - \frac{43}{6}$$

(c) Graphing gives the following plot.

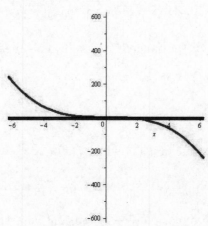

Prob. 25. We have $y = f(x) = \sqrt{x}$, and so

$$
\begin{aligned}
f'(x) &= \lim_{h\to 0} \frac{f(x+h)-f(x)}{h} = \lim_{h\to 0} \frac{\sqrt{x+h}-\sqrt{x}}{h} \\
&= \lim_{h\to 0} \left(\frac{\sqrt{x+h}-\sqrt{x}}{h} \cdot \frac{\sqrt{x+h}+\sqrt{x}}{\sqrt{x+h}+\sqrt{x}} \right) \\
&= \lim_{h\to 0} \frac{x+h-x}{h(\sqrt{x+h}+\sqrt{x})} = \lim_{h\to 0} \frac{1}{\sqrt{x+h}+\sqrt{x}} \\
&= \frac{1}{\sqrt{x}+\sqrt{x}} = \frac{1}{2\sqrt{x}}
\end{aligned}
$$

Prob. 27. $f'(x) = 6x$, $f'(1) = 6$. So we have:

$$
y - 3 = 6(x - 1)
$$

$$
y = 6x - 3.
$$

Prob. 29. We first find the value of $f'(4)$ when $f(x) = \sqrt{x}$, and we have

$$
\begin{aligned}
f'(4) &= \lim_{h\to 0} \frac{f(4+h)-f(4)}{h} = \lim_{h\to 0} \frac{\sqrt{4+h}-\sqrt{4}}{h} \\
&= \lim_{h\to 0} \left(\frac{\sqrt{4+h}-\sqrt{4}}{h} \cdot \frac{\sqrt{4+h}+\sqrt{4}}{\sqrt{4+h}+\sqrt{4}} \right) \\
&= \lim_{h\to 0} \frac{4+h-4}{h(\sqrt{4+h}+\sqrt{4})} = \lim_{h\to 0} \frac{1}{\sqrt{4+h}+\sqrt{4}} = \lim_{h\to 0} \frac{1}{\sqrt{4}+\sqrt{4}} = \frac{1}{4}.
\end{aligned}
$$

Thus we know that the slope of the tangent line is $1/4$. The equation of the tangent line is

$$
y - 2 = \frac{1}{4}(x - 4) \quad \textbf{or} \quad y = \frac{1}{4}x + 1
$$

Prob. 31. We first find the value of $f'(-1)$ when $f(x) = -3x^2$, and we have

$$
\begin{aligned}
f'(-1) &= \lim_{h\to 0} \frac{f(h-1)-f(-1)}{h} = \lim_{h\to 0} \frac{-3(h-1)^2-(-3(-1)^2)}{h} \\
&= \lim_{h\to 0} \frac{-3h^2+6h-3+3}{h} \\
&= \lim_{h\to 0} \frac{-3h^2+6h}{h} = \lim_{h\to 0}(-3h + 6) = 6.
\end{aligned}
$$

Thus we know that the slope of the tangent line is 6. This means the slope of the normal line is $m = -1/6$, and its equation is

$$
y - (-3) = -\frac{1}{6}(x - (-1)) \quad \textbf{or} \quad y = -\frac{1}{6}x - \frac{19}{6}
$$

Prob. 33. We first find the value of $f'(1)$ when $f(x) = 2x^2 - 1$, and we have

$$
\begin{aligned}
f'(1) &= \lim_{h\to 0} \frac{f(h+1)-f(1)}{h} = \lim_{h\to 0} \frac{2(h+1)^2-1-1}{h} \\
&= \lim_{h\to 0} \frac{2h^2+4h}{h} = \lim_{h\to 0}(2h + 4) = 4.
\end{aligned}
$$

Thus we know that the slope of the tangent line is 4. This means the slope of the normal line is $m = -1/4$, and its equation is

$$y - 1 = -\frac{1}{4}(x - 1) \quad \text{or} \quad y = -\frac{1}{4}x + \frac{5}{4}$$

Prob. 35. Comparing the given limit to

$$f'(a) = \lim_{h \to 0} \frac{f(a + h) - f(a)}{h}$$

we find that $f(x) = 2x^2$ and $x = a$.

Prob. 37. Comparing the given limit to

$$f'(a) = \lim_{h \to 0} \frac{f(a + h) - f(a)}{h}$$

we find that $f(x) = \frac{1}{x^2+1}$ and $a = 2$.

Prob. 39.

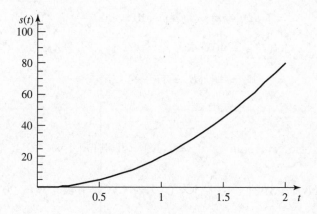

(a)

(b) The average velocity is $\frac{s(2)-s(0)}{2-0} = \frac{20(2)^2-20(0)^2}{2} = \frac{80}{2} = 40$ km/hr.

(c) At $t = 1$, the instantaneous velocity is

$$\lim_{h \to 0} \frac{s(h + 1) - s(1)}{h} = \lim_{h \to 0} \frac{20(h+1)^2-20}{h}$$
$$= \lim_{h \to 0} \frac{20h^2+40h}{h} = \lim_{h \to 0} (20h + 40) = 40.$$

Thus the instantaneous velocity at $t = 1$ is 40 km/hr.

Prob. 41.

(a) At $t = 3/4$, the car is $s(3/4) = \frac{160}{3}(3/4)^2 = 30$ km along the road. At $t = 1$, the car is $s(1) = \frac{160}{3}(1)^2 = \frac{160}{3} \approx 53.33$ km along the road.

(b) The average velocity is $\frac{s(1)-s(3/4)}{1-\frac{3}{4}} = \frac{\frac{160}{3}-30}{1/4} = \frac{280}{3} \approx 93.33$ km/hr.

(c) At $t = 3/4$, the velocity is

$$\lim_{h \to 0} \frac{s(h + \frac{3}{4}) - s(\frac{3}{4})}{h} \qquad = \lim_{h \to 0} \frac{\frac{160}{3}(h+\frac{3}{4})^2 - 30}{h}$$

$$= \lim_{h \to 0} \frac{\frac{160}{3}h^2 + 80h}{h} = \lim_{h \to 0}\left(\frac{160}{3}h + 80\right) = 80.$$

Thus the velocity at $t = 3/4$ is 80 km/hr. The speed of the car at $t = 3/4$ is also 80 km/hr.

Prob. 43. To find the value of R^*, we must solve the equation $\frac{1}{b}\frac{dB}{dt} = 0$. That is,

$$\frac{1}{b}\frac{dB}{dt} \qquad = f(R) - m = f(R) - 40 = 200\frac{R}{5+R} - 40 = 0$$

$$\Rightarrow 200\frac{R}{5+R} = 40 \quad \Rightarrow \quad \frac{5R}{5+R} = 1 \quad \Rightarrow \quad 5R = 5 + R$$

$$\Rightarrow 4R = 5 \quad \Rightarrow \quad R = \tfrac{4}{5}$$

That is, $R^* = 1.25$

Prob. 45. For equilibria, we require $\frac{dN}{dt} = f(N) = 0$. That is,

$$f(N) = 3N\left(1 - \frac{N}{20}\right) = 0 \quad \Rightarrow \quad N = 0 \text{ and } N = 20$$

Thus the points of equilibria are $N = 0$ and $N = 20$.

Prob. 47. We want $\frac{dx}{dt} = 0$, which must occur when $k(a - x)(b - x) = 0$, or when $x = a$ and $x = b$. That is, $x = 7$ or $x = 4$. However, the reaction ceases when $x = 4$ since at this point there is no more reactant B remaining.

Prob. 49. We can see that $\frac{dN}{dt} = 0$ when $N = 0$ or when $1 - \frac{N}{K} = 0$, that is when $N = K$. The value K is called the carrying capacity, because at this point the rate of change $\frac{dN}{dt}$ must be zero, and the population cannot grow any more. Thus K corresponds to the largest possible size of the population.

Prob. 51. Only **(B)** is true. See Section 4.1.3 for details.

Prob. 53. The domain is the interval $[a, b]$. The function below is differentiable at all values of x in the interval $[a, b]$ except for the value $x = c$.

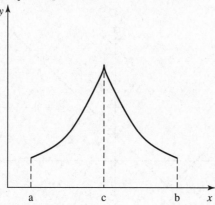

Prob. 55. Consider the following:

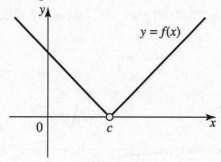

This is clearly differentiable everywhere except $X = C$, and not continuous. Hence it is not true.

Prob. 57. The function has a sharp corner at $x = -5$, and is not differentiable there.

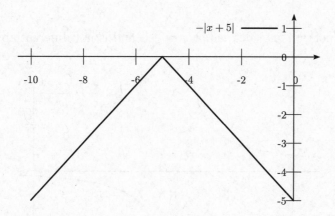

Prob. 59. The function has a sharp corner at $x = -2$, and is not differentiable there.

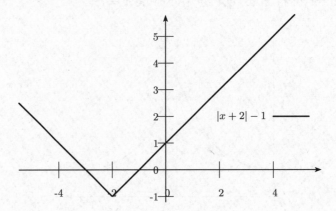

Prob. 61. The function is not defined at $x = 3$, and is not differentiable there.

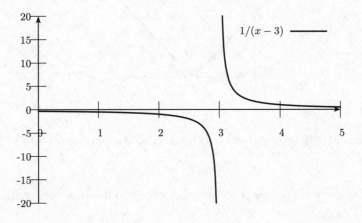

Prob. 63. The function is not defined at $x = -1$, and is not differentiable there.

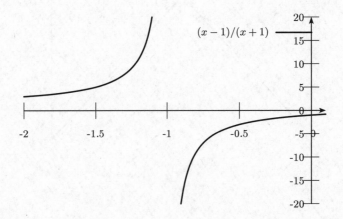

Prob. 65. The function has sharp corners at $x = \pm 1/\sqrt{2}$, and is not differentiable there.

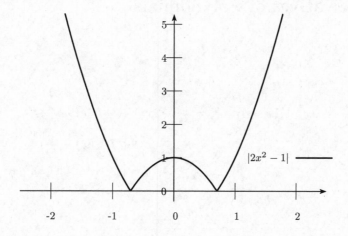

Prob. 67. The function has a discontinuity at $x = 1$, and is not differentiable there.

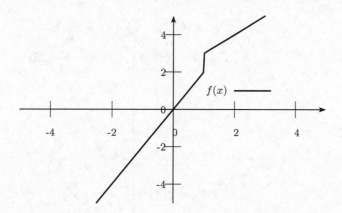

Prob. 69. The function has a discontinuity at $x = 0$, and is not differentiable there.

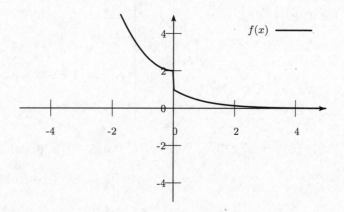

4.2   The Power Rule, the Basic Rules of Differentiation and the Derivatives of Polynomials

Prob. 1. $f'(x) = 12x^2 - 7$

Prob. 3.

$$\frac{d}{dx}(-2x^5 + 7x - 4) \quad = \frac{d}{dx}(-2x^5) + \frac{d}{dx}(7x) - \frac{d}{dx}(4)$$
$$= -2\frac{d}{dx}(x^5) + 7\frac{d}{dx}(x) - \frac{d}{dx}(4)$$
$$= -2(5x^4) + 7(1) - 0 = -10x^4 + 7$$

Prob. 5.

$$\frac{d}{dx}(3 - 4x - 5x^2) \quad = \frac{d}{dx}(3) - \frac{d}{dx}(4x) - \frac{d}{dx}(5x^2)$$
$$= \frac{d}{dx}(3) - 4\frac{d}{dx}(x) - 5\frac{d}{dx}(x^2)$$
$$= 0 - 4(1) - 5(2x) = -4 - 10x$$

Prob. 7.

$$\frac{d}{ds}(5s^7 + 2s^3 - 5s) \quad = \frac{d}{ds}(5s^7) + \frac{d}{ds}(2s^3) - \frac{d}{ds}(5s)$$
$$= 5\frac{d}{ds}(s^7) + 2\frac{d}{ds}(s^3) - 5\frac{d}{ds}(s)$$
$$= 5(7s^6) + 2(3s^2) - 5(1) = 35s^6 + 6s^2 - 5$$

Prob. 9. $h' = -\frac{1}{3} \cdot 4 \cdot t^3 + 4 \cdot 1 = -\frac{4}{3}t^3 + 4.$

Prob. 11.

$$\frac{d}{dx}(x^2 \sin(\pi/3) + \tan(\pi/4)) \quad = \frac{d}{dx}(x^2 \sin(\pi/3)) + \frac{d}{dx}(\tan(\pi/4))$$
$$= \sin(\pi/3)\frac{d}{dx}(x^2) + \frac{d}{dx}(\tan(\pi/4))$$
$$= \sin(\pi/3)(2x^1) + 0 = \sqrt{3}x$$

Prob. 13.

$$\frac{d}{dx}(-3x^4 \tan(\pi/6) - \cot(\pi/6)) \quad = \frac{d}{dx}(-3x^4 \tan(\pi/6)) - \frac{d}{dx}(\cot(\pi/6))$$
$$= -3\tan(\pi/6)\frac{d}{dx}(x^4) - \frac{d}{dx}(\cot(\pi/6))$$
$$= -3\tan(\pi/6)(4x^3) - 0 = -4\sqrt{3}x^3$$

Prob. 15.

$$\frac{d}{dt}(t^3 e^{-2} + t) = \frac{d}{dt}(t^3 e^{-2}) + \frac{d}{dt}(t)$$

$$= e^{-2}\frac{d}{dt}(t^3) + \frac{d}{dt}(t)$$

$$= e^{-2}(3t^2) + (1) = 3e^{-2}t^2 + 1$$

Prob. 17.

$$\frac{d}{ds}(s^3 e^3 + 3e) = \frac{d}{ds}(s^3 e^3) + \frac{d}{ds}(3e)$$

$$= e^3\frac{d}{ds}(s^3) + \frac{d}{ds}(3e)$$

$$= e^3(3s^2) + 0 = 3e^3 s^2$$

Prob. 19.

$$\frac{d}{dx}(20x^3 - 4x^6 + 9x^8) = \frac{d}{dx}(20x^3) - \frac{d}{dx}(4x^6) + \frac{d}{dx}(9x^8)$$

$$= 20\frac{d}{dx}(x^3) - 4\frac{d}{dx}(x^6) + 9\frac{d}{dx}(x^8)$$

$$= 20(3x^2) - 4(6x^5) + 9(8x^7) = 60x^2 - 24x^5 + 72x^7$$

Prob. 21. $f'(x) = 3\pi x^2 + \frac{1}{\pi}$

Prob. 23. We have

$$\frac{d}{dx}(ax^3) = a\frac{d}{dx}(x^3)$$

$$= a(3x^2)$$

$$= 3ax^2$$

Prob. 25. We have

$$\frac{d}{dx}(ax^2 - 2a) = \frac{d}{dx}(ax^2) - \frac{d}{dx}(2a)$$

$$= a\frac{d}{dx}(x^2) - \frac{d}{dx}(2a)$$

$$= a(2x) - 0$$

$$= 2ax$$

Prob. 27. We have

$$\frac{d}{ds}(rs^2 - r) = \frac{d}{ds}(rs^2) - \frac{d}{ds}(r)$$

$$= r\frac{d}{ds}(s^2) - \frac{d}{ds}(r)$$

$$= r(2s^1) - 0$$

$$= 2rs$$

Prob. 29. We have

$$\frac{d}{dx}(rs^2x^3 - rx + s) = \frac{d}{dx}(rs^2x^3) - \frac{d}{dx}(rx) + \frac{d}{dx}(s)$$

$$= rs^2\frac{d}{dx}(x^3) - r\frac{d}{dx}(x) + \frac{d}{dx}(s)$$

$$= rs^2(3x^2) - r(1) + 0$$

$$= 3rs^2x^2 - r$$

Prob. 31. We have

$$\frac{d}{dN}\left((b-1)N^4 - \frac{N^2}{b}\right) = \frac{d}{dN}\left((b-1)N^4\right) - \frac{d}{dN}\left(\frac{N^2}{b}\right)$$

$$= (b-1)\frac{d}{dN}\left(N^4\right) - \frac{1}{b}\frac{d}{dN}\left(N^2\right)$$

$$= (b-1)\left(4N^3\right) - \frac{1}{b}\left(2N^1\right)$$

$$= 4(b-1)N^3 - \frac{2}{b}N$$

Prob. 33. We have

$$\frac{d}{dt}(a^3t - at^3) = \frac{d}{dt}(a^3t) - \frac{d}{dt}(at^3)$$

$$= a^3\frac{d}{dt}(t) - a\frac{d}{dt}(t^3)$$

$$= a^3(1) - a(3t^2)$$

$$= a^3 - 3at^2$$

Prob. 35. We have

$$\frac{d}{dt}(V_0(1 + \gamma t)) = \frac{d}{dt}(V_0 + V_0\gamma t)$$

$$= \frac{d}{dt}(V_0) + \frac{d}{dt}(V_0\gamma t)$$

$$= \frac{d}{dt}(V_0) + V_0\gamma\frac{d}{dt}(t)$$

$$= 0 + V_0\gamma(1) = V_0\gamma$$

Prob. 37. We have

$$
\begin{aligned}
\frac{d}{dN}\left(N\left(1-\frac{N}{K}\right)\right) &= \frac{d}{dN}\left(N-\frac{N^2}{K}\right) \\
&= \frac{d}{dN}(N) - \frac{d}{dN}\left(\frac{N^2}{K}\right) \\
&= \frac{d}{dN}(N) - \frac{1}{K}\frac{d}{dN}\left(N^2\right) \\
&= 1 - \frac{1}{K}(2N) = 1 - \frac{2}{K}N
\end{aligned}
$$

Prob. 39. We have

$$
\begin{aligned}
\frac{d}{dN}\left(rN^2\left(1-\frac{N}{K}\right)\right) &= \frac{d}{dN}\left(rN^2-\frac{rN^3}{K}\right) \\
&= \frac{d}{dN}(rN^2) - \frac{d}{dN}\left(\frac{rN^3}{K}\right) \\
&= r\frac{d}{dN}(N^2) - \frac{r}{K}\frac{d}{dN}\left(N^3\right) \\
&= r(2N^1) - \frac{r}{K}(3N^2) = 2rN - \frac{3r}{K}N^2
\end{aligned}
$$

Prob. 41. We have

$$
\begin{aligned}
\frac{d}{dT}\left(\frac{2\pi^5}{15}\frac{k^4}{c^2h^3}T^4\right) &= \frac{2\pi^5}{15}\frac{k^4}{c^2h^3}\frac{d}{dT}\left(T^4\right) \\
&= \frac{2\pi^5}{15}\frac{k^4}{c^2h^3}(4T^3) = \frac{8\pi^5}{15}\frac{k^4}{c^2h^3}T^3
\end{aligned}
$$

Prob. 43. When $y = 7x^3 + 2x - 1$, we have $y' = 21x^2 + 2$. When $x = -3$ we have $y' = 191$, which means that the slope of the required tangent line is $m = 191$. Also, when $x = -3$ we have $y = -196$, and since the tangent line must pass through the point $(-3, -196)$, it's equation is

$$
y - (-196) = 191(x - (-3)) \quad \text{or} \quad y = 191x + 377.
$$

In standard form, this is $191x - y + 377 = 0$.

Prob. 45. $y' = 8x^3 - 5$, $y'(1) = 8(1)^3 - 5 = 3$. So, we have

$$
y - (-3) = 3(X - 1)
$$

$$
y = 3X - 6.
$$

Prob. 47. When $y = \frac{1}{\sqrt{2}}x^2 - \sqrt{2}$, we have $y' = \frac{2}{\sqrt{2}}x = \sqrt{2}x$. When $x = 4$ we have $y' = 4\sqrt{2}$, which means that the slope of the required tangent line is $m = 4\sqrt{2}$. Also, when

$x = 4$ we have $y = \frac{1}{\sqrt{2}}(16) - \sqrt{2} = 8\sqrt{2} - \sqrt{2} = 7\sqrt{2}$, and since the tangent line must pass through the point $(4, 7\sqrt{2})$, it's equation is

$$y - 7\sqrt{2} = 4\sqrt{2}(x - 4) \quad \text{or} \quad y = 4\sqrt{2}x - 9\sqrt{2}.$$

In standard form, this is $8x - \sqrt{2}y - 18 = 0$.

Prob. 49. $y' = 2X$, $y'(-1) = -2$. So, we have:

$$y - (3) = \frac{1}{2}(x + 1)$$

$$y = \frac{1}{2}x + \frac{7}{2}.$$

Prob. 51. When $y = \sqrt{3}x^4 - 2\sqrt{3}x^2$, we have $y' = 4\sqrt{3}x^3 - 4\sqrt{3}x = 4\sqrt{3}x(x^2 - 1)$. When $x = -\sqrt{3}$ we have $y' = -24$, which means that the slope of the tangent line is -24. Thus the slope of the normal line is $m = 1/24$. Also, when $x = -\sqrt{3}$ we have $y = 3\sqrt{3}$, and since the normal line must pass through the point $(-\sqrt{3}, 3\sqrt{3})$, it's equation is

$$y - 3\sqrt{3} = \frac{1}{24}(x - (-\sqrt{3})) \quad \text{or} \quad y = \frac{1}{24}x + \frac{\sqrt{3}}{24} + 3\sqrt{3}.$$

In standard form, this is $x - 24y + 73\sqrt{3} = 0$.

Prob. 53. $y' = 3x^2$, $y'(1) = 3$. So, we have:

$$y - (-2) = -\frac{1}{3}(x - 1)$$

$$y = -\frac{1}{3}x - \frac{5}{3}.$$

Prob. 55. When $f(x) = ax^2$, we have $f'(x) = 2ax$. When $x = 1$ we have $f'(1) = 2a$, which means that the slope of the required tangent line is $m = 2a$. Also, when $x = 1$ we have $f(1) = a$, and since the tangent line must pass through the point $(1, a)$, it's equation is

$$y - a = 2a(x - 1) \quad \text{or} \quad y = 2ax - a.$$

Prob. 57. When $f(x) = \frac{ax^2}{a^2+2} = \frac{a}{a^2+2}x^2$, we have $f'(x) = \frac{2a}{a^2+2}x$. When $x = 2$ we have $f'(2) = \frac{4a}{a^2+2}$, which means that the slope of the required tangent line is $m = \frac{4a}{a^2+2}$. Also, when $x = 2$ we have $f(2) = \frac{4a}{a^2+2}$, and since the tangent line must pass through the point $(2, 4a/(a^2 + 2))$, it's equation is

$$y - \frac{4a}{a^2 + 2} = \frac{4a}{a^2 + 2}(x - 2) \quad \text{or} \quad 4ax - (a^2 + 2)y - 4a = 0.$$

Prob. 59. When $f(x) = ax^3$, we have $f'(x) = 3ax^2$. When $x = -1$ we have $f'(-1) = 3a$, which means that the slope of the tangent line is $3a$. Thus the slope of the normal line is $m = -1/(3a)$. Also, when $x = -1$ we have $f(-1) = -a$, and since the normal line must pass through the point $(-1, -a)$, it's equation is

$$y - (-a) = -\frac{1}{3a}(x - (-1)) \quad \text{or} \quad x + 3ay + 1 + 3a^2 = 0.$$

Prob. 61. When $f(x) = \frac{ax^2}{a+1} = \frac{a}{a+1}x^2$, we have $f'(x) = \frac{2a}{a+1}x$. When $x = 2$ we have $f'(2) = \frac{4a}{a+1}$, which means that the slope of the tangent line is $\frac{4a}{a+1}$. Thus the slope of the normal line is $m = -(a+1)/(4a)$. Also, when $x = 2$ we have $f(2) = \frac{4a}{a+1}$, and since the normal line must pass through the point $(2, \frac{4a}{a+1})$, it's equation is

$$y - \frac{4a}{a+1} = -\frac{a+1}{4a}(x-2) \quad \text{or} \quad (a+1)^2 x + 4a(a+1)y - 16a^2 - 2(a+1)^2 = 0.$$

Prob. 63. A horizontal line has slope equal to zero. Thus any point $(x, f(x))$ at which there is a horizontal tangent line must satisfy $f'(x) = 0$. Here $f(x) = x^2$ and so $f'(x) = 2x$. This implies that $f'(x) = 2x = 0$ only when $x = 0$. The desired point is then $(0, 0)$.

Prob. 65. A horizontal line has slope equal to zero. Thus any point $(x, f(x))$ at which there is a horizontal tangent line must satisfy $f'(x) = 0$. Here $f(x) = 3x - x^2$ and so $f'(x) = 3 - 2x$. This implies that $f'(x) = 3 - 2x = 0$ only when $x = 3/2$. The desired point is then $(3/2, 9/4)$.

Prob. 67. A horizontal line has slope equal to zero. Thus any point $(x, f(x))$ at which there is a horizontal tangent line must satisfy $f'(x) = 0$. Here $f(x) = 3x^3 - x^2$ and so $f'(x) = 9x^2 - 2x$. This implies that $f'(x) = 9x^2 - 2x = x(9x - 2) = 0$ only when $x = 0$ and when $x = 2/9$. The desired points are $(0, 0)$ and $(2/9, -4/243)$.

Prob. 69. A horizontal line has slope equal to zero. Thus any point $(x, f(x))$ at which there is a horizontal tangent line must satisfy $f'(x) = 0$. Here $f(x) = \frac{1}{2}x^4 - \frac{7}{3}x^3 - 2x^2$ and so $f'(x) = 2x^3 - 7x^2 - 4x$. This implies that
$$f'(x) = 2x^3 - 7x^2 - 4x = x(2x^2 - 7x - 4) = x(2x + 1)(x - 4) = 0$$
only when $x = 0$, $x = 4$ and $x = -1/2$. The desired points are then $(0, 0)$, $(4, -160/3)$ and $(-1/2, -17/96)$.

Prob. 71. Suppose the point we wish to find is $(a, 4 - a^2)$. That is, the tangent line to the curve $y = 4 - x^2$ at this point is parallel to the line $y = 2$. Since $y' = -2x$, the slope of the tangent line at $(a, 4 - a^2)$ is given by $m = -2a$. Since the line $y = 2$ has slope 0 and parallel lines have the same slope, we want the tangent line to have slope 0. Thus $m = -2a = 0$ which means $a = 0$. There is only one such point, and it is $(0, 4)$.

Prob. 73. Since $y' = 4x$, $y = x$ has slope at 1, then:

$$4x = 1$$

$$x = \frac{1}{4}.$$

Hence, there is only 1 such point, namely $(]frac14, -\frac{3}{8})$.

Prob. 75. Suppose the point we wish to find is $(a, a^3 + 2a + 2)$. That is, the tangent line to the curve $y = x^3 + 2x + 2$ at this point is parallel to the line $3x - y = 2$. Since $y' = 3x^2 + 2$, the slope of the tangent line at $(a, a^3 + 2a + 2)$ is given by $m = 3a^2 + 2$. Now, the line $3x - y = 2$ can be written as $y = 3x - 2$ and so it has slope 3. Since parallel lines have the same slope, we want the tangent line to have slope 3. Thus $m = 3a^2 + 2 = 3$ which means $a = \pm 1/\sqrt{3}$. There are two such points, namely $(1/\sqrt{3}, 2 + 7/(3\sqrt{3}))$ and $(-1/\sqrt{3}, 2 - 7/(3\sqrt{3}))$.

Prob. 77. Since $y = x^2$, we have $y' = 2x$. When $x = 1$ we have $y' = 2$, which means that the slope of the tangent line at the point $(1, 1)$ is $m = 2$. The tangent line must pass through the point $(1, 1)$, and so it has equation

$$y - 1 = 2(x - 1) \quad \text{or} \quad y = 2x - 1.$$

Since $y = 2(0) - 1 = -1$ when $x = 0$, this tangent line passes through the point $(0, -1)$.

Prob. 79. Since $y = x^2$, we have $y' = 2x$. If (t, t^2) is an arbitrary point on the curve, then the slope of the tangent line at this point is $m = 2t$. The tangent line must pass through the point (t, t^2), and so it has equation

$$y - t^2 = 2t(x - t) \quad \text{or} \quad y = 2tx - t^2.$$

If this line is to pass through the point $(0, -a^2)$, then we must have $-a^2 = y = 2t(0) - t^2$, which means that $t^2 = a^2$ and so $t = \pm a$. Thus the desired points are (a, a^2) and $(-a, a^2)$, with tangent lines $y = 2ax - a^2$ and $y = -2ax - a^2$ respectively.

Prob. 81. If $P(x)$ is a polynomial of degree 4, then it can be written as $P(x) = a_4 x^4 + a_3 x^3 + a_2 x^2 + a_1 x + a_0$, where the a_i are constants for $0 \le i \le 4$ and $a_4 \ne 0$.

Differentiating, we get $P'(x) = 4a_4x^3 + 3a_3x^2 + 2a_2x + a_1$, which is a polynomial of degree 3 because $4a_4 \neq 0$.

4.3 Product Rule and Quotient Rules

Prob. 1. $f'(x) = (x+5)'(x^2-3) + (x+5)(x^2-3)'$

$$= (x^2-3) + (x+5)(2x).$$

Prob. 3. $f'(x) = (12x^3)(2x - 5x^3) + (3x^4 - 5)(2 - 15x^2)$

Prob. 5.

$\frac{d}{dx}((\frac{1}{2}x^2 - 1)(2x + 3x^2)) = x(2x + 3x^2) + (\frac{1}{2}x^2 - 1)(2 + 6x) = -2 - 6x + 3x^2 + 6x^3$.

Prob. 7. $f'(x) = \frac{1}{5}\left[2x(x^2+1) + 2x(x^2-1)\right]$.

Prob. 9. $\frac{d}{dx}(3x-1)^2 = \frac{d}{dx}(3x-1)(3x-1) = 3(3x-1) + (3x-1)3 = 6(3x-1)$.

Prob. 11. We write $f(x) = 3uv$ where $u = v = 1 - 2x$. Since the two factors of the same, $uv' = u'v = 2uu'$ and we get $f'(x) = 3(2(1-2x)(-2)) = -12(1-2x)$.

Prob. 13. $\frac{d}{ds}(2s^2 - 5s)^2 = 2(2s^2 - 5s)\frac{d}{ds}(2s^2 - 5s) = 2(2s^2 - 5s)(4s - 5)$.

Prob. 15. $\frac{d}{dt}(3(2t^2 - 5t^4)^2) = 3(2(2t^2 - 5t^4)(4t - 20t^3))$.

Prob. 17. $f'(x) = -2 - 6x + 9x^2$, so $f'(1) = 1$ and $f(1) = 0$. The tangent at 1 is given by $y = f(1) + f'(1)(x-1)$, that is, $y = x - 1$.

Prob. 19. $f'(x) = -8(-6 + 9x^2 - 16x^3 + 12x^5)$, so $f'(-1) = -56$ and $f(-1) = -8$. The tangent at -1 is given by $y = f(-1) + f'(-1)(x - (-1))$, that is, $y = -56x - 64$.

Prob. 21. $f'(x) = 3x^2 - 2x - 2$, so $f'(2) = 6$ and $f(2) = 2$. The normal at 2 is given by $y = f(2) - (1/f'(2))(x - 2)$, that is, $y = (14 - x)/6$.

Prob. 23. $f'(x) = -5(4x + 1)$, so $f'(0) = -5$ and $f(0) = 2$. The normal at 0 is $y = 2 + x/5$.

Prob. 25. $f(x) = uvw$, with $u = 2x - 1$, $v = 3x + 4$, $w = 1 - x$. We have $(uvw)' = u'(vw) + u(vw)' = u'vw + uv'w + uvw'$. Therefore $f'(x) = 2(3x+4)(1-x) + (2x-1)3(1-x) + (2x-1)(3x+4)(-1) = -18x^2 + 2x + 9$.

Prob. 27. $f'(x) = (\frac{d}{dx}(x-3))(2x^1 + 1)(1-x^2) + (x-3)(\frac{d}{dx}(2x^1 + 1))(1-x^2) + (x-3) \times (2x^1 + 1)(\frac{d}{dx}(1-x^2)) = -5 + 10x + 15x^2 - 8x^3$.

Prob. 29. Because a is a constant ("positive" is irrelevant), we can treat it as a number, taking the derivative of $(x-1)(2x-1)$ and multiplying by a. The result is
$f'(x) = a(4x-3)$.

Prob. 31. Because a is a constant ("positive" is irrelevant), the only product we need to treat using the product rule for derivatives is $x^2 - a$ times itself. We get
$f'(x) = 2a(2(x^2-a)(2x)) = 8ax(x^2-a)$.

Prob. 33. $g'(t) = 2a(at+1)$.

Prob. 35. $(fg)' = f'g + fg'$; in particular, at the point 2,
$(fg)'(2) = f'(2)g(2) + f(2)g'(2)$. Substituting the known values we get 11.

Prob. 37. $dy/dx = 2(\frac{dx}{dx}f(x) + x\frac{df(x)}{dx}) = 2(f(x) + xf'(x))$.

Prob. 39.
$dy/dx = -5(\frac{dx^3}{dx}f(x) + x^3\frac{df(x)}{dx}) - 2 = -5(3x^2f(x) + x^3f'(x)) - 2 = -5x^2(3f(x) + xf'(x)) - 2$.

Prob. 41. $dy/dx = 3(f'(x)g(x) + f(x)g'(x))$.

Prob. 43.
$dy/dx = (f'(x) + 2g'(x))g(x) + (f(x) + 2g(x))g'(x) = f'(x)g(x) + f(x)g'(x) + 4g'(x)g(x)$.

Prob. 45. The absolute growth rate is B times the specific growth rate: $Bg(B)$. Note that B is itself a function of t, but the growth rates (absolute and specific) are being thought of as functions of B alone, and t plays no real role in this problem. We are being asked to relate the slope of the graph of the absolute growth rate $Bg(B)$ versus B (at the point $B = 0$) with the specific growth rate (again at the point $B = 0$).

The slope of the graph of $Bg(B)$ is, of course, the derivative of $Bg(B)$ with respect to B, which, by the product rule, equals $g(B) + Bg'(B)$. When $B = 0$, this reduces to $g(B)$, as was to be shown.

Prob. 47. Since $f(N) = r(aN - N^2)(1 - N/K)$, we apply the product rule (treating r, a and K as constants) and get $f'(N) = r((a - 2N)(1 - N/K) + (aN - N^2)(-1/K))$.

Prob. 49. $f'(x) = \frac{3(x+1) - (3x-1)(1)}{(x+1)^2}$.

Prob. 51. $f'(x) = \frac{(6x-2)(2x+1) - (3x^2 - 2x+1)(2)}{(2x+1)^2}$

Prob. 53. $f'(x) = \frac{2x^3 - 3x^2 + 3}{(1-x)^2}$.

Prob. 55. $h'(t) = \frac{t^2 + 2t - 4}{(t+1)^2}$.

Prob. 57. $f'(x) = 2(x^2 - 2x + 2))/(1 - x)^2$.

Prob. 59. $\frac{d}{dx}(\sqrt{x}(x-1)) = (\frac{d}{dx}\sqrt{x})(x-1) + \sqrt{x}\frac{d}{dx}(x-1) = \frac{x-1}{2\sqrt{x}} + \sqrt{x} = \frac{3x-1}{2\sqrt{x}}$.

Prob. 61. $\frac{d}{dx}(\sqrt{3x}(x^2-1)) = \sqrt{3}\frac{d}{dx}(\sqrt{x}(x^2-1)) =$

$\sqrt{3}\left((\frac{d}{dx}\sqrt{x}(x^2-1)) + \sqrt{x}\frac{d}{dx}(x^2-1)\right) = \sqrt{3}(\frac{x^2-1}{2\sqrt{x}} + 2x\sqrt{x}) = \sqrt{3}\frac{5x^2-1}{2\sqrt{x}}$.

Prob. 63. Since $1/x^3 = x^{-3}$, we have $f'(x) = 3x^2 - (-3x^{-4}) = 3(x^2 - x^{-4}) = 3(x^2 + \frac{1}{x^4})$.

Prob. 65. We can treat the fraction term either as a quotient or as a product (of $3x-1$ with x^{-3}). We choose to do the latter. Then

$f'(x) = 4x - (3x^{-3} + (3x-1)(-3)x^{-4}) = (4x^5 + 6x - 3)/x^4$. For yet another approach see exercise 66.

Prob. 67. Using the fact that derivatives of $s^{1/3}$ and $s^{2/3}$ are respectively $\frac{1}{3}s^{-2/3}$ and $\frac{2}{3}s^{-2/3}$, we have

$$g'(s) = \frac{\frac{1}{3}s^{-2/3}(s^{2/3}-1) - \frac{2}{3}s^{-1/3}(s^{1/3}-1)}{(s^{2/3}-1)^2}.$$

The numerator simplifies to $-\frac{1}{3}(s^{1/3}-1)s^{-2/3}$. An even simpler expression for the derivative can be obtained if we observe that the original function is in fact equal to $1/(s^{1/3}-1)$, and treat this as a quotient of the constant function 1 by $s^{1/3}-1$. The derivative then turns out to be $\frac{1}{3}(s^{1/3}+1)^{-2}s^{-2/3}$. (This is of course equivalent to the result previously found.)

Prob. 69. We write $\sqrt{2x}$ as $\sqrt{2}x^{1/2}$, with derivative $(1/\sqrt{2})x^{-1/2}$, and we write $2/\sqrt{x}$ as $2x^{-1/2}$, with derivative $-x^{-3/2}$. Applying the product rule then gives

$f'(x) = -x^{-3/2} + (\frac{\sqrt{2}}{2} - 2)x^{-1/2} - 3\sqrt{2}x^{1/2}$.

Prob. 71. $f'(x) = -\frac{x(x^3+9x-10)}{(5+x^3)^2}$, so $f'(2) = -\frac{32}{169}$ and $f(2) = \frac{7}{13}$. The tangent at 2 is given by $y = f(2) + f'(2)(x-2)$, that is, $(155 - 32x)/169$.

Prob. 73. $f'(x) = (15 - 4x)/x^4$, so $f'(2) = \frac{7}{16}$ and $f(2) = -\frac{1}{8}$. The tangent at 2 is given by $y = f(2) + f'(2)(x-2)$, that is, $y = -1 + \frac{7}{16}x$.

Prob. 75. We treat a as a number and apply the quotient rule. The result is $3a/(x+3)^2$.

Prob. 77. An application of the quotient rule gives $f'(x) = 8ax/(4+x^2)^2$.

Prob. 79.

$$f'(R) = \frac{nR^{n-1}(K^n + R^n) - R^n(n - R^{n-1})}{(K^n + R^n)^2}$$

$$= \frac{nR^{n-1}(K^n + R^n - R^n)}{(K^n + R^n)^2} = \frac{nR^{n-1} - K^n}{(K^n + R^n)^2}.$$

Prob. 81. We write $h(t) = \sqrt{a}\sqrt{t}(t-a) + at$. Treating a as a number and applying the power rule to $\sqrt{t}$ and then the product rule, we get

$$h'(t) = \sqrt{a}t + \frac{\sqrt{a}(t-a)}{2\sqrt{t}} + a.$$

Prob. 83. We have $(f/2g)' = \frac{1}{2}(f/g)' = \frac{1}{2}(f'g - fg')/g^2$; specializing to the point 2, we get

$$(f/2g)'(2) = \frac{1}{2}(1 \cdot 3 - (-4) \cdot (-2))/3^2 = 5/18.$$

Prob. 85. We apply the quotient rule with numerator $g(x) = x^2 + 4f(x)$ and denominator $h(x) = f(x)$. The result is $(g'(x)h(x) - g(x)h'(x))/h(x)^2 = ((2x + 4f'(x))f(x) - (x^2 + 4f(x))f'(x))/f(x)^2 = (2xf(x) - x^2f'(x))/f(x)^2$. (Some time could have been saved by observing in the beginning that $y = x^2/f(x) + 4$.)

Prob. 87. We observe that $y = 1 - \frac{x}{f(x)+x}$, so

$$y' = -\frac{1(f(x)+x) - x(f'(x)+1)}{(f(x)+x)^2} = -\frac{f(x) - xf'(x)}{(f(x)+x)^2}.$$

Prob. 89. The derivative of the denominator is $2g(x)g'(x)$. Therefore

$$y' = \frac{f'(x)g(x)^2 - 2f(x)g(x)g'(x)}{g(x)^4}.$$

Prob. 91. The derivative of $f(x)g(x)$ is $f'(x)g(x) + f(x)g'(x)$. Applying the product rule again to the factors $\sqrt{x}$ and $f(x)g(x)$ we get

$y' = 1/(2\sqrt{x})f(x)g(x) + \sqrt{x}(f'(x)g(x) + f(x)g'(x))$.

Prob. 93. We write $y = c/x$ and take the derivative $y' = -c/x^2$. The tangent at the point (x_1, y_1) is given by the equation $y = y_1 + y'(x_1)(x - x_1)$, or

$$y = y_1 - \frac{c(x - x_1)}{x_1^2} = \frac{c(2x_1 - x)}{x_1^2},$$

where for the second equality we used the fact that $y_1 = c/x_1$. The intersection of the equation with the x-axis happens when $y = 0$, and solving for x gives $x = 2x_1$, which does not depend on c. (Note that each pair (x_1, y_1) belongs to only one hyperbola, and the problem does not state which variable is to be regarded as the independent variable as c varies. If y_1 were the independent variable, the x-value of the intersection would be $x = 2c/y_1$, which does depend on c.)

4.4 The Chain Rule and Higher Derivatives

Prob. 1. $f'(x) = 2(x - 3)$

Prob. 3. Let $u = 1 - 3x^2$. By the chain rule,

$$f'(x) = \frac{d}{dx} u^4 = 4u^3 \frac{du}{dx} = 4u^3 \cdot (-6x) = -24x(1 - 3x^2)^3.$$

Prob. 5. Let $u = x^2 + 3$. By the chain rule,

$$f'(x) = \frac{d}{dx} \sqrt{u} = \frac{1}{2\sqrt{u}} \frac{du}{dx} = \frac{1}{2\sqrt{u}} \cdot (2x) = \frac{x}{\sqrt{x^2 + 3}}.$$

Prob. 7. Let $u = 3 - x^3$. By the chain rule,

$$f'(x) = \frac{d}{dx} \sqrt{u} = \frac{1}{2\sqrt{u}} \frac{du}{dx} = \frac{1}{2\sqrt{u}} \cdot (-3x^2) = -\frac{3x^2}{2\sqrt{3 - x^3}}.$$

Prob. 9. Let $u = x^3 - 2$. By the chain rule,

$$f'(x) = \frac{d}{dx} \frac{1}{u^4} = -\frac{4}{u^5} \frac{du}{dx} = -\frac{4}{u^5} \cdot 3x^2 = -\frac{12x^2}{(x^3 - 2)^5}.$$

Prob. 11. Let $u = 2x^2 - 1$. Using the chain rule and the quotient rule, we have

$$
\begin{aligned}
f'(x) &= \frac{d}{dx} \frac{3x - 1}{\sqrt{u}} \\
&= \frac{\sqrt{u} \cdot 3 - (3x - 1) \cdot \frac{1}{2\sqrt{u}} \frac{du}{dx}}{(\sqrt{u})^2} \\
&= \frac{3u - \frac{1}{2}(3x - 1)(4x)}{u^{\frac{3}{2}}} \\
&= \frac{3(2x^2 - 1) - 2x(3x - 1)}{u^{\frac{3}{2}}} \\
&= \frac{2x - 3}{(2x^2 - 1)^{\frac{3}{2}}}.
\end{aligned}
$$

Prob. 13. Let $u = 2x - 1$, $v = x - 1$. Using the chain rule and the quotient rule, we have

$$
\begin{aligned}
f'(x) &= \frac{d}{dx} \frac{\sqrt{u}}{v^2} \\
&= \frac{v^2 \cdot \frac{1}{2\sqrt{u}} \frac{du}{dx} - \sqrt{u} \cdot 2v \frac{dv}{dx}}{v^4} \\
&= \frac{\frac{v}{2} \frac{du}{dx} - 2u \frac{dv}{dx}}{v^3 \sqrt{u}} \\
&= \frac{(x - 1) - 2(2x - 1)}{v^3 \sqrt{u}} \\
&= \frac{-3x + 1}{(x - 1)^3 \sqrt{2x - 1}}
\end{aligned}
$$

Prob. 15. Let $u = s + \sqrt{s}$. By the chain rule,

$$f'(s) = \frac{d}{ds}\sqrt{u} = \frac{1}{2\sqrt{u}}\frac{du}{ds} = \frac{1}{2\sqrt{s+\sqrt{s}}}\left(1 + \frac{1}{2\sqrt{s}}\right).$$

Prob. 17. Let $u = t/(t-3)$. Using the chain rule and the quotient rule, we have

$$g'(t) = \frac{d}{dt}u^3 = 3u^2\frac{du}{dt} = 3u^2 \cdot \frac{(t-3)-t}{(t-3)^2} = -\frac{9t^3}{(t-3)^4}.$$

Prob. 19. Let $u = r^2 - r$, $v = r + 3r^3$. Using the chain rule and the product rule, we have

$$
\begin{aligned}
f'(r) &= \frac{d}{dr}(u^3 v^{-4}) \\
&= 3u^2\frac{du}{dr}\cdot v^{-4} + u^3\left(-4v^{-5}\frac{dv}{dr}\right) \\
&= u^2 v^{-5}\left(3v\frac{du}{dr} - 4u\frac{dv}{dr}\right) \\
&= \frac{(r^2-r)^2}{(r+3r^3)^5}\cdot\left[3(r+3r^3)(2r-1) - 4(r^2-r)(1+9r^2)\right]
\end{aligned}
$$

Prob. 21. Let $u = 3 - x^4$. By the chain rule,

$$
\begin{aligned}
h'(x) &= \frac{d}{dx}u^{\frac{1}{5}} = \frac{1}{5}u^{-\frac{4}{5}}\frac{du}{dx} \\
&= \frac{1}{5}u^{-\frac{4}{5}}(-4x^3) \\
&= -\frac{4}{5}x^3(3-x^4)^{-\frac{4}{5}}.
\end{aligned}
$$

Prob. 23. Notice that

$$f(x) = \sqrt[7]{x^2 - 2x + 1} = \sqrt[7]{(x-1)^2}.$$

Let $u = x - 1$. By the chain rule,

$$f'(x) = \frac{d}{dx}u^{\frac{2}{7}} = \frac{2}{7}u^{-\frac{5}{7}}\frac{du}{dx} = \frac{2}{7}(x-1)^{-\frac{5}{7}}.$$

Prob. 25. $g'(s) = \frac{3}{2}(3s^7 - 7S)^{\frac{1}{2}}(21s^5 - 7)$.

Prob. 27. Let $u = 3t + 3/t$. By the chain rule,

$$
\begin{aligned}
h'(t) &= \frac{d}{dt}u^{\frac{2}{5}} = \frac{2}{5}u^{-\frac{3}{5}}\frac{du}{dt} \\
&= \frac{2}{5}u^{-\frac{3}{5}}\left(3 - \frac{3}{t^2}\right) \\
&= \frac{6}{5}\left(3t + \frac{3}{t}\right)^{-\frac{3}{5}}\left(1 - \frac{1}{t^2}\right).
\end{aligned}
$$

Prob. 29. Let $u = ax + 1$. By the chain rule,

$$
\begin{aligned}
f'(x) &= \frac{d}{dx} u^3 = 3u^2 \frac{du}{dx} \\
&= 3u^2 \cdot a = 3a(ax + 1)^2.
\end{aligned}
$$

Prob. 31. Let $u = k + N$. Then, using the chain rule and the product rule, we have

$$
\begin{aligned}
g'(N) &= \frac{d}{dN}(bNu^{-2}) \\
&= b\left(u^{-2} + N \cdot -2u^{-3} \frac{du}{dN} \right) \\
&= bu^{-3}(u - 2N) \\
&= \frac{b(k - N)}{(k + N)^3}.
\end{aligned}
$$

Prob. 33. Let $u = T_0 - T$. Then, by the chain rule,

$$
g'(T) = \frac{d}{dT} au^3 - b = 3au^2 \frac{du}{dT} = -3a(T_0 - T)^2.
$$

Prob. 35.

(a) By the chain rule,

$$
\begin{aligned}
\frac{d}{dx}[f(x^2 + 3)] &= f'(x^2 + 3)\frac{d}{dx}(x^2 + 3) \\
&= \frac{1}{x^2 + 3} \cdot 2x \\
&= \frac{2x}{x^2 + 3}.
\end{aligned}
$$

(b) By the chain rule,

$$
\begin{aligned}
\frac{d}{dx}[f(\sqrt{x - 1})] &= f'(\sqrt{x - 1})\frac{d}{dx}\sqrt{x - 1} \\
&= \frac{1}{\sqrt{x - 1}} \cdot \frac{1}{2\sqrt{x - 1}} \\
&= \frac{1}{2(x - 1)}.
\end{aligned}
$$

Prob. 37. Let $u = f(x)/g(x) + 1$. By the chain rule and the quotient rule,

$$
\begin{aligned}
\frac{d}{dx}\left(\frac{f(x)}{g(x)} + 1 \right)^2 &= \frac{d}{dx} u^2 = 2u \frac{du}{dx} \\
&= 2\left(\frac{f}{g} + 1 \right) \frac{gf' - fg'}{g^2}.
\end{aligned}
$$

Prob. 39. Let $v = g(2x) + 2x$.

$$\frac{d}{dx}\frac{[f(x)]^2}{g(2x) + 2x} = \frac{d}{dx}\frac{f^2}{v}$$

$$= \frac{v\frac{d}{dx}f^2 - f^2\frac{dv}{dx}}{v^2}$$

$$= \frac{1}{v^2}\left\{v \cdot 2f\frac{df}{dx} - f^2\frac{d}{dx}[g(2x) + 2x]\right\}$$

$$= \frac{1}{v^2}\left\{2vff' - f^2[g'(2x) \cdot 2 + 2]\right\}$$

$$= \frac{2}{[g(2x) + 2x]^2}\left\{f(x)f'(x)[g(2x) + 2x] - f(x)^2[g'(2x) + 1]\right\}.$$

Prob. 41. $\frac{dy}{dx} = 4(\sqrt{x^3 - 3x} + 3x)^3(\frac{1}{2}(x^3 - 3x)^{-\frac{1}{2}}(3x^2 - 3) + 3)$.

Prob. 43. Let $u = 3x^2 - 1$, $v = 1 + u^3 = 1 + (3x^2 - 1)^3$. By the chain rule,

$$\frac{dy}{dx} = \frac{d}{dx}v^2$$

$$= 2v\frac{d}{dx}(1 + u^3)$$

$$= 2v \cdot 3u^2\frac{d}{dx}(3x^2 - 1)$$

$$= 36x(3x^2 - 1)^2\left(1 + (3x^2 - 1)^3\right)$$

Prob. 45. Let

$$u = 2x + 1, \quad v = x^3 - 1,$$

$$w = 3v^3 - 1 = 3(x^3 - 1)^3 - 1.$$

Using the chain rule,

$$\frac{dy}{dx} = \frac{d}{dx}\left(\frac{u}{w}\right)^3 = 3\left(\frac{u}{w}\right)^2\frac{d}{dx}\frac{u}{w}$$

$$= \frac{3u^2}{w^4}\left[w\frac{d}{dx}(2x + 1) - u\frac{d}{dx}(3v^3 - 1)\right]$$

$$= \frac{3u^2}{w^4}\left[2w - 9uv^2\frac{d}{dx}(x^3 - 1)\right]$$

$$= \frac{3u^2}{w^4}(2w - 27x^2uv^2)$$

$$= 3\left(\frac{2x + 1}{3(x^3 - 1)^3 - 1}\right)^2\frac{6(x^3 - 1)^3 - 2 - 27x^2(2x - 1)(x^3 - 1)}{(3(x^3 - 1)^3 - 1)^2}$$

Prob. 47. In problems **47** to **52**, differentiate the given equations with respect to x regarding y as a function of x. Various differentiation rules (chain, product, quotient) are used when needed without further comment.

$$2x + 2y\frac{dy}{dx} = 0 \Rightarrow \frac{dy}{dx} = -\frac{x}{y}$$

Prob. 49.

$$\frac{3}{4}x^{-\frac{1}{4}} + \frac{3}{4}y^{-\frac{1}{4}}\frac{dy}{dx} = 0 \Rightarrow \frac{dy}{dx} = -\frac{x^{-\frac{1}{4}}}{y^{-\frac{1}{4}}} = -\sqrt[4]{\frac{y}{x}}$$

Prob. 51.

$$\frac{1}{2\sqrt{xy}}\left(y + x\frac{dy}{dx}\right) = 2x \Rightarrow \frac{dy}{dx} = 4\sqrt{xy} - \frac{y}{x}$$

Prob. 53. Let us first rewrite the given equation as $x^2 = y^2$. Differentiation then yields

$$2x = 2y\frac{dy}{dx} \Rightarrow \frac{dy}{dx} = -\frac{x}{y}.$$

Prob. 55. We first find dy/dx using implicit differentiation:

$$2x + 2y\frac{dy}{dx} = 0 \Rightarrow \frac{dy}{dx} = -\frac{x}{y}$$

The slope of the tangent is given by

$$\frac{dy}{dx}\bigg|_{(4,-3)} = -\frac{4}{-3} = \frac{4}{3},$$

(a) The slope of the normal is thus $-3/4$. Hence, the equation of the tangent line at $(4, -3)$ is

$$y + 3 = \frac{4}{3}(x - 4) \Rightarrow y = \frac{4}{3}x - \frac{25}{3},$$

(b) The equation for the normal line is:

$$y + 3 = -\frac{3}{4}(x - 4) \Rightarrow y = -\frac{3}{4}x.$$

Prob. 57. We first find dy/dx using implicit differentiation:

$$\frac{2x}{25} - \frac{2y}{9}\frac{dy}{dx} = 0 \Rightarrow \frac{dy}{dx} = \frac{9x}{25y}$$

The slope of the tangent is given by

$$\frac{dy}{dx}\bigg|_{(\frac{25}{3},4)} = \frac{9 \cdot \frac{25}{3}}{25 \cdot 4} = \frac{3}{4},$$

and the slope of the normal is thus $-4/3$.

(a) Hence, the equation of the tangent line at $(25/3, 4)$ is

$$y - 4 = \frac{3}{4}\left(x - \frac{25}{3}\right) \Rightarrow y = \frac{3}{4}x - \frac{9}{4},$$

(b) and that of the normal line is

$$y - 4 = -\frac{4}{3}\left(x - \frac{25}{3}\right) \Rightarrow y = -\frac{4}{3}x + \frac{136}{9}.$$

Prob. 59.

(a) Differentiating the equation $x^{2/3} + y^{2/3} = 4$ with respect to x yields

$$\frac{2}{3}x^{-\frac{1}{3}} + \frac{2}{3}y^{-\frac{1}{3}}\frac{dy}{dx} = 0 \Rightarrow \frac{dy}{dx} = -\sqrt[3]{\frac{y}{x}}.$$

Thus, the value of dy/dx at $(-1, 3\sqrt{3})$ is

$$-\sqrt[3]{\frac{3\sqrt{3}}{-1}} = \sqrt{3}.$$

(b) The curve $x^{2/3} + y^{2/3} = 4$:

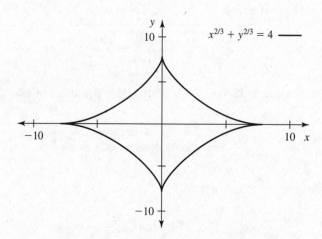

Prob. 61. To simplify notations, in problems **61** to **64**, we use $x', y', \ldots$ to denote derivatives with respect to t. Differentiating $x^2 + y^2 = 1$ with respect to t yields

$$2xx' + 2yy' = 0 \Rightarrow y' = -\frac{xx'}{y}.$$

We already have $x = 1/2$ and $x' = 2$, so we only need to find y. From $x^2 + y^2 = 1$ and the assumption that $y > 0$, we obtain

$$\left(\frac{1}{2}\right)^2 + y^2 = 1 \Rightarrow y = \frac{\sqrt{3}}{2}.$$

Therefore, we have

$$y' = -\frac{\frac{1}{2} \cdot 2}{\frac{\sqrt{3}}{2}} = -\frac{2}{\sqrt{3}}.$$

Prob. 63. Since $x^2 y = 1$ and $x = 2$, we have $y = 1/4$. Differentiating $x^2 y = 1$ with respect to t yields

$$2xx'y + x^2 y' = 0 \Rightarrow y' = -\frac{2x'y}{x} = -\frac{2 \cdot 3 \cdot \frac{1}{4}}{2} = -\frac{3}{4}.$$

Prob. 65. The chain rule gives us

$$\frac{dV}{dt} = \frac{d(x^3)}{dt} = 3x^2 \frac{dx}{dt}.$$

Prob. 67. The chain rule gives us

$$\frac{dS}{dt} = \frac{d(4\pi r^2)}{dt} = 8\pi r \frac{dr}{dt}.$$

Prob. 69. Let us express all volumes in m^2. In mathematical terms, the statement that water is drained at a rate of 250 liters per minute means $dV/dt = -0.25$, and the quantity of interest now is dh/dt. Since $V = 25\pi h$, by differentiating with respect to t, we have

$$\frac{dV}{dt} = 25\pi \frac{dh}{dt},$$

and thus,

$$\frac{dh}{dt} = \frac{-0.25}{25\pi} \approx -0.0032.$$

In words, the height is dropping at a rate of about 3.2cm per minute.

Prob. 71. Let x be the distance traveled by the eastbound biker in miles and y be that of the southbound biker. The distance between is $\sqrt{x^2 + y^2}$. The rate at which this distance is changing is given by

$$\frac{d}{dt}\sqrt{x^2 + y^2} = \frac{1}{2\sqrt{x^2 + y^2}}\left(2x\frac{dx}{dt} + 2y\frac{dy}{dt}\right)$$
$$= \frac{1}{\sqrt{x^2 + y^2}}\left(x\frac{dx}{dt} + y\frac{dy}{dt}\right).$$

After 20 minutes, $x = 5$ and $y = 6$. Also, we always have $dx/dt = 15$ and $dy/dt = 18$. Thus, the above expression gives us the first answer

$$\frac{1}{\sqrt{5^2 + 6^2}}(5 \cdot 15 + 6 \cdot 18) \approx 23.4.$$

After 40 minutes, $x = 10$ and $y = 12$, so the second answer is

$$\frac{1}{\sqrt{10^2 + 12^2}}(10 \cdot 15 + 12 \cdot 18) \approx 23.4$$

In fact, the distance between the two bikers is increasing at a constant rate.

Prob. 73.

$$f'(x) = 3x^2 - 6x, \quad f''(x) = 6x - 6$$

Prob. 75. Note that

$$\frac{x-1}{x+1} = \frac{x+1-2}{x+1} = 1 - \frac{2}{x+1}.$$

Hence we have

$$g'(x) = \frac{2}{(x+1)^2}$$
$$g''(x) = -\frac{4}{(x+1)^3}.$$

Prob. 77.

$$g'(t) = \frac{1}{2\sqrt{3t^3 + 2t}} \cdot (9t^2 + 2)$$
$$= \frac{1}{2}(3t^3 + 2t)^{-\frac{1}{2}}(9t^2 + 2)$$
$$g''(t) = -\frac{1}{4}(3t^3 + 2t)^{-\frac{3}{2}}(9t^2 + 2)^2 + \frac{1}{2}(3t^3 + 2t)^{-\frac{1}{2}} \cdot 18t$$
$$= \frac{1}{4}(3t^3 + 2t)^{-\frac{3}{2}}\left[-(9t^2 + 2)^2 + 36t(3t^3 + 2t)\right]$$
$$= \frac{1}{4}(3t^3 + 2t)^{-\frac{3}{2}} \cdot (27t^4 + 36t^2 - 4)$$

Prob. 79.

$$f'(s) = \frac{1}{2\sqrt{s^{\frac{3}{2}} - 1}} \cdot \frac{3}{2}s^{\frac{1}{2}} = \frac{3}{4}\left(\frac{s}{s^{\frac{3}{2}} - 1}\right)^{\frac{1}{2}}$$
$$f''(s) = \frac{3}{8}\left(\frac{s}{s^{\frac{3}{2}} - 1}\right)^{-\frac{1}{2}} \cdot \frac{(s^{\frac{3}{2}} - 1) - s \cdot \frac{3}{2}s^{\frac{1}{2}}}{(s^{\frac{3}{2}} - 1)^2} = -\frac{3(\frac{1}{2}s^{\frac{3}{2}} + 1)}{8\sqrt{s}(s^{\frac{3}{2}} - 1)^{\frac{3}{2}}}$$

Prob. 81.

$$g'(t) = -\frac{5}{2}t^{-7/2} - \frac{1}{2}t^{-1/2}$$
$$g''(t) = \frac{35}{4}t^{-9/2} + \frac{1}{4}t^{-3/2}$$

Prob. 83. By repeated differentiation, the first ten derivatives of x^5 are:

$$f'(x) = 5x^4, \quad f''(x) = 20x^3, \quad f'''(x) = 60x^2, \quad f^{(4)}(x) = 120x,$$

$$f^{(5)}(x) = 120, \quad f^{(6)}(x) = \cdots = f^{(10)}(x) = 0.$$

Prob. 85. Let $p(x) = ax^2 + bx + c$. Its first two derivatives are $p'(x) = 2ax + b$ and $p''(x) = 2a$. Thus we have

$$p(0) = c, \quad p'(0) = b, \quad p''(0) = 2a.$$

The condition $p(0) = 3$, $p'(0) = 2$ and $p''(0) = 6$ means

$$c = 3, \quad b = 2, \quad a = 3,$$

and hence the polynomial is $p(x) = 3x^2 + 2x + 3$.

Prob. 87.

(a) The velocity is
$$v(t) = h'(t) = v_0 - gt,$$

while the acceleration is
$$a(t) = v'(t) = -g.$$

(b) By the result in (a), $v(t) = v_0 - gt = 0$ when

$$t = \frac{v_0}{g}.$$

Before this moment, i.e. when $t < v_0/g$, $v(t)$ is positive, so the object is traveling upward. After this moment, i.e. when $t > v_0/g$, $v(t)$ is negative, so the object is traveling downward.

4.5 Derivatives of Trigonometric Functions

Prob. 1. $2\cos x + \sin x$

Prob. 3. $3\cos x - 5\sin x - 2\sec x \tan x$

Prob. 5. $\sec^2 x + \csc^2 x$

Prob. 7. $3\cos(3x)$

Prob. 9. $6\cos(3x+1)$

Prob. 11. $4\sec^2 x$

Prob. 13. $4\sec(1+2x)\tan(1+2x)$

Prob. 15. $6x\cos(x^2)$

Prob. 17. $f'(x) = 3\sin^2(x^2-3)\cdot\cos(x^2-3)\cdot 2x$

Prob. 19. $f'(x) = 6\sin x^2\cdot\cos x^2\cdot 2x = 12x\sin x^2\cos x^2$.

Prob. 21. $f'(x) = -4\sin x^2\cdot(2x)+4\cos x\cdot\sin x$.

Prob. 23. $f'(x) = 4\cdot 2\cos x\cdot(-\sin x)+2\cdot(-\sin x^4)\cdot 4x^3$

Prob. 25. $-4x\sec^2(1-x^2)$

Prob. 27. $-18\tan^2(3x-1)\sec^2(3x-1)$

Prob. 29. $f'(x) = \frac{1}{2}(\sin(2x^2-1))^{-\frac{1}{2}}\cdot\cos(2x^2-1)\cdot 4x$

Prob. 31. $g'(s) = -\frac{1}{2}(\cos s)^{-\frac{1}{2}}\cdot\sin s+\sin(\sqrt{xs})\cdot\frac{1}{2}xs^{-\frac{1}{2}}$.

Prob. 33.
$$\frac{2\cos 2t(\cos 6t-1)+6\sin 6t(\sin 2t+1)}{(\cos 6t-1)^2}$$

Prob. 35. Note that
$$f(x) = \frac{\sin(x^2+1)}{\cos(x^2+1)}.$$

Hence we have

$$f'(x) = \frac{2x}{\cos^2(x^2-1)}\Big[\cos(x^2-1)\cos(x^2+1)+\sin(x^2-1)\sin(x^2+1)\Big]$$

Prob. 37. $2\cos(2x-1)\cos(3x+1)-3\sin(2x-1)\sin(3x+1)$

Prob. 39. $6x[\sec^2(3x^2-1)\cot(3x^2+1)-\tan(3x^2-1)\csc^2+1]$

Prob. 41. Note that $f(x) = \tan x$. Hence $f'(x) = \sec^2 x$.

Prob. 43. $f'(x) = 0$ since $\sin^2 x-\sin^2 x = -1$.

Prob. 45.

$$-\frac{6x\cos(3x^2-1)}{\sin^2(3x^2-1)}$$

Prob. 47. $g(x) = \frac{1}{\csc^3(1-5x^2)} = \sin^3(1-5x^2)$. So,

$$g'(x) = 3\sin^2(1-5x^2)\cdot\cos(1-5x^2)\cdot(-10x).$$

Prob. 49.

$$-\frac{3(2\sec^2 2x - 1)}{(\tan 2x - x)^2}$$

Prob. 51. $h'(s) = 3\sin^2 s \cdot \cos s - 3\cos^2 s - \sin s.$

Prob. 53. $\frac{2(1+x^2)\cos 2x - 2x\sin 2x}{(1+x^2)^2}$

Prob. 55. $-\frac{1}{x^2}\sec^2 \frac{1}{x}.$

Prob. 57. Rewrite $f(x)$ as $\cos^2 x / \cos x^2$. Then we have

$$f'(x) = \frac{1}{\cos^2 x^2}\left[-2\cos x^2 \sin x \cdot \cos x + 2x\sin x^2 \cos^2 x\right].$$

Prob. 59. First, compute

$$\frac{dy}{dx} = \frac{\pi}{3}\cos\left(\frac{\pi}{3}x\right).$$

The tangent is horizontal when $dy/dx = 0$, which is, $\frac{\pi}{3}x = n\pi + \frac{\pi}{2}$, or

$$x = 3n + \frac{3}{2},$$

where n is any integer.

Prob. 61. By definition,

$$\begin{aligned}
\frac{d}{dx}\cos x &= \lim_{h \to 0}\frac{\cos(x+h) - \cos x}{h} \\
&= \lim_{h \to 0}\frac{1}{h}\left[\cos x \cos h - \sin x \sin h - \cos x\right] \\
&= \cos x \lim_{h \to 0}\frac{\cos h - 1}{h} - \sin x \lim_{h \to 0}\frac{\sin h}{h}.
\end{aligned}$$

The limit in the second term is 1, as is discussed earlier in the book. To evaluate the other limit, consider the trick

$$\begin{aligned}
\frac{\cos h - 1}{h} &= \frac{\cos h - 1}{h} \cdot \frac{\cos h + 1}{\cos h + 1} \\
&= \frac{\cos^2 h - 1}{h(\cosh +1)} = -\frac{\sin^2 h}{h(\cos h + 1)} \\
&= -\left(\frac{\sin h}{h}\right) \cdot \sin h \cdot \left(\frac{1}{\cos h + 1}\right).
\end{aligned}$$

As $h \to 0$, the first factor approaches 1, as mentioned just above, while the second and the third approach 0 and 1/2 respectively. Hence, the limit of the whole things is 0. Going back to the calculation we start with, this implies we have

$$\frac{d}{dx}\cos x = \cos x \cdot 0 - \sin x \cdot 1 = -\sin x.$$

Prob. 63. By the chain rule, we have

$$
\begin{aligned}
\frac{d}{dx}\sec x &= \frac{d}{dx}\frac{1}{\cos x} \\
&= -\frac{1}{\cos^2 x}\cdot(-\sin x) \\
&= \frac{1}{\cos x}\cdot\frac{\sin x}{\cos x} \\
&= \sec x \tan x.
\end{aligned}
$$

Prob. 65.

$$
\frac{x}{\sqrt{x^2+1}}\cos\sqrt{x^2+1}
$$

Prob. 67.

$$
\frac{9x^2+3}{2\sqrt{3x^2+3x}}\cos\sqrt{3x^3+3x}
$$

Prob. 69. $4x\sin(x^2-1)\cos(x^2-1)$

Prob. 71. $27x^2\tan^2(3x^3-3)\sec^2(3x^3-3)$

Prob. 73.

(a) $\frac{dc}{dt}=\frac{\pi}{2}\cos(\frac{\pi}{2}t)$

(b) The graph of $c(t)$ and dc/dt:

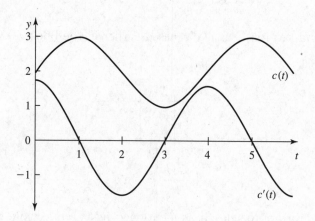

(c) (i) Whenever $c(t)$ reaches a maximum, $dc/dt = 0$.

(ii) When dc/dt is positive, $c(t)$ is increasing.

(iii) In this case, when $dc/dt = 0$, $c(t)$ either reaches a maximum or a minimum. More generally speaking, the value of $c(t)$ is stationary/ the tangent of the graph is horizontal.

4.6 Derivatives of Exponential Functions

Prob. 1. $f'(x) = 3e^{3x}$

Prob. 3. $-12e^{1-3x}$

Prob. 5. $(-4x + 3)e^{-2x^2+3x-1}$

Prob. 7. $f'(x) = e^{7(x^7+1)^2} \cdot 7 \cdot 2(x^7 + 1) \cdot 2x.$

Prob. 9. $(1 + x)e^x$

Prob. 11. $(2x - x^2)e^{-x}$

Prob. 13. Using the quotient rule and after simplifications, we have for $f'(x)$:

$$\frac{(1 - x)^2 e^x - 2x}{(1 + x^2)^2}$$

Prob. 15. Using the quotient rule and after simplifications, we have for $f'(x)$:

$$\frac{2(e^x - 1 - e^{-x})}{(2 + x^{-x})^2}$$

Prob. 17. $f'(x) = e^{\sin(3x)} \cdot \cos(3x) \cdot 3.$

Prob. 19. $2x \cos(x^2 - 1)e^{\sin(x^2-1)}$

Prob. 21. $e^x \cos(e^x)$

Prob. 23. $(2e^{2x} + 1) \cos(e^{2x} + x)$

Prob. 25. $(1 - \cos x)e^{x-\sin x}$

Prob. 27. $2s \sec s^2 \tan s^2 e^{\sec s^2}$

Prob. 29. $(\sin x + x \cos x)e^{x \sin x}$

Prob. 31. $-3(2x + \sec^2 x)e^{x^2+\tan x}$

Prob. 33. $(\ln 2)2^x$

Prob. 35. $(\ln 2)2^{x+1}$

Prob. 37. $f'(x) = \ln 5 \cdot 5^{\sqrt{2x-1}} \cdot \frac{1}{2}(2x - 1)^{-\frac{1}{2}} \cdot 2.$

Prob. 39. $2(\ln 2)x2^{x^2+1}$

Prob. 41. $2(\ln 2)t2^{t^2-1}$

Prob. 43. $\frac{\ln 2}{2\sqrt{x}} 2^{\sqrt{x}}$

Prob. 45. $\frac{x \ln 2}{\sqrt{x^2-1}} 2^{\sqrt{x^2-1}}$

Prob. 47. $\frac{\ln 5}{2\sqrt{t}} \cdot 5^{\sqrt{t}}$

Prob. 49. $g'(x) = -2(\ln 2)(\sin x) 2^{2\cos x}$.

Prob. 51. $\frac{\ln 3}{5} r^{-4/5} 3^{r^{1/5}}$

Prob. 53.

$$\lim_{h \to 0} \frac{e^{2h} - 1}{h} = \lim_{h \to 0} \frac{e^{2h} - e^0}{h} = \frac{d}{dx} e^{2x} \bigg|_{x=0} = 2e^{2x} \big|_{x=0} = 2$$

Prob. 55.

$$\lim_{h \to 0^+} \frac{e^h - 1}{\sqrt{h}} = \lim_{h \to 0^+} \frac{e^h - e^0}{h} \cdot \sqrt{h} = \left(\lim_{h \to 0^+} \frac{e^h - e^0}{h} \right) \cdot \left(\lim_{h \to 0^+} \sqrt{h} \right) = \frac{d}{dx} e^x \bigg|_{x=0} \cdot 0 = 0$$

Prob. 57. Let c be the subtangent. We have

$$\frac{2}{c} = \frac{dy}{dx} \bigg|_{x=1} = \frac{d(2^x)}{dx} \bigg|_{x=1} = (\ln 2) 2^x |_{x=1} = 2 \ln 2.$$

Hence, $c = 1/\ln 2$.

Prob. 59.

(a) $N(0) = 1$

(b)

$$\frac{dN}{dt} = \frac{d(e^{2t})}{dt} = 2e^{2t} = 2N$$

Prob. 61. The rate of growth is

$$\frac{d}{dt}[N(0)2^t] = N(0) \ln 2 \cdot 2^t = (\ln 2) N(t),$$

which is proportional to the population size.

Prob. 63.

(a) Let $a = \frac{K}{N(0)} - 1$. We have

$$\frac{dN}{dt} = \frac{d}{dt} \frac{K}{1 + ae^{-rt}} = \frac{Kare^{-rt}}{(1 + ae^{-rt})^2}$$

(b) Compute

$$rN\left(1 - \frac{N}{K}\right) = r \cdot \frac{K}{1 + ae^{-rt}} \quad \cdot \left(1 - \frac{1}{1 + ae^{-rt}}\right) = \frac{rK}{1 + ae^{-rt}} \cdot \frac{ae^{-rt}}{1 + ae^{-rt}}.$$

From the result in (a), we see that

$$\frac{dN}{dt} = rN\left(1 - \frac{N}{K}\right).$$

(c) By the result in (b), the per capita growth rate is

$$\frac{1}{N}\frac{dN}{dt} = r\left(1 - \frac{N}{K}\right).$$

Its graph as a function of N is as below

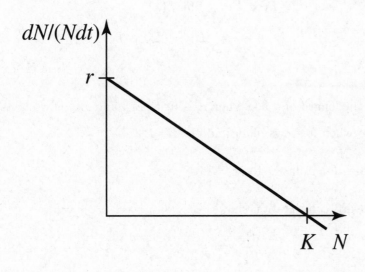

Prob. 65.

(a) The graphs of $L(x) = 10 - 9e^{-x}$ and $L(x) = 10 - 9e^{-0.1x}$:

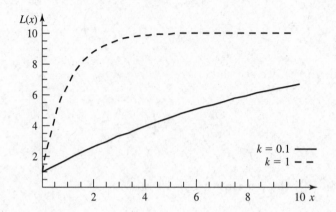

(b) Consider

$$L(0) = L_\infty - (L_\infty - L_0) \cdot 1$$

$$= L_0,$$

$$L(\infty) = \lim_{x \to \infty} L(x)$$

$$= L_\infty - (L_\infty - L_0) \cdot 0$$

$$= L_\infty.$$

Hence, L_0 is the length of a fish when it is just born and L_∞ is what its length approaches to when it grows older and older.

(c) The fish reaches $L = 5$ faster for $k = 1.0$.

(d) A straightforward computation shows

$$\frac{dL}{dx} = k(L_\infty - L_0)e^{-kx}$$

$$= k(L_\infty - L(x)).$$

As a fish grows up, its length keeps increasing, thus the quantity on the right hand side is decreasing. Therefore, the rate of growth is also decreasing with age.

Prob. 67. The differential equation is

$$\frac{dW}{dt} = -4w(t).$$

Prob. 69. We follow along the lines of the solution to **Prob. 68**. $W(t)$ satisfies the differential equation

$$W'(t) = -\frac{\ln 2}{5} W(t).$$

Prob. 71.

(a) The differential equation implies that $W(t)$ is an exponential function. Its exact form is $W(t) = 6e^{-3t}$. Check that it in deed satisfies the differential equation and the initial condition. Thus, at $t = 4$, the amount of the material left is

$$W(4) = 6e^{-12} \approx 2.5 \times 10^{-3}.$$

(b) The half life is τ if and only if $W(\tau)/W(0) = 1/2$. Since we have

$$\frac{1}{2} = \frac{W(\tau)}{W(0)} = \frac{6e^{-3\tau}}{6} = e^{-3\tau}.$$

Taking logarithm of both sides yields

$$-\ln 2 = -3\tau \quad \Rightarrow \quad \tau = \frac{\ln 2}{3}.$$

Prob. 73.

(a) After every unit time, the value of $W(t)$ drops to $2/5$ of its original value. Since $W(0) = 5$, we have for general t

$$W(t) = 5 \left(\frac{2}{5} \right)^t.$$

Differentiation gives

$$W'(t) = 5 \left(\ln \frac{2}{5} \right) \left(\frac{2}{5} \right)^t.$$

Hence, we have the differential equation $W'(t) = (\ln \frac{2}{5}) W(t)$.

(b) By the result in (a), $W(3) = 5(\frac{2}{5})^3 = 0.32$.

(c) The half life τ is determined by

$$\frac{1}{2} = \frac{W(\tau)}{W(0)} = \frac{5(2/5)^\tau}{5} = \left(\frac{2}{5} \right)^\tau.$$

Taking logarithm of both sides gives

$$\ln \frac{1}{2} = \tau \ln \frac{2}{5},$$

i.e. $\tau = \ln \frac{1}{2} / \ln \frac{2}{5} \approx 0.76$.

4.7   Derivatives of Inverse and Logarithmic Functions

Prob. 1. To find the inverse:

$$
\begin{aligned}
x &= \sqrt{2y+1} \\
x^2 &= 2y+1 \\
y &= \frac{1}{2}(x^2-1).
\end{aligned}
$$

(i) $y' = \frac{1}{2}(2x) = x$.

(ii) by (4.12) we have $\dfrac{1}{\frac{1}{2}(2(\frac{1}{2}(x^2-1)+1)^{-\frac{1}{2}}} = \dfrac{1}{\frac{1}{2}(x^2)^{-\frac{1}{2}}} = x$.

Prob. 3. To find the inverse:

$$
\begin{aligned}
x &= 2y^2-1 \\
\frac{1}{2}(x+1) &= y^2 \\
y &= \sqrt{\frac{1}{2}(x+1)}.
\end{aligned}
$$

(i) $y' = \sqrt{\frac{1}{2}} \cdot \frac{1}{2}(x+1)^{-\frac{1}{2}}$.

(ii) by (4.12) we have:

$$
\frac{1}{4\left(\sqrt{\frac{1}{2}(x+1)}\right)} = \frac{1}{2} \cdot \sqrt{\frac{1}{2}}(x+1)^{-\frac{1}{2}}.
$$

Prob. 5. $x = f^{-1}(y) = \sqrt[3]{1-\frac{y}{2}}$

(i) $(f^{-1})'(y) = -\frac{1}{6}(1-\frac{y}{2})^{-2/3}$

(ii) $f'(x) = -6x^2$. Hence, $(f^{-1})'(y) = 1/f'(x) = -1/6x^2 = -\frac{1}{6}(1-\frac{y}{2})^{-2/3}$.

Prob. 7. We have $f'(x) = 4x$, hence $(f^{-1})'(f(x)) = 1/f'(x) = 1/4x$. If we put $x = 1$, we have $(f^{-1})'(0) = (f^{-1})'(f(1)) = 1/4$.

Prob. 9. We have $f'(x) = \frac{1}{2\sqrt{x+1}}$, hence $(f^{-1})'(f(x)) = 1/f'(x) = 2\sqrt{x+1}$. If we put $x = 3$, we have $(f^{-1})'(2) = (f^{-1})'(f(3)) = 4$.

Prob. 11. We have $f'(x) = 1+e^x$, hence $(f^{-1})'(f(x)) = 1/f'(x) = \frac{1}{1+e^x}$. If we put $x = 0$, we have $(f^{-1})'(1) = (f^{-1})'(f(0)) = 1/2$.

Prob. 13. We have $f'(x) = 1 - \cos x$, hence $(f^{-1})'(f(x)) = 1/f'(x) = \frac{1}{1-\cos x}$. If we put $x = \pi$, we have $(f^{-1})'(\pi) = (f^{-1})'(f(\pi)) = 1/2$.

Prob. 15. We have $f'(x) = 2x + \sec^2 x$, hence $(f^{-1})'(f(x)) = 1/f'(x) = \frac{1}{2x+\sec^2 x}$. If we put $x = 0$, we have $(f^{-1})'(0) = (f^{-1})'(f(0)) = 1$.

Prob. 17. $f'(x) = \frac{1}{\sin x} \cdot \cos x$.

$$\frac{d}{dx} f^{-1}(x)\Big|_{x=-\ln 2} \frac{1}{f'(f^{-1}(\ln 2))} = \frac{1}{\frac{1}{\sin \frac{\pi}{6}} \cdot \cos \frac{\pi}{6}} = \frac{1}{\sqrt{3}}.$$

Prob. 19. We have $f'(x) = 5x^4 + 1$, hence $(f^{-1})'(f(x)) = 1/f'(x) = \frac{1}{5x^4+1}$. If we put $x = 0$, we have $(f^{-1})'(1) = (f^{-1})'(f(0)) = 1$.

Prob. 21. We have $f'(x) = -xe^{-x^2/2} + 2$, hence $(f^{-1})'(f(x)) = 1/f'(x) = \frac{1}{2-xe^{-x^2/2}}$. If we put $x = 0$, we get $(f^{-1})'(1) = (f^{-1})'(f(0)) = 1/2$.

Prob. 23. $1/(x+1)$

Prob. 25. $-2/(1-2x)$

Prob. 27. $2/x$

Prob. 29. $f'(x) = \frac{1}{2x^3-x} \cdot (6x^2 - 1)$.

Prob. 31. $2\ln x/x$

Prob. 33. $8\ln x/x$

Prob. 35. Note that $f(x) = \frac{1}{2}\ln(1 + x^2)$. Its derivative is $x/(x^2 + 1)$.

Prob. 37. Note that $f(x) = \ln x - \ln(x + 1)$. Its derivative is $\frac{1}{x} - \frac{1}{x+1} = \frac{1}{x(x+1)}$.

Prob. 39. Note that $f(x) = \ln(1 - x) - \ln(1 + 2x)$. Its derivative is

$\frac{1}{1-x} - \frac{2}{1+2x} = \frac{4x-1}{(1-x)(1+2x)}$.

Prob. 41. $(1 - \frac{1}{x})e^{x-\ln x} = (1 - \frac{1}{x})\frac{e^x}{x}$

Prob. 43. $\cot x$

Prob. 45. $\frac{2x\sec^2 x^2}{\tan x^2} = \frac{2x}{\sin x^2 \cos x^2}$

Prob. 47. $\ln x + 1$

Prob. 49. $(1 - \ln x)/x^2$

Prob. 51. $\cos(\ln 3t)/t$

Prob. 53. $2x/(x^2 - 3)$

Prob. 55. $\frac{-2x}{\ln 10(1-x^2)}$

Prob. 57. $\frac{1}{\ln 10} \cdot \frac{3x^2-3}{x^3-3x}$

Prob. 59. $\frac{1}{\ln 3} \cdot \frac{4u^3}{3+u^4}$

Prob. 61.

(a) Let $f(x) = \ln x$. By definition of the derivative, we have

$$f'(1) = \lim_{h\to 0} \frac{\ln(1+h) - \ln 1}{h} = \lim_{h\to 0} \frac{\ln(1+h)}{h},$$

since $\ln 1 = 0$.

(b) We know that $f'(x) = 1/x$ and hence $f'(1) = 1$. Also, using the property of logarithmic functions, we have $\frac{\ln(1+h)}{h} = \ln(1+h)^{1/h}$. Thus the result in (a) becomes

$$1 = \lim_{h\to 0} \ln(1+h)^{1/h} = \ln\left(\lim_{h\to 0}(1+h)^{1/h}\right),$$

where the second equality comes from the fact ln is a continuous function.

(c) If we let $h = 1/n$, then the limit $h \to 0$ corresponds to $n \to \infty$, and thus the result in (b) becomes

$$\ln\left[\lim_{n\to\infty}\left(1 + \frac{1}{n}\right)^n\right] = 1.$$

Applying the function exp to both side will yield the desired result.

Prob. 63.

$$\ln f(x) \;=\; \ln(2x^2)$$

$$\ln f(x) \;=\; \ln 2 + x \ln x$$

Taking the derivative with respect to x:

$$\frac{1}{f(x)} \cdot f'(x) = \ln x + 1$$

$$\texttt{therefore} f'(x) = (\ln x + 1)(2x^x).$$

Prob. 65. We have $\ln y = x \ln(\ln x)$, and differentiation yields

$$\frac{y'}{y} = \ln(\ln x) + \frac{1}{\ln x},$$

or $y' = (\ln x)^{x-1}[\ln x \ln(\ln x) + 1]$.

Prob. 67. We have $\ln y = (\ln x)^2$, and differentiation yields

$$\frac{y'}{y} = \frac{2 \ln x}{x},$$

or $y' = 2(\ln x)x^{\ln x - 1}$.

Prob. 69. We have $\ln y = \ln x / x$, and differentiation yields

$$\frac{y'}{y} = \frac{1 - \ln x}{x^2},$$

or $y' = x^{\frac{1}{x} - 2}(1 - \ln x)$.

Prob. 71. We have $\ln y = x^x \ln x$, and thus

$$\ln(\ln y) = x \ln x + \ln(\ln x).$$

Differentiation with respect to x then gives

$$\frac{y'}{y \ln y} = \ln x + 1 + \frac{1}{x \ln x}.$$

Finally, we rewrite y and $\ln y$ in terms of x:

$$y' = x^{x^x + x} \ln x \left(\ln x + 1 + \frac{1}{x \ln x} \right).$$

Prob. 73. We have $\ln y = \cos x \ln x$, and differentiation yields

$$\frac{y'}{y} = -\sin x \ln x + \frac{\cos x}{x}.$$

Thus, $y' = x^{\cos x}(-\sin x \ln x + \frac{\cos x}{x})$.

Prob. 75. First take the logarithm of both sides:

$$\ln y = 2x + 3 \ln(9x - 2) - \frac{1}{4}\Big[\ln(x^2 + 1) + \ln(3x^3 - 7) \Big].$$

Then differentiate both sides with respect to x:

$$\frac{y'}{y} = 2 + \frac{27}{9x - 2} - \frac{x}{2(x^2 + 1)} - \frac{9x^2}{4(3x^3 - 7)}.$$

Finally, multiply both sides by y and rewrite y in terms of x. Since the final expression is not particularly enlightening, let us leave it here.

4.8   Approximation and Local Linearity

Prob. 1. $f'(x) = 1/2\sqrt{x}$, hence we have

$$
\begin{aligned}
\sqrt{65} &\approx \sqrt{64} + \frac{1}{2\sqrt{64}}(65 - 64) \\
&= 8 + 1/16 \approx 8.06,
\end{aligned}
$$

while $\sqrt{65} = 8.0622\ldots$.

Prob. 3. Let $f(x) = \sqrt[3]{x}$. Then, $f'(x) = \frac{1}{3}x^{-2/3}$, hence we have

$$
\sqrt[3]{124} \approx \sqrt[3]{125} + \frac{1}{2}125^{-2/3}(124 - 125) = 5 - 1/50 = 4.98,
$$

while $\sqrt[3]{124} = 4.9866\ldots$.

Prob. 5. Let $f(x) = x^2 5$. Then, $f'(x) = 25x^4$, hence we have

$$
0.99^2 5 \approx 1^2 5 + 25 \cdot 1^4(0.99 - 1) = 1 - 0.25 = 0.75
$$

while $0.99^2 5 = 0.7778\ldots$.

Prob. 7. Let $f(x) = \sin x$. Then, $f'(x) = \cos x$, hence we have

$$
\sin\left(\frac{\pi}{2} + 0.02\right) \approx \sin\frac{\pi}{2} + \cos\frac{\pi}{2} \cdot 0.02 = 1 + 0 = 1
$$

while $\sin(\frac{\pi}{2} + 0.02) = 0.99998\ldots$.

Prob. 9. Let $f(x) = \ln x$. Then, $f'(x) = 1/x$, hence we have

$$
\ln 1.01 \approx \ln 1 + \frac{1}{1}(1.01 - 1) = 0 + 0.01 = 0.01
$$

while $\ln 1.01 = 0.009950\ldots$.

Prob. 11. $\frac{1}{1+x} \approx 1 - x$ near $x = 0$

Prob. 13. $f'(x) = -2(1 + x)^{-2}$. Now

$$
\begin{aligned}
L(x) &= f(a) + f'(a)(x - a) \\
&= 1 + (-\frac{1}{2}(x - 1) = 1 - \frac{1}{2}x + \frac{1}{2} = -\frac{1}{2}x + \frac{3}{2}.
\end{aligned}
$$

Prob. 15. $\frac{1}{(1+x)^2} \approx 1 - 2x$ near $x = 0$

Prob. 17. $\ln(1 + x) \approx x$ near $x = 0$

Prob. 19. $f'(x) = \frac{1}{\ln 10} \cdot \frac{1}{x}$. Now

$$L(x) = f(a) + f'(a)(x-a) = \log 1 + \frac{1}{\ln 10}(x-1) = \frac{x}{\ln(1)} - \frac{1}{\ln(10)}.$$

Prob. 21. $e^x \approx 1 + x$ near $x = 0$

Prob. 23. $e^{-x} \approx 1 - x$ near $x = 0$

Prob. 25. $e^{x-1} \approx 1 + (x-1)$ near $x = 1$

Prob. 27. $(1+x)^{-n} \approx 1 - nx$ near $x = 0$

Prob. 29. $f'(x) = \frac{1}{2}(1+x^2)^{-\frac{1}{2}} \cdot 2x = x \cdot (1+x^2)^{-\frac{1}{2}}$. Now,

$$L(x) = f(a) + f'(a)(x-a) = 1 + 1(x-0) = x + 1.$$

Prob. 31. Since $N' = 0.03N$, $N'(4) = 0.03N(4) = 0.03 \times 100 = 3$ and thus

$$
\begin{aligned}
N(4.1) &\approx N(4) + N'(4)(4.1-4) \\
&= 100 + 3 \times 0.1 = 100.3.
\end{aligned}
$$

Prob. 33. Since $B' = 0.01B$, $B'(1) = 0.01B(1) = 0.01 \times 5 = 0.05$ and thus

$$
\begin{aligned}
B(1.1) &\approx B(1) + B'(1)(1.1-1) \\
&= 5 + 0.05 \times 0.1 = 5.005.
\end{aligned}
$$

Prob. 35. 2 ± 0.2

Prob. 37. 12 ± 1.2

Prob. 39. 7.4 ± 1.5

Prob. 41. The error in x is $\Delta x = 0.02x = 0.03$, hence the error $f(x)$ is $\Delta f = |f'(1.5)|\Delta x = 0.81$ and its percentage error is 6%.

Prob. 43. The error in x is $\Delta x = 0.02x = 0.4$, hence the error $f(x)$ is $\Delta f = |f'(20)|\Delta x = 0.02$ and its percentage error is 0.67%.

Prob. 45. The assumption means $\Delta r/r = 3\%$. Taking logarithm yields

$$\ln V = \ln(\frac{4}{3}\pi) + 3\ln r.$$

A small change on both sides is

$$\frac{\Delta V}{V} = \Delta(\ln V) = 3\Delta(\ln r) = 3\left(\frac{\Delta r}{r}\right).$$

Hence the accuracy in V is $3 \times 3\% = 9\%$.

Prob. 47. We have $N = kL^{2.11}$ for some constant k. Taking logarithm yields

$$\ln N = \ln k + 2.11 \ln L.$$

A small change on both sides is

$$\frac{\Delta N}{N} = \Delta(\ln N) = 2.11\Delta(\ln L) = 2.11 \left(\frac{\Delta L}{L} \right).$$

In order that $\Delta N/N = 5\%$, we must have $\Delta L/L = \frac{1}{2.11} \times 5\% \approx 2.4\%$.

Prob. 49. To make use of the fact that $\Delta(\ln R) = \Delta R/R$, let us first take the logarithm of R:

$$\ln R = \ln k + \ln(a - x) + \ln(b - x).$$

A small change in both sides is

$$
\begin{aligned}
\frac{\Delta R}{R} &= \Delta(\ln R) \\
&= \Delta \ln(a - x) + \Delta \ln(b - x) \\
&= -\frac{\Delta x}{a - x} - \frac{\Delta x}{b - x} \\
&= -x \left(\frac{1}{a - x} + \frac{1}{b - x} \right) \cdot \frac{\Delta x}{x} \\
&= -\frac{x(a + b - 2x)}{(a - x)(b - x)} \cdot \frac{\Delta x}{x}.
\end{aligned}
$$

This gives the relation between the percentage error in R and the percentage error in x.

4.10   Review Problems

Prob. 1. $-12x^3 - x^{-3/2}$

Prob. 3. $-\frac{2}{3}(1 - t)^{-2/3}(1 + t)^{-4/3}$

Prob. 5. $e^{2x}\left(2 \sin \frac{\pi x}{2} + \frac{\pi}{2} \cos \frac{\pi x}{2}\right)$

Prob. 7. Note that

$$f(x) = \frac{\ln(x + 1)}{\ln x}.$$

Hence, an application of the quotient rule gives us:

$$f'(x) = \frac{\frac{1}{x+1} \ln x - \frac{1}{x} \ln(x + 1)}{(\ln x)^2}.$$

Prob. 9. $f'(x) = -xe^{-x^2/2}$

$f''(x) = (x^2 - 1)e^{-x^2/2}$

Prob. 11. $h'(x) = 1/(x+1)^2$

$h''(x) - 2/(x+1)^3$

Prob. 13. In problems **13** to **16**, differentiate the given equation with respect to x and collect terms to express $y' = dy/dx$ in x and y.

$$2xy + x^2 y' - 2yy'x - y^2 = \cos x \Rightarrow y' = \frac{\cos x + y^2 - 2xy}{x^2 - 2xy}$$

Prob. 15. $\frac{1-y'}{x-y} = 2 \Rightarrow y' = 1 - 2x + 2y$

Prob. 17. Differentiating both sides with respect to x yields $2x + 2yy' = 0 \Rightarrow y' = -\frac{x}{y}$. The quotient rule then gives us $y'' = \frac{xy'-y}{y^2}$

Prob. 19. Differentiating both sides with respect to x yields

$$y'e^y = \frac{1}{x} \quad \Rightarrow \quad y' = \frac{e^{-y}}{x}.$$

The quotient rule then gives us

$$y'' = -\frac{(xy' + 1)e^{-y}}{x^2}$$

Prob. 21. Let x be the distance the birds have flown from when they were directly overhead, and y be our distance from the birds. We have $dx/dt = 6$ at all times and $y^2 = 100^2 + x^2$. Differentiating this equation with respect to t yields

$$2y\frac{dy}{dt} = 2x\frac{dx}{dt} \Rightarrow \frac{dy}{dt} = \frac{x}{y}\frac{dx}{dt}.$$

When $y = 320$, $x = \sqrt{320^2 - 100^2} \approx 304$, and thus $dy/dt = \frac{304}{320} \cdot 6 = 5.7$.

Prob. 23.

(a) $f'(x)e^{f(x)}$

(b) $f'(x)/f(x)$

(c) $2f(x)f'(x)$

Prob. 25.

(a) The graph of $y = \frac{x^2}{1+x^2}$:

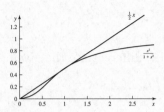

(b) On the one hand, the slope of the line is $f'(c)$. On the other hand, since this line connects $(0,0)$ and $(c, f(c))$, its slope is $f(c)/c$. Thus we have

$$f'(c) = \frac{f(c)}{c},$$

or

$$\frac{2c}{(1+c^2)^2} = \frac{c}{1+c^2}.$$

Solving the equation, we will get $c^2 = 1$, or $c = 1$, since it is assumed that $c > 0$.

This line is also plotted in the above graph.

Prob. 27. $y' = -e^{-x^2}(2x \cos x + \sin x)$. When $x = \pi/3$, it is equal to $-e^{-\pi^2/9}(\frac{\pi}{3} + \frac{\sqrt{3}}{2}) \approx -0.64$. The tangent line is thus

$$y = e^{-\pi^2/9} - 0.64\left(x - \frac{\pi}{3}\right).$$

Prob. 29. Implicit differentiation yields:

$$\ln y + \frac{xy'}{y} = y' \ln x + \frac{y}{x} \Rightarrow y' = \frac{\ln y - y/x}{\ln x - x/y}.$$

When $x = 1$, $y = 1$ and thus the above result tells us $y' = 1$. The tangent line then has the equation

$$y = 1 + (x - 1) = x.$$

Prob. 31. Since $p'(x) = 2ax + b$, $p''(x) = 2a$, the given conditions are

$$6 = p(-1) = a - b + c$$

$$8 = p'(1) = 2a + b$$

$$4 = p''(0) = 2a,$$

and the system has the unique solution

$$a = 2, \quad b = 4, \quad c = 8.$$

Prob. 33.

(a) The graph of $s(t)$:

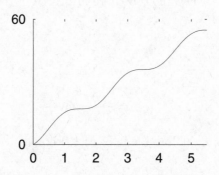

We didn't backtrack because we can see from the graph that $s(t)$ keeps increasing.

The distance between the two towns is given by $s(5.5) \approx 17.3$.

(b) $v(t) = s'(t) = 3\pi(1 + \sin \pi t) \; a(t) = v'(t) = 3\pi^2 \cos \pi t$

(c) The graph of $s(t)$ is shown above. Here are the graphs of $v(t)$

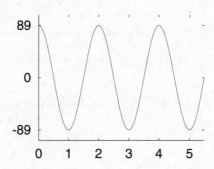

and $a(t)$

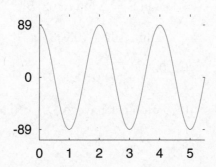

No backtracking means $s(t)$ being an increasing function. This is also equivalent to $v(t) \geq 0$, which is clear from the graph.

(d) At each peak, we switch from slowing down ($a < 0$) to speeding up ($a > 0$). This happens two times, so there are two peaks on the road. Similarly, a changes from positive to negative at each valley, so there are three of them.

Prob. 35.

(a) For the given $N(t)$,

$$\frac{dN}{dt} = -\frac{A\pi}{2T} \sin \frac{\pi t}{2T}.$$

On the other hand, the right hand side of (4.13) equals

$$\frac{\pi}{2T} \left[K - \left(K + A \cos \frac{\pi(t-T)}{2T} \right) \right] = -\frac{\pi}{2T} \cdot A \cos \left(\frac{\pi t}{2T} - \frac{\pi}{2} \right) = -\frac{A\pi}{2T} \sin \frac{\pi t}{2T}.$$

The two computations above show that $N(t)$ satisfies the given differential equation.

(b) The graph of $N(t) = 100 + 50 \cos \frac{\pi t}{2}$:

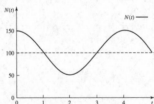

(c) The population grows and shrinks periodically between $K + A$ and $K - A$, with period $4T$.

Prob. 37. The relative errors in S and B are related as follows:

$$
\begin{aligned}
\frac{\Delta S}{S} &= \Delta \ln S = \Delta \ln[(1.162)B^{0.933}] \\
&= \Delta[\ln 1.162 + (0.933) \ln B] \\
&= 0.933 \cdot \frac{\Delta B}{B}.
\end{aligned}
$$

Therefore, if we want $\Delta B/B < 10\%$, we must have $\Delta S/S < 0.933 \times 10\% = 0.33\%$.

Chapter 5

Applications of Differentiation

5.1 Extrema and the Mean Value Theorem

Prob. 1. $y = 2x - 1$, $x \in [0, 1]$:

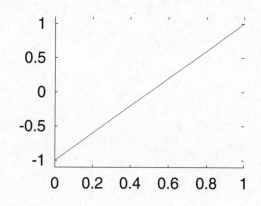

global maximum: $(1, 1)$

global minimum: $(0, -1)$

Prob. 3. $f(x) = \sin(2x)$, $0 \leq x \leq \pi$.

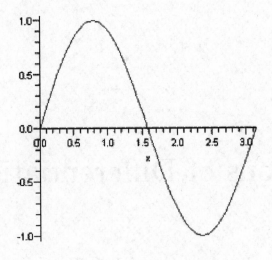

Hence, global minimum: $x = \frac{3\pi}{4}$, $\ y = -1$

global maximum at: $x = \frac{\pi}{4}$ $\ y = 1$

Prob. 5. $y = |x|$, $x \in [-1, 1]$:

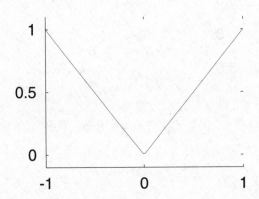

global maxima: $(\pm 1, 1)$

global minimum: $(0, 0)$

Prob. 7. $f(x) = e^{-|x|}$, $-1 \le x \le 1$

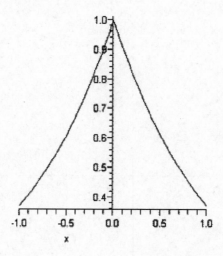

Hence global minima at: $x = \pm 1$, $y = \frac{1}{c}$

global maxima: $x = 0$, $Y = 1$

Prob. 9. An example of a function with the desired properties is $f(x) = \cos(\pi x) - x$:

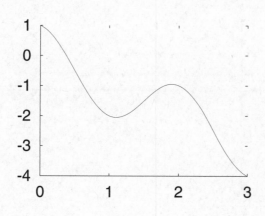

Prob. 11. An example of a function with the desired properties is $f(x) = x^2$:

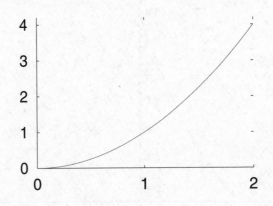

Prob. 13. $f(x) = 3 - x$, $x \in [-1, 3)$

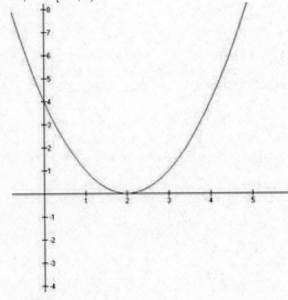

global maximum: $x = -1$ $y = 4$

no minima

Prob. 15. $y = x^2 - 2$, $x \in [-1, 1]$:

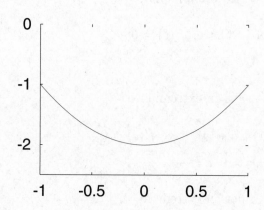

global maxima: $(\pm 1, -1)$

global minimum: $(0, -2)$

Prob. 17. $y = -x^2 + 1$, $x \in [-2, 1]$:

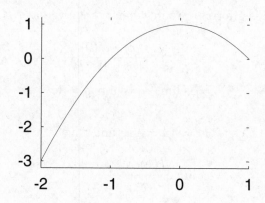

global maximum: $(0, 1)$

global minimum: $(-2, -3)$

Prob. 19. $f(x) = x^2$. Need to have $f'(c) = 0$: $f'(x) = 2x$, hence $2c = 0$, $c = 0$.

Testing:

x	-1	1
$f'(x)$	$-$	$+$

, global minimum.

Prob. 21. $f(x) = -x^2$, need to have $f'(c) = 0$:

$f'(x) = -2x$, $2c = 0$, $c = 0$.

Testing: $\dfrac{\begin{array}{c|cc} x & -1 & 1 \\ \hline f'(x) & + & - \end{array}}{}$, global maximum.

Prob. 23. $f(x) = x^3$, need to have: $f'(c) = 0$:

$f'(x) = 3x^2$, hence $3c^2 = 0$, $c = 0$

Testing: $\dfrac{\begin{array}{c|cc} x & -1 & 1 \\ \hline f'(x) & + & + \end{array}}{}$, not on extremum.

Prob. 25. $f(x) = (x + 1)^3$, need to have $f'(c) = 0$:

$f'(x) = 3(x + 1)^2$, $3(c + 1)^2 = 0$, $c = -1$.

Testing: $\dfrac{\begin{array}{c|cc} x & -2 & 0 \\ \hline f'(x) & + & + \end{array}}{}$, not an extremum.

Prob. 27. The absolute value of any number is nonnegative, so we have

$$f(x) = |x| \geq 0 = f(0),$$

i.e. $x = 0$ is a global, and thus local, minimum.

The graph of $f(x) = |x|$ has an angle at $x = 0$, corresponding to the fact that f is not differentiable there. Here is a formal proof: when $x > 0$, we have $f(x) = |x| = x$, and thus

$$\lim_{x \to 0^+} \frac{f(x) - f(0)}{x - 0}$$

$$= \lim_{x \to 0^+} \frac{x}{x} = \lim_{x \to 0^+} 1 = 1.$$

However, when $x < 0$, we have $f(x) = |x| = -x$, and thus

$$\lim_{x \to 0^-} \frac{f(x) - f(0)}{x - 0}$$

$$= \lim_{x \to 0^-} \frac{-x}{x} = \lim_{x \to 0^-} -1 = -1.$$

Therefore, the derivative

$$f'(0) = \lim_{x \to 0} \frac{f(x) - f(0)}{x - 0}$$

doesn't exist, and f is not differentiable at $x = 0$.

Prob. 29. The absolute value of any number is nonnegative, so we have

$$f(x) = |x^2 - 1| \geq 0 = f(1) = f(-1),$$

i.e. $x = \pm 1$ are global, and thus local, minima.

Let us first consider what happens near $x = 1$. When $x > 1$, $x^2 - 1 > 0$ and thus $f(x) = |x^2 - 1| = x^2 - 1$. Then we have

$$
\begin{aligned}
\lim_{x \to 1^+} \frac{f(x) - f(1)}{x - 1} &= \lim_{x \to 1^+} \frac{x^2 - 1}{x - 1} \\
&= \lim_{x \to 1^+} (x + 1) \\
&= 2.
\end{aligned}
$$

When $-1 < x < 1$, $x^2 - 1 = (x + 1)(x - 1) < 0$ and thus $f(x) = |x^2 - 1| = -x^2 + 1$. So we have

$$
\begin{aligned}
\lim_{x \to 1^-} \frac{f(x) - f(1)}{x - 1} &= \lim_{x \to 1^-} \frac{-x^2 + 1}{x - 1} \\
&= \lim_{x \to 1^-} -(x + 1) \\
&= -2.
\end{aligned}
$$

Therefore, the derivative

$$
f'(1) = \lim_{x \to 1} \frac{f(x) - f(1)}{x - 1}
$$

doesn't exist, and f is not differentiable at $x = 1$.

The situation is similar near $x = -1$. When $-1 < x < 1$, $f(x) = -x^2 + 1$ and

$$
\lim_{x \to -1^+} \frac{f(x) - f(-1)}{x + 1} = \lim_{x \to -1^+} \frac{-x^2 + 1}{x + 1} = 2.
$$

When $x < -1$, $f(x) = x^2 - 1$ and

$$
\lim_{x \to -1^-} \frac{f(x) - f(-1)}{x + 1} = \lim_{x \to -1^-} \frac{x^2 - 1}{x + 1} = -2.
$$

Therefore, f is not differentiable at $x = -1$.

Prob. 31. The graph of $y = |1 - |x||$ looks like:

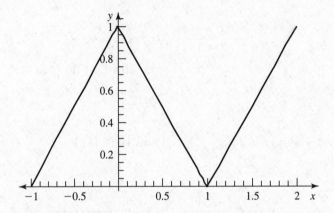

The local maxima are $(0, 1), (2, 1)$ and they are both global. The local minima are $(\pm 1, 0)$ and also, they are both global.

Prob. 33.

(a) The graph of $y = dN/dt = 2N(1 - N/100)$:

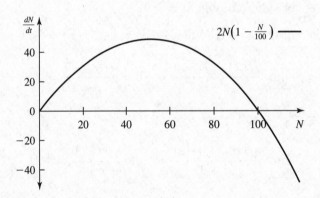

From the graph, we see that the growth rate is maximal when $N = 50$.

(b) Since $f(N)$ is a polynomial in N, it is differentiable throughout any open set in which it is defined. In particular, it is differentiable when $N > 0$. Expanding $f(N) = r(N - \frac{N^2}{K})$, we get

$$f'(N) = r\left(1 - \frac{2}{K}N\right).$$

(c) From the result in (b), we have $f'(N) = 0$ if and only if $N = K/2$. If $K = 100$, this happens at $N = 50$, the same number we obtain in (a).

Prob. 35.

(a) The slope of the line connecting $(0,0)$ and $(2,4)$ is
$$\frac{4-0}{2-0} = 2.$$

(b) Notice that the two points in (a) are $(0, f(0))$ and $(2, f(2))$. Since f is continuous on $[0,2]$ and differentiable throughout $(0,2)$, by the MVT, there must exist a number $c \in (0,2)$ such that $f'(c)$ equals the above slope. To find c, we let $f'(c) = 2$. Since $f'(x) = 2x$, we have $2c = 2$, and thus $c = 1$.

Prob. 37. Since $f(x) = x^2$ is differentiable (and continuous) everywhere, and $f(1) = 1 = f(-1)$, by Rolle's theorem, there exists some $c \in (-1, 1)$ such that $f'(c) == 0$, or, the graph of $f(x)$ has a horizontal tangent at $x = c$. Since $f'(x) = 2x$, we have $2c = 0$, or $c = 0$.

Prob. 39. Since $f(x) = x(1 - x)$ is differentiable (and continuous) everywhere, we may apply Rolle's theorem to f, i.e. if $f(a) = f(b)$ for some $a < b$, then there exists $c \in (a, b)$ with $f'(c) = 0$. For example, since $f(0) = 0 = f(1)$, f' vanishes somewhere in $[0, 1]$.

Prob. 41. Since $f(x) = -x^2 + 2$ is differentiable (and continuous) everywhere, by the MVT, there exists some $c \in (-1, 2)$ such that
$$f'(c) = \frac{f(2) - f(-1)}{2 - (-1)} = \frac{-2 - 1}{3}$$
$$= -1.$$

Prob. 43. An example of such a function is $f(x) = x^2$:

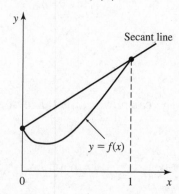

The slope of the secant connecting $(0,0)$ and $(1,1)$ is equal to the slope of the tangent of the graph at exactly one point, namely $(1/2, 1/4)$. The existence of such a point inside the interval $[0,1]$ is predicted by the MVT.

Prob. 45.

(a) The slope of the line connecting $(a, f(a))$ and $(b, f(b))$ is

$$\frac{f(b) - f(a)}{b - a} = \frac{b^2 - a^2}{b - a}$$
$$= \frac{(a+b)(b-a)}{b-a} = a + b.$$

(b) Since f is differentiable (and continuous) everywhere, by the mean value theorem, there must exist a number $c \in (a, b)$ such that $f'(c)$ equals the above slope. To find c, let $f'(c) = a + b$. Since $f'(x) = 2x$, we have $2c = a + b$, or $c = (a+b)/2$, i.e. the midpoint in between a and b.

Prob. 47. Since f is not constant on $[a, b]$ and $f(a) = f(b) = 0$, we must have $f(c) \neq 0$ for some $c \in (a, b)$. Assume $f(c) > 0$. (The case $f(c) < 0$ is completely similar.) Note that the MVT also applies to the subintervals $[a, c]$ and $[c, b]$. For the subinterval $[a, c]$, we thus have some $c_1 \in (a, c) \subset (a, b)$ with

$$f'(c_1) = \frac{f(c) - f(a)}{c - a} = \frac{f(c)}{c - a} > 0,$$

since $f(a) = 0$, $f(c) > 0$ and $a < c$. For the subinterval $[c, b]$, we have some $c_2 \in (c, b) \subset (a, b)$ with

$$f'(c_2) = \frac{f(b) - f(c)}{b - c} = \frac{-f(c)}{b - c} > 0,$$

since $f(b) = 0$, $f(c) > 0$ and $b > c$.

Prob. 49.

(a) Between $t = 0$ and $t = 5$, the car moves a distance of $s(5) - s(0) = 1.25$ (meters). Thus, the average velocity is

$$\frac{\text{distance traveled}}{\text{time elapsed}} = \frac{1.25 \text{ meters}}{5 \text{ secs}} = 1.25 \text{ meter per sec.}$$

(b) The instantaneous velocity at time t is given by

$$s'(t) = \frac{3}{100}t^2.$$

(c) From (a) and (b), the instantaneous velocity is equal to the average velocity when $3t^2/100 = 1.25$, or $t = \sqrt{125/3} \approx 6.45$.

Prob. 51. The assumption $|dB/dt| \leq 1$, or

$$-1 \leq B'(t) \leq 1, \qquad 0 \leq t \leq 3,$$

together with Corollary 1 of the MVT, tells us that

$$-1 \cdot (3 - 0) \leq B(3) - B(0) \leq 1 \cdot (3 - 0).$$

Since $B(0) = 3$, this means

$$0 \leq B(3) \leq 6.$$

Prob. 53. Consider any number $x(\neq 2)$. Since f' vanishes identically between 2 and x, Corollary 2 of the MVT implies that f is constant in the interval between 2 and x. Hence, $f(x) = f(2) = 3$.

Prob. 55. The assumption $|f(x) - f(y)| \leq |x - y|^2$ implies

$$0 \leq \left| \frac{f(x) - f(y)}{x - y} \right|$$
$$\leq \frac{|x - y|^2}{|x - y|} = |x - y|.$$

Regard x as a variable and let it approach y. By the sandwich principle, we have

$$\lim_{x \to y} \left| \frac{f(x) - f(y)}{x - y} \right| = 0,$$

which implies

$$f'(y) = \lim_{x \to y} \frac{f(x) - f(y)}{x - y} = 0.$$

(Fact: $|g(t)| \to 0$ implies $g(t) \to 0$. Proof: use $-|g(t)| \leq g(t) \leq |g(t)|$ and the sandwich principle.) Since y is arbitrary, f' vanishes identically. By Corollary 2 of the MVT, f is a constant function.

5.2   Monotonicity and Concavity

Prob. 1. In problems **1** to **20**, the range of x over which f is increasing (decreasing) is given by solving $f' > 0$ ($f' < 0$), and the range over which f is concave up (down) is given by solving $f'' > 0$ ($f'' < 0$). $f(x) = 3x - x^2$ $f'(x) = 3 - 2x$, $f''(x) = -2$. So f is increasing for $x < 3/2$, decreasing for $x > 3/2$ and concave down for all $x \in \mathbb{R}$

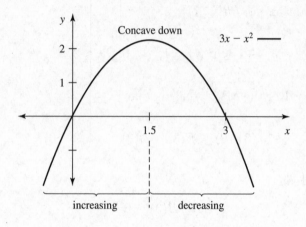

Prob. 3. $f(x) = x^2 + x - 4$ so $f'(x) = 2x + 1$ and $f''(x) = 2$. So f is increasing for $x > -1/2$, decreasing for $x < -1/2$ and concave up for all $x \in \mathbb{R}$

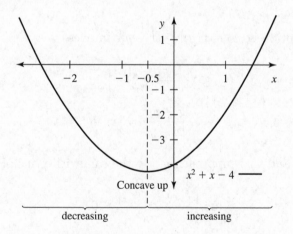

Prob. 5. $f(x) = -\frac{2}{3}x^3 + \frac{7}{2}x^2 - 3x + 4$, so $f'(x) = -2x^2 + 7x - 3 = -(2x-1)(x-3)$ and $f''(x) = -4x + 7$. So f is increasing for $1/2 < x < 3$, decreasing for $x < 1/2$, $x > 3$, concave up: $x < 7/4$ and concave down: $x > 7/4$

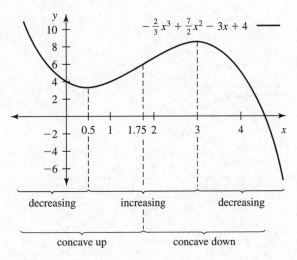

Prob. 7. $f(x) = \sqrt{x+1}$, $x \geq -1$, so $f'(x) = \frac{1}{2}(x+1)^{-\frac{1}{2}}$, $x > 1$ and $f''(x) = -\frac{1}{4}(x+1)^{-\frac{3}{2}}$ So f is increasing for all $x > -1$ and concave down for all $x > -1$

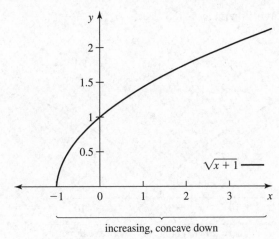

Prob. 9. $f(x) = \frac{1}{x}$, $x \neq 0$, so $f'(x) = -\frac{1}{x^2}$ and $f''(x) = \frac{2}{x^3}$. So f is decreasing: all $x \neq 0$, concave up for $x > 0$ and concave down for $x < 0$.

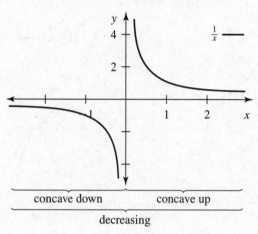

Prob. 11. $f(x) = (x^2 + 1)^{\frac{1}{3}}$, $x \in \mathbb{R}$ so $f'(x) = \frac{2}{3}x(x^2 + 1)^{-\frac{2}{3}}$ and

$f''(x) = -\frac{2}{9}(x^2 - 3)(x^2 + 1)^{-\frac{5}{3}}$. So f is increasing: $x > 0$, decreasing: $x < 0$, concave up: $-\sqrt{3} < x < \sqrt{3}$ and concave down: $x > \sqrt{3}$, $x < -\sqrt{3}$

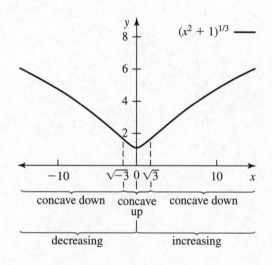

Prob. 13. $f(x) = \frac{1}{(1+x)^2}$, $x \neq -1$ so $f'(x) = -\frac{2}{(1+x)^3}$ and $f''(x) = \frac{6}{(1+x)^4}$i. So f is increasing: $x < -1$, decreasing: $x > -1$ and concave up: all $x \neq -1$

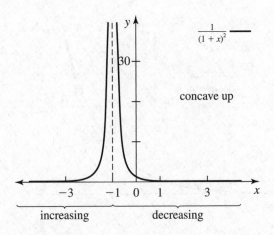

Prob. 15. $f(x) = \sin x$, $0 \leq x \leq 2\pi$ so $f'(x) = \cos x$ and $f''(x) = -\sin x$. So f is increasing: $\left(0, \frac{\pi}{2}\right) \cup \left(\frac{3\pi}{2}, 2\pi\right)$, decreasing: $\left(\frac{\pi}{2}, \frac{3\pi}{2}\right)$, concave up: $(\pi, 2\pi)$ and concave down: $(0, \pi)$

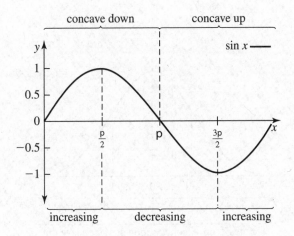

Prob. 17. $f(x) = e^x$ so $f'(x) = f''(x) = e^x$ and f is increasing for all $x \in \mathbb{R}$ and concave up for all $x \in \mathbb{R}$

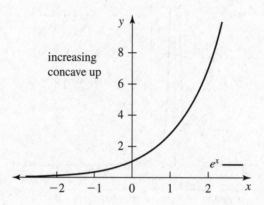

Prob. 19. $f(x) = e^{-x^2/2}$ so $f'(x) = -xe^{-x^2/2}$ and $f''(x) = (x^2 - 1)e^{-x^2/2}$. So f is increasing for $x < 0$, decreasing for $x > 0$, concave up for $x > 1$, $x < -1$ and concave down: $-1 < x < 1$

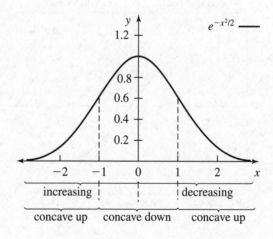

Prob. 21.

(a) The graph of a function increasing at an accelerating rate:

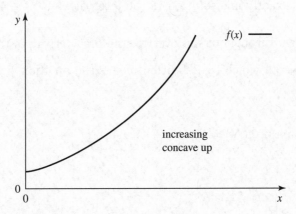

(b) The graph of a function increasing at an decelerating rate:

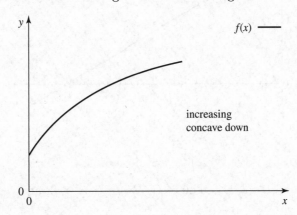

(c) In the situation of (a), the function must have positive first and second derivatives. The graph is concave up. For (b), the function must have a positive first derivative and a negative second derivative. The graph is concave down.

Prob. 23.

(a) If f' is strictly positive (negative) throughout (a, b), the graph of f in between a and b is always climbing up (down), and thus can intersect the y-axis at most once. In other words, $f(x) = 0$ for at most one x in (a, b). Since we already know there exists at least one such x, there is exactly one. (Equivalently, if $f(x) = 0$ has two solutions, Rolle's theorem says f' vanishes somewhere in (a, b), contradicting the assumption.)

(b) Let $f(x) = x^3 - 4x + 1$. Since $f(-1) = 4 > 0$ and $f(1) = -2 < 0$, $f(x) = 0$ has a solution in $(-1, 1)$. But $f'(x) = 3x^2 - 4 < 3(1) - 4 < 0$ in $(-1, 1)$. Part (a) then tells us $f(x) = 0$ has exactly one solution in the interval.

Prob. 25. If $f''(x) < 0$ throughout an interval, the first derivative test tells us f' is decreasing in this interval, which by definition means the function f is concave down.

Prob. 27.

(a) Below is the graph of $g(N) = 3\left(1 - \frac{N}{10}\right)$ versus N:

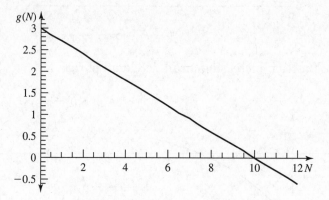

(b) We have $g(N) = r(1 - N/K)$. Thus, $g'(N) = -r/K < 0$, and $g(N)$ is decreasing for all $N > 0$.

Prob. 29. Compute the derivative

$$f'(N) = \left[1 - \left(\frac{N}{K}\right)^\theta\right] - N \cdot \frac{\theta N^{\theta-1}}{K^\theta}$$

$$= 1 - (1 + \theta)\left(\frac{N}{K}\right)^\theta .$$

It is easy to deduce that the expression is positive, i.e. the growth rate is increasing, when

$$(0 <) N < \frac{K}{(1+\theta)^{\frac{1}{\theta}}},$$

and negative otherwise.

Prob. 31. Since $f'(P) = -ae^{-aP}$ is always negative, f decreases with P.

Prob. 33.

(a) Since

$$y'(x) = \frac{1170}{x^2}e^{-10/x}$$

is always positive, the height y increases with age x. For $x > 0$, we have $-10/x < 0$ and thus $e^{-10/x} < 1$. Therefore, the height $y = 117e^{-10/x}$ never exceeds 117. As the tree ages ($x \to \infty$), $-10/x$ approaches zero and thus $y(x)$ approaches 117.

(b) Compute

$$y''(x) = 1170e^{-10/x}\left(\frac{10 - 2x}{x^4}\right).$$

Hence, the graph is concave up (down), i.e. $y'' > 0$ ($y'' < 0$), when $x < 5$ ($x > 5$).

(c) The graph of $y = 117e^{-10/x}$:

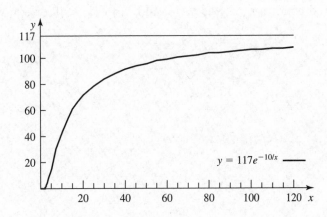

(d) The greatest rate of growth corresponds to the largest slope. Looking at the graph, we notice that it is the steepest at $x = 5$ (as expected from the calculation in (b)), and it is exactly where the graph changes from being concave up to being concave down.

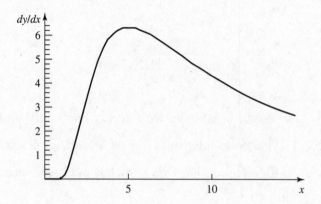

Prob. 35. In terms of the function $X(F)$, we are looking for a condition such that $X'(F) > 0$ and $X''(F) < 0$ for all $F \geq 0$. Since

$$X'(F) = c\gamma F^{\gamma-1},$$
$$X''(F) = c\gamma(\gamma-1)F^{\gamma-2},$$

and $c > 0$, the desired condition on γ is $\gamma > 0$ and $\gamma(\gamma - 1) < 0$, or equivalently,

$$0 < \gamma < 1.$$

Prob. 37.

(a) At equilibrium, i.e. $dN/dt = 0$, we have

$$0 = Ne^{-aN} - N^2 = N(e^{-aN} - N).$$

The nontrivial equilibrium $(N^* \neq 0)$ thus satisfies $e^{-aN^*} = N^*$.

(b) If we differentiate

$$e^{-aN^*} = N^*$$

with respect to a, regarding N^* as a function of a, we obtain

$$-[a(N^*)' + N^*]e^{-aN^*} = (N^*)',$$

which after rearrangement becomes

$$(N^*)' = -\frac{N^* e^{-aN^*}}{1 + ae^{-aN^*}} < 0.$$

In other words, N^* decreases with a.

Prob. 39.

(a) Compute

$$\frac{d}{dN}\frac{A(N)}{N} = \frac{d}{dN}\frac{S}{1 + (aN)^b}$$

$$= -\frac{Sab(aN)^{b-1}}{[1 + (aN)^b]^2}.$$

Since the right hand side is negative, $A(N)/N$ is a decreasing function of N.

(b) Using the fact that $\log(x/y) = \log x - \log y$, we have

$$k = \log NS - \log\left[\frac{NS}{1 + (aN)^b}\right]$$

$$= \log(NS) - \left[\log(NS) - \log(1 + (aN)^b)\right]$$

$$= \log[1 + (aN)^b].$$

(i) Recall that

$$\log N = \log_{10} N = \frac{\ln N}{\ln 10}.$$

Hence we have

$$\frac{d\log N}{dN} = \frac{d}{dN}\left(\frac{\ln N}{\ln 10}\right)$$

$$= \frac{1}{N\ln 10}.$$

(ii) Using the chain rule and the result of (i), we have

$$\frac{dk}{d\log N} = \frac{dk/dN}{d\log N/dN}$$

$$= (\ln 10)N\frac{dk}{dN}.$$

Using the first result in (b), we compute

$$\frac{dk}{dN} = \frac{d}{dN}\frac{\ln[1 + (aN)^b]}{\ln 10}$$

$$= \frac{1}{\ln 10} \cdot \frac{ab(aN)^{b-1}}{1 + (aN)^b}.$$

Hence, we have

$$
\begin{aligned}
\frac{dk}{d\log N} &= (\ln 10)N \cdot \frac{1}{\ln 10} \cdot \frac{ab(aN)^{b-1}}{1+(aN)^b} \\
&= \frac{b(aN)^b}{1+(aN)^b} \\
&= \frac{b}{(aN)^{-b}+1}.
\end{aligned}
$$

(iii) Since $a, b > 0$, as $N \to \infty$, $(aN)^{-b} \to 0$ and thus

$$
\lim_{N\to\infty}\frac{dk}{d\log N} = \lim_{N\to\infty}\frac{b}{1+(aN)^{-b}} = b.
$$

(iv) Let us consider the relation between $dk/d\log N$ and $A'(N) = dA/dN$. By definition,

$$
k = \log(NS) - \log A(N) = \log N - \log S - \log A(N).
$$

Using the chain rule, we have

$$
\begin{aligned}
\frac{dk}{d\log N} &= 1 - \frac{d\log A(N)/dN}{d\log N/dN} \\
&= 1 - \frac{A'(N)}{A(N)\ln 10}\bigg/ \frac{1}{N\ln 10} \\
&= 1 - \frac{NA'(N)}{A(N)}.
\end{aligned}
$$

Therefore, if $dk/d\log N > 1\,(< 1)$, then $A'(N)$ is negative (positive), i.e. $A(N)$ is decreasing (increasing). (Note: N and $A(N)$ are positive.)

Since the initial density of seeds is by definition NS and S is a constant, a higher density means increasing N. From the result in (ii), we observe that $dk/d\log N$ increases with N. $(N\uparrow \Rightarrow (aN)^{-b}\downarrow \Rightarrow 1/[1+(aN)^{-b}]\uparrow)$ And from what we have just deduced, if $dk/d\log N$ increases beyond 1, the number of surviving plants switches from growing to dropping. Therefore, to summarize, if the initial density of seeds is higher (lower) than some fixed value, there will be less (more) plants in the following year than this year.

(v) From the result in (iv), the case $dk/d\log N = 1$ corresponds to $A'(N) = 0$. Thus, over the range where $dk/d\log N = 1$ holds, the number of plants is in equilibrium.

Prob. 41.

(a) Since

$$\frac{dY}{dX} = \frac{d(bX^a)}{dX} = abX^{a-1},$$

$$\frac{d}{dX}\frac{Y}{X} = \frac{d(bX^{a-1})}{dX}$$

$$= (a-1)bX^{a-2},$$

the condition we are looking for is $ab > 0$ and $(a-1)b < 0$. Since b is positive, we conclude $0 < a < 1$.

In this case, $Y'' = a(a-1)bX^{a-2}$ is negative, i.e. Y as a function of X is concave down.

(b) Now X is the body length, Y is the skull length and $0 < a < 1$. From (a), we know

$$\frac{d}{dX}\frac{Y}{X} < 0.$$

As a vertebrate grow up (X increases), the size of the skull Y becomes smaller and smaller *compared to the size of the body X*.

Prob. 43. Differentiating both sides of the given equation $y' = ky/x$ with respect to x yields

$$y'' = k\left(\frac{xy' - y}{x}\right).$$

If we use the equation $y' = ky/x$, or $xy' = ky$, again, we have

$$y'' = k\left(\frac{ky - y}{x}\right) = k(k-1)\frac{y}{x}.$$

Since x and y are positive, y as a function of x is concave up, i.e. $y'' > 0$, if and only if $k(k-1) > 0$, or $k > 1$. (k is assumed to be positive.)

5.3 Extrema, Inflection Points and Graphing

Prob. 1. $y = (2-x)^2 = 4 - 4x + x^2$

$y' = -4 + 2x = 2(-2 + x) = 0$, $x = 2$.

Testing: $\dfrac{\begin{array}{c|c|c} x & 1 & 3 \\ \hline y' & - & + \end{array}}{}$, hence y is increasing: $(2,3)$ y is decreasing: $(-2, 1)$.

Also, $x = 2$ is a minimum. Also checking endpoints:

$$f(-2) = 16 \leftarrow \texttt{global max}$$

$$f(2) = 0 \leftarrow \texttt{global min}$$

$$f(3) = 1$$

Prob. 3. $y = \ln(2x - 1)$, $1 \le x \le 2$.

$y' = \frac{2}{2x-1}$, undefined at $x = \frac{1}{2}$, not in the domain for $1 \le x \le 2$ $y' > 0$, hence always increasing. Need to check endpoints: $f(1) = \ln(1) = 0 \leftarrow$ global min $f(2) = \ln(3) \leftarrow$ global max.

Prob. 5. $y = xe^{-x}$, $0 \le x \le 1$. $y' = e^{-x} = e^{-x}(1 - x) = 0$ gives $x = 1$. It is already an endpoint, hence:

$$f(0) = 0 \leftarrow \texttt{global min}.$$

$$f(1) = \frac{1}{2} \leftarrow \texttt{global max}.$$

Prob. 7. Given, $y = (x - 1)^3 + 1$ and $x \in \mathbf{R}$. To find the maxima or minima, we first find the derivative of the above function,

$$y' = 3(x - 1)^2$$

Equating the above derivative to zero, we get

$$y' = 0,$$

$$3(x - 1)^2 = 0$$

$$x = 1$$

Finding the second derivative of the function we have

$$y'' = 6(x - 1)$$

$$y'' \begin{cases} > 0 & \texttt{for} \ \ > 1; \\ < 0 & \texttt{for} \ \ x < 1; \\ = 0 & \texttt{for} \ \ x = 1. \end{cases}$$

Thus the function continuously increases for $x >$ and decreases for $x < 1$. Thus the function has no maxima and minima.

Prob. 9. $y = \cos(\pi x^2)$, $-1 \leq x \leq 1$

$y' = \sin(\pi x^2)x = 0$, $x = \pm 1$, $x = 0$.

Testing: $\dfrac{x \;\; \begin{array}{c|c} -\frac{1}{2} & \frac{1}{2} \end{array}}{y' \;\; \begin{array}{c|c} + & - \end{array}}$, hence y increases on $(-1,0)$ y decreases on $(0,1)$.

$f(-1) = -1 \leftarrow$ global min

$f(0) = 1 \leftarrow$ global max

Prob. 11. Given, $y = e^{-|x|}$ and $x \in \mathbf{R}$. We redefine the function y as

$$y = \begin{cases} e^{-x} & \text{for} \quad x > 0; \\ e^{x} & \text{for} \quad x < 0. \end{cases}$$

To find the maxima or minima, we first find the derivative of the above function,

$$y' = \begin{cases} -e^{-x} & \text{for} \quad x > 0; \\ e^{x} & \text{for} \quad x < 0. \end{cases}$$

Equating the above derivative to zero, we get

$$y' = 0,$$

So, no value of x satisfies the equation and hence we there is no maxima and minima in the above interval. Evaluating the function t $x = 0$, we find that

$$y(0) = 1$$

which is,in fact a global maxima. Thus, we have one global maxima, which is at $x = 0$.

$$x = 0 \Rightarrow \texttt{Global maxima}$$

Prob. 13. Given, $y = \frac{x^3}{3} + \frac{x^2}{2} - 6x + 2$ and $x \in \mathbf{R}$. To find the maxima or minima, we first find the derivative of the above function, $y' = x^2 + x - 6$ Equating the above derivative to zero, we get

$$y' = 0,$$

$$x^2 + x - 6 = 0$$

$$(x+3)(x-2)=0$$

$$x=-3,2$$

Finding the second derivative of the above function we have,

$$y''=2x+1$$

$$y'' = \begin{cases} <0 & \text{for} \quad x < -\frac{1}{2}; \\ >0 & \text{for} \quad x > \frac{1}{2}. \end{cases}$$

Since -3 lies in the 1st interval and 2 lies in the second interval we have, local maxima and minima at these points.

$$x=-3 \Rightarrow \texttt{Local maxima}$$

$$x=2 \Rightarrow \texttt{Local minima}$$

Prob. 15. $y=(x-1)^{1/3}$, $y'=\frac{1}{3}\cdot(x-1)^{-\frac{2}{3}}$, undefined at $x=1$, testing:

x	0	2
y'	+	+

always increasing, no extrema.

Prob. 17. We have to prove that the $f'(x)=0$ is not a sufficient condition for the extrema to exist. We start by letting the function to be,

$$f(x)=x^3$$

So finding its derivative,

$$f'=3x^2$$

Equating it to zero we have,

$$f'=0$$

$$3x^2=0, x=0$$

So, we should have a extrema here. But, we see that the derivative doesn't change sign at $x=0$.

$$\lim_{x\to+0} f' = \lim_{x\to+0} 3x^2$$

$$=0$$

and similarly,

$$\lim_{x\to-0} f' = \lim_{x\to+0} 3x^2$$

$$= 0$$

Thus f' doesn't change its sign and hence $x = 0$ is not a local maxima or minima.

Prob. 19. Given the function as, $f(x) = x^3 - 2$ and $x \in \mathbf{R}$ Since, we have to find the points of inflection, we need to find the first derivative as well as second derivative and check whether the second derivative changes sign at the inflection point. For the point to be inflection, second derivative should change sign .

$$f'(x) = 3x^2$$

$$f''(x) = 6x$$

Thus

$$6x = 0, x = 0$$

Hence we get $x = 0$ as the candidate inflection point. Also, the second derivative changes sign at $x = 0$ as

$$f''(x) = \begin{cases} > 0 & \text{for} \quad x > 0; \\ < 0 & \text{for} \quad x < 0. \end{cases}$$

Thus it is, indeed a point of inflection.

Prob. 21. Given the function as,

$$f(x) = e^{-x^2}$$

and

$$x \geq 0$$

Since, we have to find the points of inflection, we need to find the first derivative as well as second derivative and check whether the second derivative changes sign at the inflection point. For the point to be inflection, second derivative should change sign .

$$f'(x) = -2xe^{-x^2}$$

$$f''(x) = -2e^{-x^2}(1 - 2x^2)$$

Thus

$$-2e^{-x^2}(1 - 2x^2) = 0$$

$$1 - 2x^2 = 0$$

$$x = \frac{1}{\sqrt{2}}, -\frac{1}{\sqrt{2}}$$

Hence we get $x = \pm\frac{1}{\sqrt{2}}$ as the candidate inflection point. Also, the second derivative changes sign at $x = 0$ as

$$f''(x) = \begin{cases} > 0 & \text{for} \quad 0x > \frac{1}{\sqrt{2}}x < -\frac{1}{\sqrt{2}}; \\ < 0 & \text{for} \quad -\frac{1}{\sqrt{2}} < x < \frac{1}{\sqrt{2}}. \end{cases}$$

Thus it is, indeed a point of inflection.

Prob. 23. Given the function as,

$$f(x) = \tan x$$

and

$$-\frac{\pi}{2} \le x \le \frac{\pi}{2}$$

Since, we have to find the points of inflection, we need to find the first derivative as well as second derivative and check whether the second derivative changes sign at the inflection point. For the point to be inflection, second derivative should change sign .

$$f'(x) = \sec^2 x$$

$$f''(x) = 2\sec^2 x \tan x$$

Thus

$$2\sec^2 x \tan x = 0, x = 0$$

Hence we get $x = 0$ as the candidate inflection point. Also, the second derivative changes sign at $x = 0$ as

$$f''(x) = \begin{cases} > 0 & \text{for} \quad 0 < x < \frac{\pi}{2}; \\ < 0 & \text{for} \quad -\frac{\pi}{2} < x < 0. \end{cases}$$

Thus it is, indeed a point of inflection.

Prob. 25. Given the function as, $f(x) = x^4$ and $x \in \mathbf{R}$ Since, we have to find the points of inflection, we need to find the first derivative as well as second derivative and check whether the second derivative changes sign at the inflection point. For the point to be inflection, second derivative should change sign .

$$f'(x) = 4x^3$$

$$f''(x) = 12x^2$$

Thus

$$12x^2 = 0, x = 0$$

Hence we get $x = 0$ as the candidate inflection point. But it is not a inflection point as the 2nd derivative does not change sign around $x = 0$. Thus it is, not a point of inflection.

Prob. 27. Given the function as

$$y = \frac{2x^3}{3} - 2x^2 - 6x + 2$$

where,

$$-2 \leq x \leq 5$$

To find the local maxima and minima we have to find the first and second derivative.

First, we find the first derivative,

$$y' = 2x^2 - 4x - 6$$

Equating it to zero, we have

$$y' = 0$$

$$2x^2 - 4x - 6 = 0$$

$$2(x - 3)(x + 1) = 0$$

$$x = 3, -1$$

Next, we find the second derivative,

$$y'' = 4x - 4$$

From, above we can say that,

$$y'' = \begin{cases} > 0 & \text{for } x \geq 1, \\ < 0 & \text{for } x \leq 1; \end{cases}$$

Thus we can say that the function has local maxima and minima.

$$x = 3 \Rightarrow \text{Local minima}$$

$$x = -1 \Rightarrow \texttt{Local maxima}$$

Also, to find the point of inflection, we equate second derivative to 0,

$$y'' = 0$$

$$4x - 4 = 0$$

$$x = 1$$

From above, we see that the function does change sign at $x = 1$ and hence it is the point of inflection.

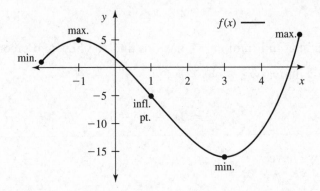

Prob. 29. Given the function as

$$y = |x^2 - 9|$$

where,

$$-4 \le x \le 5$$

We can redefine the function as

$$y = \begin{cases} x^2 - 9 & -4 \le x \le -3, \ 3 \le x \le 5; \\ 9 - x^2 & -3 < x < 3. \end{cases}$$

To find the local maxima and minima we have to find the first and second derivative.

First, we find the first derivative,

$$y' = \{2x -4 \le x \le -3, \ 3 \le x \le 5; -2x -3 < x < 3.$$

Thus equating each of them to zero we find that, the second part is only zero for $x = 0$

Next, we find the second derivative,

$$y'' = \begin{cases} 2 & -4 \le x \le -3, \quad 3 \le x \le 5; \\ -2 & -3 < x < 3. \end{cases}$$

Thus the function has a local maxima at $x = 0$.

$$x = 0 \Rightarrow \texttt{Local maxima}$$

As the second derivative is not equal to zero for any x, we do not have any point of inflection.

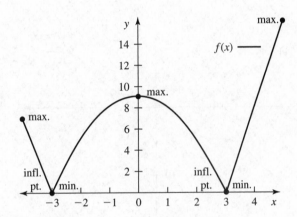

Prob. 31. Given the function as $y = x + \cos x$ where, $x \in \mathbf{R}$ To find the local maxima and minima we have to find the first and second derivative. First, we find the first derivative,

$$y' = 1 - \sin x$$

Equating it to zero, we have

$$y' = 0$$

$$1 - \sin x = 0$$

$$x = \frac{n\pi}{2}, \texttt{where } n = 1, 5, 9, \cdots$$

Next we find the second derivative of the function,

$$y'' = -\cos x$$

$$y'' = \begin{cases} > 0 & \text{for} \quad \frac{n\pi}{2} < x < \frac{3n\pi}{2}; \\ < 0 & \text{for} \quad -\frac{n\pi}{2} < x < \frac{n\pi}{2}. \end{cases}$$

where $n = 1, 2, \ldots$ To find the point of inflection we equate second derivative to 0

$$y'' = 0$$

$$-\cos x = 0$$

$$x = \frac{n\pi}{2}, where, n = 1, 5, 9, \cdots$$

Thus the above represents, the point of inflection as we see that the 2nd derivative changes sign at them.

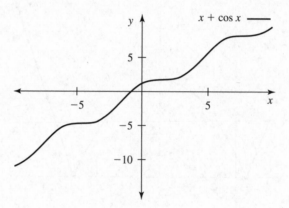

Prob. 33. $y = \frac{x^2-1}{x^2+1}$ where, $x \in \mathbf{R}$ To find the local maxima and minima we have to find the first and second derivative. First, we find the first derivative,

$$y' = \frac{4x}{(x^2 + 1)^2}$$

Equating it to zero, we have

$$y' = 0$$

$$\frac{4x}{(x^2 + 1)^2} = 0$$

$$x = 0$$

Next, we find the second derivative,

$$y'' = 4\frac{1 - 3x^4 - 2x^2}{(x^2 + 1)^4}$$

From, above we can say that,

$$y'' = \begin{cases} > 0 & \text{for} \quad x \geq \frac{1}{3} \text{ and } -1 \leq x \leq -\frac{1}{3}, \\ < 0 & \text{for} \quad -\frac{1}{3} \leq x \leq \frac{1}{3} \text{ and } x < 01; \end{cases}$$

Thus the function has a local maxima at $x = 0$ Also the point of inflection are

$$x = \pm\frac{1}{3}, -1$$

as the function changes sign as it passes through it.

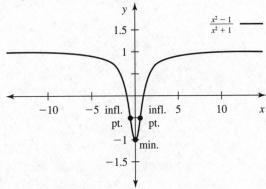

Prob. 35. Given the function,

$$f(x) = \frac{x}{x-1}$$

and

$$x \neq 1$$

(a)

$$\lim_{x \to +\infty} f(x) = \lim_{x \to +\infty} \frac{x}{x-1}$$

$$= \lim_{x \to +\infty} \frac{1}{1 - \frac{1}{x}}$$

$$= \frac{1}{1-0} = 1$$

And,

$$\lim_{x \to -\infty} f(x) = \lim_{x \to -\infty} \frac{x}{x-1}$$

$$= \lim_{x \to -\infty} \frac{1}{1 - \frac{1}{x}}$$

$$= \frac{1}{1-0} = 1$$

Thus we have

$$\lim_{x \to +\infty} f(x) = \lim_{x \to -\infty} f(x) = 1$$

(b)

$$\lim_{x \to 1^+} f(x) = \lim_{x \to 1^+} \frac{x}{x-1}$$

$$= \lim_{x \to 1^+} \frac{1}{1 - \frac{1}{x}}$$

$$= \frac{1}{1 - (1 - \epsilon)} = +\infty$$

And,

$$\lim_{x \to 1^-} f(x) = \lim_{x \to 1^-} \frac{x}{x-1}$$

$$= \lim_{x \to 1^-} \frac{1}{1 - \frac{1}{x}}$$

$$= \frac{1}{1 - (1 + \epsilon)} = -\infty$$

Thus $x = 1$ is vertical asmyptote.

(c) We first compute y' and the y'',

$$y' = -\frac{1}{(x-1)^2}$$

$$y'' = 2\frac{1}{(x-1)^3}$$

The function is decreasing in the interval $x > 1$ No, the function has no local extrema.

(d) The function is concave up in the interval $x > 1$ and is concave down in the interval $x < 1$.

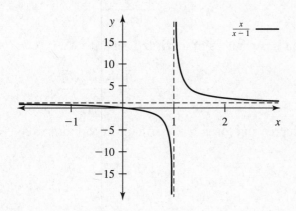

(e)

Prob. 37. Given the function,

$$f(x) = \frac{2x^2 - 5}{x + 2}$$

and

$$x \neq 2, -2$$

(a)

$$\lim_{x \to -2^+} f(x) = \lim_{x \to -2^+} \frac{2x^2 - 5}{x + 2}$$

$$= \lim_{x \to -2^+} \frac{2(-2 + \epsilon)^2 - 5}{(-2 + \epsilon) + 2}$$

$$= \infty$$

And,

$$\lim_{x \to -2^-} f(x) = \lim_{x \to -2^-} \frac{2x^2 - 5}{x + 2}$$

$$= \lim_{x \to -2^-} \frac{2(-2 - \epsilon)^2 - 5}{(-2 - \epsilon) + 2}$$

$$= -\infty$$

Thus $x = -2$ is vertical asmyptote.

(b) Next, we find f' and f''.

$$f' = \frac{2x^2 + 8x + 5}{(x+2)^2}$$

Thus the function has chances of local extrema at

$$x = \frac{-4 + \sqrt{6}}{2}, \frac{-4 - \sqrt{6}}{2}$$

The function is increasing for $x > -2$. Again in the interval

$$x < -2$$

the function has local minima and it is at

$$x = \frac{-4 - \sqrt{6}}{2}$$

(c) It is concave up for

$$\frac{-4 - \sqrt{6}}{2} < x < -2$$

(d) The oblique assymptote willa line with slope $m = 2$ which is the $\lim_{x \to \infty} f(x)/x$

(e)

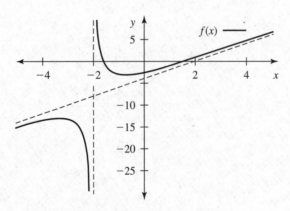

Prob. 39. Given the function, $f(x) = \frac{x^2}{x^2+1}$ and $x \in \mathbf{R}$

(a) First we find the derivative of the above function.

$$f'(x) = \frac{2x}{(1+x^2)^2}$$

Thus the function is increasing for $x > 0$ and $x < 0$, as the function is even.

(b) For concavity, we compute $f''(x)$

$$f'' = 2\frac{1 - 2x^2 - 4x^4}{(1 + x^2)^4}$$

To compute

$$f'' = 0$$

we have

$$2\frac{(3x^2 - 1)(x^2 + 1)}{(1 + x^2)^4} = 0$$

$$3x^2 - 1 = 0$$

$$x = \pm\frac{1}{\sqrt{3}}$$

Thus we have

$$y'' = \begin{cases} > 0 & \text{for } x > \frac{1}{\sqrt{3}} \text{ and } x < -\frac{1}{\sqrt{3}}; \\ < 0 & \text{for } -\frac{1}{\sqrt{3}} < x < \frac{1}{\sqrt{3}}. \end{cases}$$

Thus the points of inflection are

$$x = \pm\frac{1}{\sqrt{3}}$$

The graph is concave up for the portion $x > \frac{1}{\sqrt{3}}$.

(c) The graph has a horizontal asymptote at $y = 1$.

$$\lim_{x \to +\infty} f(x) = \lim_{x \to +\infty} \frac{x^2}{x^2 + 1}$$

$$= \lim_{x \to +\infty} \frac{1}{1 + \frac{1}{x^2}}$$

$$= 1$$

And,

$$\lim_{x \to -\infty} f(x) = \lim_{x \to -\infty} \frac{x^2}{x^2 + 1}$$

$$= \lim_{x \to -\infty} \frac{1}{1 + \frac{1}{x^2}}$$

$$= 1$$

Thus we have

$$\lim_{x \to +\infty} f(x) = \lim_{x \to -\infty} f(x) = 1$$

Hence $y = 1$ is a horizontal asymptote.

(d)

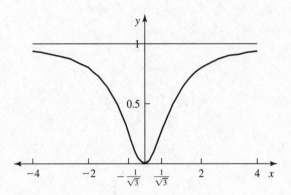

Prob. 41. Given the function, $f(x) = \frac{x}{x+a}$ and $x \in \mathbf{R}$

(a) First we find the derivative of the above function.

$$f'(x) = \frac{a}{(a+x)^2}$$

Thus the function is decreasing for $x > -a$. There is no local extrema.

(b) For concavity, we compute $f''(x)$

$$f'' = -\frac{2a}{(a+x)^3}$$

Thus we have

$$y'' = \begin{cases} > 0 & \text{for} \quad x < -a; \\ < 0 & \text{for} \quad x > -a. \end{cases}$$

There are no inflection points. Its concave up for $x > -a$.

(c) The graph has a horizontal asymptote at $y = 0$.

$$\lim_{x \to +\infty} f(x) = \lim_{x \to +\infty} \frac{x}{x+a}$$

$$= \lim_{x \to +\infty} \frac{1}{1 + \frac{a}{x}}$$

$$= 1$$

And,

$$\lim_{x \to -\infty} f(x) = \lim_{x \to -\infty} \frac{x}{x+a}$$

$$= \lim_{x \to -\infty} \frac{1}{1+\frac{a}{x}}$$

$$= 1$$

Thus we have

$$\lim_{x \to +\infty} f(x) = \lim_{x \to -\infty} f(x) = 1$$

Hence $y = 1$ is a horizontal asymptote .

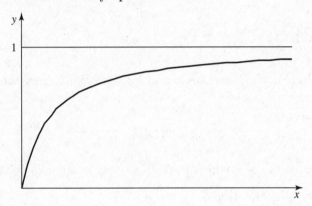

(d)

Prob. 43. Given the growth rate of population as

$$f(N) = N(1 - (\frac{N}{K})^\theta)$$

and

$$N \geq 0$$

So, we have to find N such that $f(N)$ is maximal. Hence we find 1st derivative and evaluate it to zero.

$$f'(N) = 1 - (\frac{N}{K})^\theta - \frac{N}{K}\theta(\frac{N}{K})^{\theta-1}$$

$$= 1 - (\frac{N}{K})^\theta(1+\theta)$$

Hence equating it to zero, we have

$$1 - (\frac{N}{K})^\theta(1+\theta) = 0$$

$$\frac{N}{K})^{\theta} = \frac{1}{1 + \theta}$$

$$N = K(\frac{1}{1 + \theta})^{\frac{1}{\theta}}$$

Thus we have the maximal value at the above population. We, confirm this by taking the second derivative of the function and by noting that the

$$f''(K(\frac{1}{1 + \theta})^{\frac{1}{\theta}}) < 0$$

and thus has a maxima.

5.4   Optimization

Prob. 1. If l is the length of the rectangle and b is its height, then the perimeter p is given by

$$p = 2(l + b).$$

The area of the rectangle is $l \times b$, which is given to be 25. Since $lb = 25$, we have $l = \frac{25}{b}$, and substituting this in the equation for p we get

$$p = 2\left(\frac{25}{b} + b\right)$$

Thus we have p as a function of b alone. We now differentiate p to get the extreme values of the perimeter. We have

$$\frac{dp}{db} = 2\left(-\frac{25}{b^2} + 1\right)$$

Equating the above equation to zero, we have

$$2\left(-\frac{25}{b^2} + 1\right) = 0 \qquad \Rightarrow \qquad \frac{25}{b^2} = 1 \qquad \Rightarrow \qquad b^2 = 25$$

Since b represents a distance, it must be positive and thus $b = 5$. We must now check whether this value of b corresponds to a maximum or a minimum. To do this we calculate

$$\frac{d^2p}{db^2} = \frac{100}{b^3},$$

which is positive for all $b > 0$. Thus the graph of p is concave up at $b = 5$, and we have found an absolute minimum. To find the length of the rectangle, we use the fact that

$bl = 25$ to obtain $l = 5$ as well. Thus the rectangle with smallest perimeter is in fact a square of side 5in, with a perimeter of 20in.

Prob. 3. Let the upper right corner of the rectangle be the point (x, y) on the parabola (see the figure provided with the question). Then the area of the rectangle is $A = 2xy$. Since the point (x, y) is on the parabola, we know that $y = 3 - x^2$, and so we get $A = 2x(3 - x^2)$. This gives us A as a function of x only. To get the extreme values of A, we first differentiate the function $A(x)$, to get

$$A'(x) = 2(3 - x^2) + 2x(-2x) = 6 - 6x^2.$$

Setting $A'(x) = 0$, we have $6 - 6x^2 = 0 \Rightarrow x^2 = 1 \Rightarrow x = \pm 1$. Since (x, y) is in the first quadrant, we select $x = 1$. Now, $A''(x) = -12x$ and so the function $A(x)$ is concave down for all $x > 0$. Thus $x = 1$ gives us a maximum value of A, and from this we have $y = 3 - x^2 = 3 - 1 = 2$. The height of the rectangle is 2 and its length is 2, with a maximal area of 4.

Prob. 5. Let l denote the length of the rectangular area, and b its height. Since the field is bounded by a river on one side, the perimeter p is given by $p = 2l + b$, which is given to equal 320 ft. Thus, $b = 320 - 2l$ and substituting for b in the equation for the area A of the rectangle, we get $A = lb = l(320 - 2l) = 320l - 2l^2$. To optimize the area, we differentiate A with respect to l.

$$\frac{dA}{dl} = 320 - 4l$$

Setting $dA/dl = 0$, we have

$$320 - 4l = 0 \quad \Rightarrow \quad 320 = 4l \quad \Rightarrow \quad l = 80.$$

To confirm that we have found a maximum, we check that the graph of A is concave down by calculating

$$\frac{d^2 A}{dl^2} = -4$$

and noticing that this expression is always negative. Thus the maximum value of A is 12800 ft^2, which occurs when the length is $l = 80$ ft and the height is $b = 320 - 2l = 320 - 160 = 160$ ft.

Prob. 7. We are given a right triangle whose sides have length a and b. Further, the hypotenuse of the triangle has length 5. Thus, the perimeter p of the triangle is given by

$$p = a + b + 5.$$

Since the triangle is a right triangle, we have that $25 = a^2 + b^2$, and so we have $a = \sqrt{25 - b^2}$. Substituting for a in the equation for the perimeter, we have

$$p = \sqrt{25 - b^2} + b + 5$$

To optimize the perimeter, we differentiate the function p to get

$$\frac{dp}{db} = -\frac{b}{\sqrt{25 - b^2}} + 1$$

Setting this derivative to equal zero, we have

$$-\frac{b}{\sqrt{25 - b^2}} + 1 = 0 \quad \Rightarrow \quad b = \sqrt{25 - b^2} \quad \Rightarrow \quad 2b^2 = 25 \quad \Rightarrow \quad b = \frac{5\sqrt{2}}{2}$$

We also note that dp/db does not exist when $b = 5$, but the base of the triangle cannot equal its hypotenuse and so we reject this value of b. Now, by calculating that

$$\frac{d^2 p}{db^2} = -\frac{25 + 2b^2}{(25 - b^2)^{3/2}}$$

and noting that this second derivative is negative for all $0 < b < 5$, we find that the graph of p is concave down and hence the value $b = \frac{5\sqrt{2}}{2}$ corresponds to a maximum, not a minimum. Since

$$\lim_{b \to 0^+} \left(\sqrt{25 - b^2} + b + 5 \right) = 10 = \lim_{b \to 5^-} \left(\sqrt{25 - b^2} + b + 5 \right)$$

we can find right triangles that have perimeters as close to 10 (and greater) as we wish, but none that actually have a perimeter equal to 10. Thus, the function p has no minimum when b is in the interval $(0, 5)$. However, if we allow the base of the triangle to equal either 0 or 5, then the perimeter of this degenerate triangle is exactly 10 cm.

Prob. 9. Since the lower left corner of the rectangle is at $(0, 0)$ and the upper right corner is at the point $(x, 1/x)$, we know that the perimeter p is given by

$$p = \frac{2}{x} + 2x.$$

To optimize the perimeter, we first differentiate the above equation to get

$$\frac{dp}{dx} = -\frac{2}{x^2} + 2$$

Setting the derivative equal to zero, we have

$$-\frac{2}{x^2} + 2 = 0 \quad \Rightarrow \quad 2x^2 = 2 \quad \Rightarrow \quad x = \pm 1.$$

Since the upper right corner has a positive x-coordinate, we select $x = 1$. To confirm that this value of x corresponds to a minimum for p, we calculate the second derivative

$$\frac{d^2p}{dx^2} = \frac{4}{x^3}$$

and note that it is always positive for $x > 0$. This means that the graph of p is concave up, and hence the optimal value must be a minimum. When $x = 1$, the minimum perimeter of the rectangle is $p = \frac{2}{x} + 2x = 4$.

Prob. 11. Suppose that the coordinates of any point on the line are (x, y), where y is given by $y = 4 - 3x$.

(a) The distance D between any two points (x_1, y_1) and (x_2, y_2) is given by

$$D = \sqrt{(x_2 - x_1)^2 + (y_2 - y_1)^2}$$

In our case, the two points are $(0, 0)$ and (x, y), and so

$$D = \sqrt{(x - 0)^2 + (y - 0)^2} = \sqrt{x^2 + y^2} = \sqrt{x^2 + (4 - 3x)^2}$$

If f denotes this distance then since f depends on x alone, we can write

$$f(x) = \sqrt{x^2 + (4 - 3x)^2}.$$

(b) To find a point on the line which is closest to the origin, we must optimize the function $f(x)$. To do this, we differentiate the above equation with respect to x, to get

$$f'(x) = \frac{10x - 12}{\sqrt{x^2 + (4 - 3x)^2}}$$

Setting the above equation to zero, we have

$$\frac{10x - 12}{\sqrt{x^2 + (4 - 3x)^2}} = 0 \quad \Rightarrow \quad 10x - 12 = 0 \quad \Rightarrow \quad x = 6/5.$$

We note that since the discriminant of $x^2 + (4 - 3x)^2 = 10x^2 - 24x + 16$ is

$24^2 - 4(10)(16) = -64 < 0$, the denominator of f' cannot be zero. Hence the only

critical number is $x = 6/5$. To show that this corresponds to a minimum, we note

that $f'(x) < 0$ for $x < 6/5$ and $f'(x) > 0$ for $x > 6/5$. (Note that we could also

deduce this geometrically; there can be no maximal distance between the point (x, y)

on the line and the origin, since the line is unbounded.) With $x = 6/5$ we get

$y = 4 - 3x = 4 - 3\frac{6}{5} = \frac{2}{5}$, and so the coordinates of the closest point are $(6/5, 2/5)$.

(c) The square of the distance is clearly found by simply squaring f, namely

$$g(x) = [f(x)]^2 = x^2 + (4 - 3x)^2$$

Again, to optimize $g(x)$, we differentiate to get

$$g'(x) = 20x - 24$$

and when we set this derivative equal to zero, we have

$$20x - 24 = 0 \quad \Rightarrow \quad x = \frac{24}{20} = \frac{6}{5}.$$

Now, again we can confirm that this value corresponds to a minimum, since

$g'' = 20 > 0$ and so the graph of g is concave up everywhere. When $x = 6/5$, we have

$y = 2/5$ as before, and thus we get the same coordinates as in part (b).

Prob. 13. Suppose the coordinates of any point on the curve $y = 1/x$ are given by (x, y).

We want to minimize the distance between (x, y) and $(0, 0)$. As in Question 11, we can

minimize the square of the distance instead. This squared distance D is given by

$$D = (x - 0)^2 + (y - 0)^2 = x^2 + y^2$$

Substituting the value of y from above, we get

$$D(x) = x^2 + \frac{1}{x^2}$$

To optimize D, we first differentiate to get

$$D'(x) = 2x - \frac{2}{x^3}$$

and then we set this derivative equal to zero and solve:

$$2x - \frac{2}{x^3} = 0 \quad \Rightarrow \quad x^4 - 1 = 0 \quad \Rightarrow \quad x = \pm 1.$$

We also notice that when $x = 0$, the derivative D' does not exist. However, the curve $y = 1/x$ has no point on it corresponding to $x = 0$, and so we reject this value. Thus there are two critical numbers, $x = 1$ and $x = -1$. Since

$$D''(x) = 2 + \frac{6}{x^4} > 0$$

for all $x \neq 0$, the graph of D is concave up everywhere on it's domain, and so both critical numbers must correspond to minima. When $x = 1$ we get $y = 1$, and when $x = -1$ we get $y = -1$. In both cases, the minimal distance is $\sqrt{D} = \sqrt{1 + 1} = \sqrt{2}$.

Prob. 15. Suppose that $f(x)$ is a positive differentiable function that has a local minimum at $x = c$. Then we know that $f'(c) = 0$, and that $f'(x) < 0$ for $x < c$ and x close to c and $f'(x) > 0$ for $x > c$ and x close to c. Let us define the function $g(x) = [f(x)]^2$. Then we have $g'(x) = 2f(x)f'(x)$. Thus $g'(c) = 2f(c)f'(c) = 0$. Also $g'(x) < 0$ for $x < c$ and x close to c since $f(x)$ is positive and $f'(x)$ is negative. Similarly $g'(x) > 0$ for $x > c$ and x close to c since $f(x)$ and $f'(x)$ are positive. Hence $g(x)$ also has a local minimum at $x = c$.

Prob. 17. Let l denote the height of the cylinder and r the radius of the circular base. The volume V of the cylinder is given by $V = \pi r^2 l$ and since we know that $V = 1000$ cm^3, we have

$$l = \frac{1000}{\pi r^2}$$

The amount of material used corresponds to the total surface area P, which is given by $P = 2\pi r^2 + 2\pi r l$. Thus we have to minimize the total surface area P. Substituting for l from above we have

$$P(r) = 2\pi r^2 + \frac{2000}{r}$$

and then differentiating, we get

$$P'(r) = 4\pi r - \frac{2000}{r^2}$$

To optimize P, we set this derivative equal to zero.

$$4\pi r - \frac{2000}{r^2} = 0 \quad \Rightarrow \quad 4\pi r^3 = 2000 \quad \Rightarrow \quad r = \sqrt[3]{500/\pi}$$

We note that the derivative $P'(r)$ does not exist when $r = 0$, but this is not a critical number since we cannot construct a cylinder of radius 0. Now, the graph of P is concave up since

$$P''(r) = 4\pi + \frac{4000}{r^3} > 0$$

for all positive values of r. Thus $r = \sqrt[3]{500/\pi}$ must correspond to a minimum. The radius of the can must be $\sqrt[3]{500/\pi}$ cm, and the height should be

$$l = \frac{1000}{\pi r^2} = \frac{1000}{\pi(\sqrt[3]{500/\pi})^2} = 2\sqrt[3]{500/\pi} \text{ cm,}$$

which is twice the radius.

Prob. 19. Please refer to Figure 5.59 in the textbook. The area A of the sector is given by $A = \frac{1}{2}r^2\theta$, and the perimeter p is given by $p = r\theta + 2r$. Solving the first of these equations for θ gives us $\theta = 2A/r^2$, and then we can substitute this expression into the formula for p to get

$$p = r\theta + 2r = \frac{2A}{r} + 2r$$

To optimize p we calculate $p'(r)$ and set it equal to zero.

$$p'(r) = 2 - \frac{2A}{r^2} \quad \text{and so} \quad p'(r) = 0 \quad \Rightarrow \quad 2 = \frac{2A}{r^2} \quad \Rightarrow \quad r = \sqrt{A}$$

since r must be positive. Now $p''(r) = 4A/r^3 > 0$ for any positive A and r. This means that the graph of p is concave up, and hence the critical number $r = \sqrt{A}$ corresponds to a minimum.

(a) When $A = 2$, the minimizing value of $r = \sqrt{2}$ and so $\theta = 2A/r^2 = 4/2 = 2$ radians.

(b) When $A = 10$, the minimizing value of $r = \sqrt{10}$ and so $\theta = 2A/r^2 = 20/10 = 2$ radians. The angle is the same no matter what the value of A, since
$$\theta = 2A/r^2 = 2A/A = 2.$$

Prob. 21. Please look at Example 4 in the textbook; we shall use the same variables here. The height of the cylinder is h and r is the radius of the circular base. The volume V of the cylinder is given by $V = \pi r^2 l$, and given that $V = 355$ cm^3 we have $l = \frac{355}{\pi r^2}$. Now, since the top of the can is three times as thick as the base, the total surface area P is given by

$$P = \pi r^2 + 3\pi r^2 + 2\pi rl = 4\pi r^2 + 2\pi rl = 4\pi r^2 + \frac{710}{r}.$$

To minimize P over the interval $(0, \infty)$ we differentiate and get

$$\frac{dP}{dr} = 8\pi r - \frac{710}{r^2}$$

Setting this derivative equal to zero we have

$$8\pi r - \frac{710}{r^2} = 0 \quad \Rightarrow \quad 8\pi r^3 = 710 \quad \Rightarrow \quad r = \sqrt[3]{\frac{710}{8\pi}} = \frac{1}{2}\sqrt[3]{\frac{710}{\pi}}$$

Now the second derivative is

$$\frac{d^2 P}{dr^2} = 8\pi + \frac{1420}{r^3}$$

which is positive for all $r > 0$, and thus the graph of p is concave up. This means that $r = \sqrt[3]{710/(8\pi)}$ cm corresponds to a minimum. The height of the cylinder is

$$l = \frac{355}{\pi r^2} = \frac{355(8\pi)^{2/3}}{\pi 710^{2/3}} = 2\sqrt[3]{\frac{710}{\pi}} \text{ cm}.$$

The approximate values are $r \approx 3$ cm and $l \approx 12.2$ cm. These are closer to actual values for real soda cans.

Prob. 23. Let $f(a) = ab = a(a - 4) = a^2 + 4a$.

$f'(a) = 2a + 4 = 0$, $a = -2$, $b = -6$.

$a \cdot b = 12$.

Prob. 25. Here

$$w(t) = \frac{f(t)}{C + t}$$

where we assume that $f(t)$ is concave down for $t \geq 0$, and also that $f(0) = 0$ and $0 \leq f \leq 1$.

(a) To find the optimal brooding time, we first differentiate w to get

$$\frac{dw}{dt} = \frac{f'(t)(C + t) - f(t)}{(C + t)^2}$$

Setting this derivative equal to zero we have

$$\frac{f'(t)(C + t) - f(t)}{(C + t)^2} = 0 \quad \Rightarrow \quad f'(t)(C + t) - f(t) = 0 \quad \Rightarrow \quad f(t) = f'(t)(C + t)$$

Now suppose the line through the point $(-C, 0)$ is tangential to the curve f, and the point of tangency is $(t, f(t))$. The point-slope equation of this tangent line would be $f(t) - 0 = f'(t)(t - (-C))$, or $f(t) = f'(t)(t + C)$, which is precisely when the derivative dw/dt would be zero. Since we have assumed that f is concave down, $f'' < 0$ and so at the point of tangency

$$
\begin{aligned}
\frac{d^2 w}{dt^2} &= \frac{f''(t)(C + t)^2 - 2f'(t)(C + t) + 2f(t)}{(C + t)^3} \\
&= \frac{f''(t)(C + t)^2 - 2f'(t)(C + t) + 2f'(t)(C + t)}{(C + t)^3} \\
&= \frac{f''(t)}{(C + t)} < 0
\end{aligned}
$$

which means that w is maximized.

(b) When $C = 2$ and

$$
f(t) = \frac{t}{1 + t}
$$

we have

$$
f(t) = f'(t)(C + t) \;\Rightarrow\; \frac{t}{1 + t} = \frac{1}{(1 + t)^2}(2 + t) \;\Rightarrow\; t(1 + t) = (2 + t) \;\Rightarrow\; t^2 = 2
$$

and since $t \geq 0$ we have $t = \sqrt{2}$. Thus, for maximum $w(t)$ we should have $t = \sqrt{2}$.

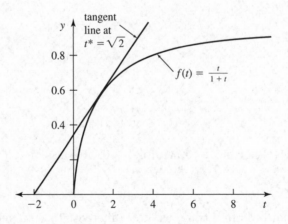

Prob. 27. We are given that the function $r(x)$ satisfies the equation

$$
\frac{e^{-x(r(x) + L)}(1 - e^{-kx})^3 c}{1 - e^{-(r(x) + L)}} = 1
$$

where k, L and c are positive constants.

(a) We have to find the $\frac{dr}{dx}$. To do this, we apply logarithms on both sides of the equation above to get

$$\ln\left(\frac{e^{-x(r(x)+L)}(1-e^{-kx})^3 c}{1-e^{-(r(x)+L)}}\right) = \ln 1$$

$$\ln\left(e^{-x(r(x)+L)}\right) + 3\ln\left(1-e^{-kx}\right) + \ln c - \ln\left(1-e^{-(r(x)+L)}\right) = 0$$

$$-x(r(x)+L) + 3\ln\left(1-e^{-kx}\right) + \ln c = \ln\left(1-e^{-(r(x)+L)}\right)$$

Now we differentiate with respect to x. We have

$$-r(x) - L - xr'(x) + \frac{3ke^{-kx}}{1-e^{-kx}} = \frac{r'(x)e^{-(r(x)+L)}}{1-e^{-(r(x)+L)}}$$

$$r'(x)\left(x + \frac{e^{-(r(x)+L)}}{1-e^{-(r(x)+L)}}\right) = -r(x) - L + \frac{3ke^{-kx}}{1-e^{-kx}}$$

$$r'(x)\left(\frac{x-(x-1)e^{-(r(x)+L)}}{1-e^{-(r(x)+L)}}\right) = \frac{[r(x)+L+3k]e^{-kx} - r(x) - L}{1-e^{-kx}}$$

$$r'(x) = \frac{[r(x)+L+3k]e^{-kx} - r(x) - L}{1-e^{-kx}} \cdot \frac{1-e^{-(r(x)+L)}}{x-(x-1)e^{-(r(x)+L)}}$$

This is an equation for dr/dx.

(b) Setting $dr/dx = 0$ and considering only the numerator, we have that either $e^{r(x)+L} = 1$ or

$$[r(x) + L + 3k]e^{-kx} - r(x) - L = 0$$

We can reject the first case, since it leads to the solution $r(x) = -L$, which is a negative rate of increase. Working with the latter equation, we have

$$[L + 3k]e^{-kx} - L = r(x) - r(x)e^{-kx}$$

$$3ke^{-kx} - L[1 - e^{-kx}] = r(x)[1 - e^{-kx}]$$

$$r(x) = \frac{3ke^{-kx}}{1-e^{-kx}} - L$$

This is what we were meant to show.

5.5   L'Hospital's Rule

Prob. 1. L'Hospital's rule gives:

$$\lim_{x \to 5} \frac{2x}{1} = 10.$$

Prob. 3. L'Hospital's rule gives:

$$\lim_{x \to -2} \frac{6x - 5}{1} = -17.$$

Prob. 5. Applying L'Hospital's rule we have, a substituting the value, we have

$$y = \lim_{x \to 0} \frac{1}{\sqrt{2x + 4}} = \frac{1}{2}.$$

Prob. 7. Applying L'Hospital's rule we have, and substituting the value, we have

$$y = \lim_{x \to 0} \frac{\cos x}{\cos x - x \sin x} = 1.$$

Prob. 9. Applying L'Hospital's rule we have,

$$y = \lim_{x \to 0} \frac{\sin x}{\tan x + x \sec^2 x}$$

Differentiating again, and substituting the value, we have

$$y = \lim_{x \to 0} \frac{\cos x}{\sec^2 x + \sec^2 x + 2x \sec^2 x \tan x} = \frac{1}{2}.$$

Prob. 11. Applying L'Hospital's rule we have,

$$y = \lim_{x \to 0^+} \frac{x + 1}{2\sqrt{x}} = \infty$$

Prob. 13. L'Hospital's rule gives:

$$\lim_{x \to \infty} \frac{\frac{1}{\ln(x)} \cdot \frac{1}{x}}{1} = 0.$$

Prob. 15. Applying L'Hospital's rule we have, and substituting the value, we have

$$y = \lim_{x \to 0} \frac{(\ln 2)2^x}{(\ln 3)3^x} = \frac{\ln 2}{\ln 3}.$$

Prob. 17. L'Hospital's rule gives:

$$\lim_{x \to 0} \frac{-\ln 3 \cdot 3^{-x}}{\ln 2 \cdot 2^x} = -\frac{\ln 3}{\ln 2}.$$

Prob. 19. Applying L'Hospital's rule we have,

$$y = \lim_{x \to 0} \frac{e^x - 1}{2x}$$

Differentiating again,

$$y = \lim_{x \to 0} \frac{e^x}{2}$$

Substituting the value, we have

$$= \frac{1}{2}.$$

Prob. 21. Applying L'Hospital's rule we have,

$$y = \lim_{x \to \infty} \frac{2\ln x}{x2x}$$

Differentiating again,

$$y = \lim_{x \to \infty} \frac{1}{2x^2}$$

Substituting the value, we have

$$= 0.$$

Prob. 23. Applying L'Hospital's rule we have, and substituting the value, we have

$$y = \lim_{x \to (\frac{\pi}{2})^-} \frac{\sec^2 x}{2\sec^2 x \tan x} = \lim_{x \to (\frac{\pi}{2})^-} \frac{1}{2\tan x} = 0.$$

Prob. 25. Converting into form so that L'Hospital's rule can be applied,

$$y = \lim_{x \to \infty} \frac{x}{e^x}$$

Applying L'Hospital's rule we have, and substituting the value, we have

$$y = \lim_{x \to \infty} \frac{1}{e^x} = 0.$$

Prob. 27. Re-write as: $\lim_{x \to \infty} \frac{x^5}{e^x}$, successive applications of the rule give: $\lim_{x \to \infty} \frac{5 \cdot 4 \cdot 3 \cdot 2 \cdot 1}{e^x} = 0$.

Prob. 29. Converting into form so that L'Hospital's rule can be applied,

$$y = \lim_{x \to 0} \frac{\ln x}{\frac{1}{\sqrt{x}}}$$

Applying L'Hospital's rule we have, and substituting the value, we have

$$y = \lim_{x \to 0} -2x^{\frac{1}{2}} = 0.$$

Prob. 31. Re-write as $\lim\limits_{x \to 0^+} \frac{\ln x}{\frac{1}{x^5}}$, L'Hospital rule gives:

$$\lim_{x \to 0^+} \frac{\frac{1}{x}}{\frac{1}{x5}} = \lim_{x \to 0^+} x^4 = 0$$

Prob. 33. Given,

$$y = \lim_{x \to (\frac{\pi}{2})^-} \left(\frac{\pi}{2} - x\right) \sec x$$

Converting into form so that L'Hospital's rule can be applied,

$$y = \lim_{x \to (\frac{\pi}{2})^-} \frac{\left(\frac{\pi}{2} - x\right)}{\cos x}$$

Applying L'Hospital's rule we have, and substituting the value, we have

$$y = \lim_{x \to (\frac{\pi}{2})^-} \frac{-1}{-\sin x} = 1.$$

Prob. 35. Converting into form so that L'Hospital's rule can be applied, Substituting

$$t = \frac{1}{x}$$

and hence when

$$x \to \infty, t \to 0$$

$$y = \lim_{t \to 0} \frac{\sin t}{\sqrt{t}}$$

Applying L'Hospital's rule we have, and substituting the value, we have

$$y = \lim_{t \to 0} \frac{2 \cos t \sqrt{t}}{1} = 0.$$

Prob. 37. Converting into form so that L'Hospital's rule can be applied,

$$y = \lim_{x \to 0^+} \frac{\cos x - 1}{\sin x}$$

Applying L'Hospital's rule we have, and substituting the value, we have

$$y = \lim_{x \to 0^+} \frac{-\sin x}{\cos x} = 0.$$

Prob. 39. Converting into form so that L'Hospital's rule can be applied,

$$y = \lim_{x \to 0^+} \frac{x - \sin x}{x \sin x}$$

Applying L'Hospital's rule we have,

$$y = \lim_{x \to 0^+} \frac{1 - \cos x}{\sin x + x \cos x}$$

Differentiating again and substituting the value, we have

$$y = \lim_{x \to 0^+} \frac{\sin x}{2 \cos x - x \sin x} = 0.$$

Prob. 41. Converting into form so that L'Hospital's rule can be applied . Taking ln on both sides,

$$\ln y = \lim_{x \to 0^+} \frac{2 \ln x}{\frac{1}{x}}$$

Applying L'Hospital's rule we have, and substituting the value, we have

$$\ln y = \lim_{x \to 0^+} \frac{-2x^2}{x} \ln y = 0. \Rightarrow y = 1$$

Prob. 43. Converting into form so that L'Hospital's rule can be applied . Taking ln on both sides,

$$\ln y = \lim_{x \to \infty} \frac{\ln x}{x}$$

Applying L'Hospital's rule we have, and substituting the value, we have

$$\ln y = \lim_{x \to \infty} \frac{1}{x} = 0 \Rightarrow y = 1$$

Prob. 45. Converting into form so that L'Hospital's rule can be applied . Taking ln on both sides,

$$\ln y = \lim_{x \to \infty} \frac{\ln(1 + \frac{3}{x})}{\frac{1}{x}}$$

Substituting

$$t = \frac{1}{x}$$

such that

$$x \to \infty, t \to 0$$

Thus we have,

$$\ln y = \lim_{t \to 0} \frac{\ln(1 + 3t)}{t}$$

Applying L'Hospital's rule we have, and substituting the value, we have

$$\ln y = \lim_{t \to 0} \frac{3}{1 + t} = 3 \Rightarrow y = e^3$$

Prob. 47. Let $y = (1 - \frac{2}{x})^x$, taking ln of both sides:

$$\ln y = x \cdot \ln(1 - \frac{2}{x}) = \frac{\ln(1 - \frac{2}{x})}{\frac{1}{x}}.$$

Letting $t = \frac{1}{x}$, $\lim\limits_{t \to 0} \frac{\ln(1-2t)}{t}$.

Now, L'Hospital rule gives:

$$\lim_{t \to 0} \frac{-2}{1 - 2t} = -2 \text{ hence } y = e^{-2}.$$

Prob. 49. Converting into form so that L'Hospital's rule can be applied . Taking ln on both sides,

$$\ln y = \lim_{x \to \infty} \frac{\ln(\frac{x}{1+x})}{\frac{1}{x}}$$

Substituting

$$t = \frac{1}{x}$$

such that

$$x \to \infty, t \to 0$$

Thus we have,

$$\ln y = \lim_{t \to 0} -\frac{\ln(1 + t)}{t}$$

Applying L'Hospital's rule we have, and substituting the value, we have

$$\ln y = \lim_{t \to 0} -\frac{1}{1 + t} = -1 \Rightarrow y = \frac{1}{e}$$

Prob. 51. Given,

$$y = \lim_{x \to 0^+} x e^x$$

Now the above function is not in L'Hospital rule format, and hence here we can directly get the limit by substituting the value of x in above function . So,

$$y = \lim_{x \to 0^+} x e^x$$

$$y = 0$$

Prob. 53. Given

$$y = \lim_{x \to (\frac{\pi}{2})^+} \tan x + \sec x$$

We rewrite above function as,

$$y = \lim_{x \to (\frac{\pi}{2})^+} \frac{1 + \sin x}{\cos x}$$

Here we can not apply L'Hospital Rule. By directly substituting the value of $x = \frac{\pi}{2}$ we get the limit. Thus we have,

$$y = \infty$$

Prob. 55. $\lim_{x \to 1} \frac{x^2 - 1}{x + 1}$, L'Hospital does not apply, hence

$$\lim_{x \to 1} \frac{(x + 1)(x - 1)}{x + 1} = \lim_{x \to 1} (x - 1) = 0.$$

Prob. 57. Making the substitution $y = -x$ we can rewrite the limit as

$$\lim_{x \to -\infty} x e^x = \lim_{y \to \infty} -y e^{-y} = -\lim_{y \to \infty} \frac{y}{e^y} = -\lim_{x \to \infty} \frac{1}{e^x} = 0$$

where we applied L'Hospital on the last calculation.

Prob. 59. Converting into form so that L'Hospital's rule can be applied . Taking ln on both sides,

$$\ln y = \lim_{x \to 0^+} \frac{3 \ln x}{\frac{1}{x}}$$

Applying L'Hospital's rule we have,

$$\ln y = \lim_{x \to 0^+} \frac{-3x^2}{x}$$

Substituting the value, we have

$$\ln y = 0 \Rightarrow y = 1$$

Prob. 61. Given

$$y = \lim_{x \to 0} \frac{a^x - 1}{b^x - 1}$$

We can apply L'Hospital rule to above limit. Applying L'Hospital's rule we have,

$$y = \lim_{x \to 0} \frac{a^x \ln a}{b^x \ln b}$$

$$y = \frac{\ln a}{\ln b}$$

Prob. 63. Given,

$$y = \lim_{x \to \infty} (1 + \frac{c}{x^p})^x$$

Converting into form so that L'Hospital's rule can be applied . Taking ln on both sides,

$$\ln y = \lim_{x \to \infty} \frac{\ln(1 + \frac{c}{x^p})}{\frac{1}{x}}$$

Substituting

$$t = \frac{1}{x}$$

such that

$$x \to \infty, t \to 0$$

Thus we have,

$$\ln y = \lim_{t \to 0} \frac{\ln(1 + ct^p)}{t}$$

Now we have different case depending upon $p > 1$, $p < 0$ $p = 1$. when $p > 1$, Applying L'Hospital's rule we have,

$$\ln y = \lim_{t \to 0} \frac{pct^{p-1}}{1 + ct^p}$$

Substituting the value, we have

$$\ln y = 0$$

$$y = 1$$

when $p = 1$, Applying L'Hospital's rule we have,

$$\ln y = \lim_{t \to 0} \frac{c}{1 + ct}$$

Substituting the value, we have

$$\ln y = c$$

$$y = e^c$$

when $p < 0$, Applying L'Hospital's rule we have,

$$\ln y = \lim_{t \to 0} \frac{pct^{p-1}}{1 + ct^p}$$

Substituting the value, we have

$$\ln y = \infty$$

$$y = \infty$$

Thus we can summarize as

$$y = \begin{cases} 1 & \text{for} \quad p > 1 \\ e^c & \text{for} \quad p = 1 \\ \infty & \text{for} \quad p < 0. \end{cases}$$

Prob. 65. Given

$$y = \lim_{x \to \infty} \frac{\ln x}{x^p}$$

Applying L'Hospital Rule,we have

$$y = \lim_{x \to \infty} \frac{1}{px^p}$$

Thus substituting the value, we get,

$$y = 0$$

as denominator tends to ∞.

Prob. 67. Given,

$$y = 121 e^{\frac{-17}{x}}$$

where y is the height in feet and x is the age of the tree in years.

(a) Rate of Growth is given by,

$$\frac{dy}{dx} = 2057 \frac{e^{\frac{-17}{x}}}{x^2}$$

Thus for $x \to 0^+$ we have

$$\frac{dy}{dx} = \lim_{x \to 0^+} \frac{2057}{x^2} e^{\frac{-17}{x}}$$

Let $t = \frac{1}{x}$ Rewriting we have,

$$= \lim_{t \to \infty} 2057(e^{-17t} t^2)$$

Applying L'Hospital's Rule we have and substituting we have

$$y = 0$$

Similarly for $x \to \infty$, we have

$$\frac{dy}{dx} = \lim_{x \to \infty} \frac{2057}{x^2} e^{\frac{-17}{x}}$$

Applying L'Hospital's Rule we have and substituting we have

$$y = 0$$

Thus the rate of growth is zero at end points.

(b) To find x such that the growth is maximal, we differentiate the below equation, and equate it to zero,

$$Growth\,Rate = 2057\frac{e^{\frac{-17}{x}}}{x^2}$$

Differentiating we have

$$y' = 2057(\frac{17e^{\frac{-17}{x}} - 2xe^{\frac{-17}{x}}}{x^4})$$

Hence we have

$$x = \frac{17}{2}$$

(c) For $x < \frac{17}{2}$, the function is increasing and above that it is decreasing.

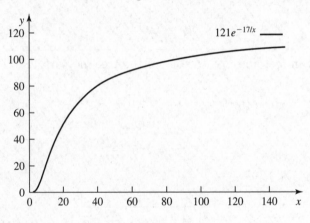

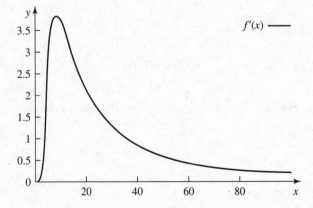

(d)

5.6 Difference Equations: Stability

Prob. 1.

(a) Given $N_{t+1} = (1.03)N_t$. This can be written as,

$$N_{t+1} = (1.03)((1.03)N_{t-1})$$

$$\Rightarrow N_{t+1} = (1.03)^{t+1}N_0$$

Given, $N_0 = 10$, Population when the generation, $t = 5$ is

$$N5 = (1.03)^5 * 10$$

$$= 11.59$$

(b) If $N_x = 2N_0$, then $x = ?$

$\Rightarrow N_x = 2(10) = 20$. We have $N_{t+1} = (1.03)^{t+1}N_0$.

Place $(t + 1)$ as x,

$$\Rightarrow \quad N_x = (1.03)^x N_0$$

$$\Rightarrow \quad 20 = (1.03)^x N_0$$

$$\Rightarrow \quad 20 = (1.03)^x 10$$

$$\Rightarrow \quad \frac{20}{10} = (1.03)^x$$

applying log on both sides,

$$\Rightarrow \quad \log(2) = log((1.03)^x x)$$

$$\Rightarrow \quad \log(2) = x \cdot \log(1.03)$$

$$\Rightarrow \quad x = \log(2)/\log(1.03)$$

$$\Rightarrow \quad x = 23.45$$

by round-off, $x = 24$. I.e., It will be in 24th generation, the population size will be twice that of N_0.

Prob. 3.

(a) Given population model is, $N_{t+1} = bN_t$. Also given, population increases by 2% each generation. Therefore, the N term at $t = 1$ is

$$N_1 = bN_0$$

$$\Rightarrow \quad N_0 + (0.02)N_0 = bN_0$$

$$\Rightarrow \quad (1.02)N_0 = bN_0$$

$$\text{therefore} \quad b = 1.02.$$

(b) If $N_0 = 20$, then $N_t = ?$ when $t = 10$. Given population model can be written as $N_t = b^t N_0$

$$\Rightarrow \quad N_{10} = (1.02)^{10} 20$$

$$\Rightarrow \quad N_{10} = 24.37$$

(c) If $N_t = 2N_0$, then $t = ?$

$$\Rightarrow \quad 2N_0 = (1.02)tN_0$$

$$\Rightarrow \quad 2 = (1.02)^t$$

applying log on both sides,

$$\Rightarrow \quad \log(2) = \log((1.02)^t)$$

$$\Rightarrow \quad \log(2) = t\log(1.02)$$

$$\Rightarrow \quad t = \log(2)/\log(1.02)$$

$$\Rightarrow \quad t = 35$$

Prob. 5.

(a) Given population model is, $N_{t+1} = bN_t$. Also given, population increases by $x\%$ each generation. Therefore, the N term at $t = 1$ is

$$N_1 = bN_0$$

$$\Rightarrow \qquad N_0 + (x/100)N_0 = bN_0$$

$$\Rightarrow \qquad (1 + x/100)N_0 = bN_0$$

therefore $\quad b = (100 + x)/100.$

(b) If $N_t = 2N_0$, then determine t.

$$\Rightarrow \qquad 2N_0 = ((100 + x)/100)^t N_0$$

$$\Rightarrow \qquad 2 = ((100 + x)/100)^t$$

applying log on both sides,

$$\Rightarrow \qquad \log(2) = \log(((100 + x)/100)^t)$$

$$\Rightarrow \qquad \log(2) = t \log((100 + x)/100)$$

$$\Rightarrow \qquad t = \log(2)/\log((100 + x)/100)$$

Now, for $x = 0.1$,

$$t \quad = \log(2)/\log((100 + 0.1)/100)$$

$$t \qquad\qquad = 693.49$$

For $x = 0.5$,

$$t \quad = \log(2)/\log((100 + 0.5)/100)$$

$$t \qquad\qquad = 138.97$$

For $x = 1$,

$$t \quad = \log(2)/\log((100 + 1)/100)$$

$$t \qquad\qquad = 69.66$$

For $x = 2$,

$$t \quad = \log(2)/\log((100 + 2)/100)$$

$$t \qquad\qquad = 35.00$$

For $x = 5$,

$$t = \log(2)/\log((100+5)/100)$$

$$t = 14.20$$

For $x = 10$,

$$t = \log(2)/\log((100+10)/100)$$

$$t = 7.27$$

Prob. 7.

(a) Given population model is,

$$N_{t+1} = 0.9N_t$$

The equilibrium can be obtained by solving $N = RN$. But the only solution for this is $N^* = 0$. As $0 < R < 1$, N_t will return to equilibrium $N^* = 0$ if $N_0 > 0$. Therefore as $0 < R < 1$, this population model is stable.

(b) The fixed points are found graphically, where the graphs of $N_{t+1} = 1.3N_t$ and $N_{t+1} = N_t$ intersects.

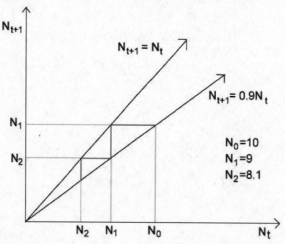

For any N_0, the population growth is converging towards equilibrium $N^* = 0$. Therefore, given population model is stable.

Prob. 9. Given model is, $x_{t+1} = \frac{2}{3} - \frac{2}{3}x_t^2$. To find the equilibrium, we need to solve $x = f(x)$ i.e., to solve $x = \frac{2}{3} - \frac{2}{3}x^2$

$$\Rightarrow \quad \frac{2}{3}x^2 + x - \frac{2}{3} = 0$$

$$\Rightarrow \quad 2x^2 + 3x - 2 = 0$$

The left hand side can be factored into $(2x - 1)(x + 2)$, and we find that

$(2x - 1)(x + 2) = 0$,

$$\texttt{therefore} \quad x = \frac{1}{2} \quad \texttt{or} \quad x = -2$$

To determine stability, we need to evaluate the derivative of $f(x) = \frac{2}{3} - \frac{2}{3}x^2$ at the equilibrium. Now, $f'(x) = -\frac{4}{3}x$ if $x = \frac{1}{2}$, then

$$\left| f'\left(\frac{1}{2}\right) \right| = \left| -\frac{2}{3} \right| = \frac{2}{3} < 1$$

and if $x = -2$, then

$$|f'(-2)| = \left| \frac{8}{3} \right| = \frac{8}{3} > 1$$

Thus $x = \frac{1}{2}$ is locally stable and $x = -2$ is unstable. As $f'\left(\frac{1}{2}\right) = -\frac{2}{3} > 0$, the equilibrium is approached with oscillations.

Prob. 11. Given model is $x_{t+1} = \frac{x_t}{0.5 + x_t}$. To find the equilibrium, we need to solve $x = f(x)$ i.e., to solve $x = \frac{x}{0.5 + x}$

$$\Rightarrow \quad x(0.5 + x) = x$$

$$\Rightarrow \quad x(x - 0.5) = 0$$

Therefore $x = 0$ or $x = 0.5$.

Now,

$$f'(x) = \frac{0.5 + x - x}{(0.5 + x)^2} = \frac{0.5}{(0.5 + x)^2}.$$

Since $f'(0) = \frac{1}{0.5} = 2 > 1$, we conclude that $x^* = 0$ is unstable. Since $f'(0.5) = 0.5 \in (0, 1)$, we conclude that $x^* = 0.5$ is locally stable and is approached without oscillations.

Prob. 13.

(a) Given model is $x_{t+1} = \frac{5x_t^2}{4 + x_t^2}$. To find the equilibrium, we need to solve $x = f(x)$ i.e., to solve $x = \frac{5x^2}{4 + x^2}$

$$\Rightarrow \quad x(x^2 - 5x + 4) = 0.$$

Therefore $x = 0$ or $x = 4$ or $x = 1$.

Now,

$$f'(x) = \frac{40x}{(4 + x^2)^2}$$

Since $f'(0) = 0$, we conclude that $x^* = 0$ is locally stable and is approached without oscillations. Since $f'(4) = \frac{2}{5} \in (0,1)$, we conclude that $x^* = 4$ is locally stable and is approached without oscillations. Since $f'(1) = \frac{8}{5} > 1$, we conclude that $x^* = 1$ is unstable.

(b) (i) Considering $x_0 = 0.5$

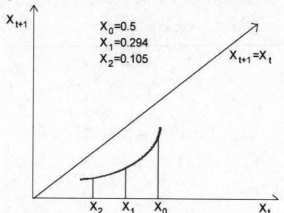

The subsequent terms at $t = 0$, $t = 1$, $t = 2$ shows that the value is converging towards closest fixed point $x^* = 0$

(ii) Considering $x_0 = 2$. The subsequent values at $t = 0$, $t = 1$ and $t = 2$ shows that value x value is reaching towards fixed point $x^* = 4$.

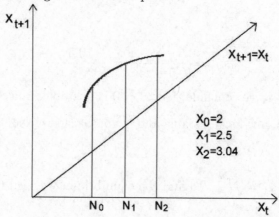

Prob. 15. (a) We see that if $P = 0$, then $R(0) = 0$. Also, since this function is exponential in nature, P is the only factor determining sign. Hence, for $P > 0$, $R(P) > 0$.

(b) 0, (c) $1/\beta$, (d) $P = 2/\beta$ is an inflection point,

(e)

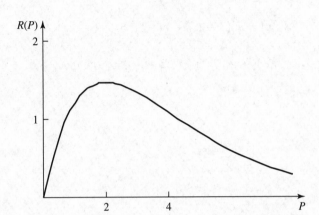

Prob. 17. (a)

t	N_t	t	N_t
0	100	11	366.8
1	367.9	12	93.4
2	92.8	13	366.8
3	366.7	14	93.4
4	93.5	15	366.8
5	366.9	16	93.4
6	93.4	17	366.8
7	366.8	18	93.4
8	93.4	19	366.8
9	336.8	20	93.4
10	93.4	21	366.8

(b) To find equilibria, solve $N = 10Ne^{-0.01N}$; $N = 0$ is another equilibrium.

(c) Oscillations seem to appear and the system does not seem to converge to the nontrivial equilibrium.

Prob. 19. (a) Need to solve: $N = N(1 + 1.5(1 - \frac{N}{100})$ $N = 0$, $1 = 1 + 1.5 - \frac{1.5N}{100}$ $N = 100$.

(b) for $N_0 = 10$, starting at 10, will go to 100.

Prob. 21. Let $r > 1$, need to solve: $X = rX(1 - x)$ $x = 0$, $1 = r(1 - x)\frac{1}{4} = 1 - x$,

$x = 1 - \frac{1}{r}$. Now,

$$f'(x) = r[(1 - x) + x(-1)] = r[1 - 2x].$$

$$f'(1 - \frac{1}{r}) = r(1 - 2 + \frac{2}{4}) = -r + 2$$

To $2e$ stable $|-r + 2| < 1$, $1 < r < 3$.

Prob. 23. Need to solve:

$$
\begin{aligned}
N &= rN^{1-\gamma}N \\
1 &= rN^{1-\gamma} \\
\frac{1}{r} &= N^{1-\gamma}, \quad N = r^{\frac{1}{\gamma-1}}.
\end{aligned}
$$

Now, $f'(r\frac{1}{\gamma-1}(= r(2 - \gamma)(r^{-1}) = 2 - \gamma$

To be stable we must have: $|2 - \gamma| < 1$ $1 < \gamma < 3$.

Prob. 25. Need to solve:

$$
\begin{aligned}
N &= e^{r(1-\frac{N}{K})}N \\
1 &= e^{r(1-\frac{N}{K})}, \quad N = K.
\end{aligned}
$$

Now,

$$
\begin{aligned}
f'(N) &= e^{r(1-\frac{r}{K})} + Ne^{r(1-\frac{N}{K})}(-\frac{r}{K}) \\
f'(K) &= e^0 + ke^0(-\frac{r}{K}) = 1 - 4.
\end{aligned}
$$

To be stable: $\left|1 - r\right| < 1$, $0 < r < 1$.

5.7   Numerical Methods

Prob. 1. $\sqrt{7} = 2.645751$

Prob. 3. 0.6529186

Prob. 5. 1.895494

Prob. 7. (a) $|x_n| = 2^n x_0$, (b) ∞

Prob. 9. (a) $x_0 = 3, x_1 = 4.166667, x_2 = 4.003333, x_3 = 4.000001,$

(b) $x_0 = x_1 = x_2 = \ldots = 4$

5.8 Antiderivatives

Prob. 1. $F(x) = \frac{4x^3}{3} - \frac{x^2}{2} + C$

Prob. 3. $F(x) = \frac{x^3}{3} + \frac{3x^2}{2} + 3x + C$

Prob. 5. Given the function as,

$$f(x) = x^4 - 3x^2 + 1$$

Integrating, we get

$$\int f(x) = F(x) = \frac{x^5}{5} - x^3 + x + C.$$

where C is a constant.

Prob. 7. Given the function as,

$$f(x) = 4x^3 - 2x + 3$$

Integrating, we get

$$\int f(x) = F(x) = x^4 - x^2 + 3x + C.$$

where C is a constant.

Prob. 9. $F(x) = x + \ln|x| - \frac{1}{x} + C$

Prob. 11. Given the function as,

$$f(x) = 1 - \frac{1}{x^2}$$

Integrating, we get

$$\int f(x) = F(x) = x + \frac{1}{x} + C.$$

where C is a constant.

Prob. 13. Given the function as,

$$f(x) = \frac{1}{1+x}.$$

Integrating, we get

$$\int f(x) = F(x) = \ln(1+x) + C.$$

where C is a constant.

Prob. 15. Given the function as,

$$f(x) = 5x^4 + \frac{5}{x^4}$$

Integrating, we get

$$\int f(x) = F(x) = x^5 - \frac{5}{3x^3} + C.$$

where C is a constant.

Prob. 17. $F(x) = \frac{1}{2}\ln|1 + 2x| + C$

Prob. 19. Given the function as,

$$f(x) = e^{-3x}$$

Integrating, we get

$$\int f(x) = F(x) = -\frac{e^{-3x}}{3} + C.$$

where C is a constant.

Prob. 21. Given the function as,

$$f(x) = 2e^{2x}$$

Integrating, we get

$$\int f(x) = F(x) = e^{2x} + C.$$

where C is a constant.

Prob. 23. Given the function as,

$$f(x) = \frac{1}{e^{2x}}$$

Integrating, we get

$$\int f(x) = F(x) = -\frac{1}{e^{2x}} + C.$$

where C is a constant.

Prob. 25. Given the function as,

$$f(x) = \sin(2x)$$

Integrating, we get

$$\int f(x) = F(x) = -\frac{\cos(2x)}{2} + C.$$

where C is a constant.

Prob. 27. $F(x) = -3\cos(\frac{x}{3}) + 3\sin(\frac{x}{3}) + C.$

Prob. 29. Given the function as,

$$f(x) = 2\sin(\frac{\pi x}{2}) - 3\cos(\frac{\pi x}{2})$$

Integrating, we get

$$\int f(x) = F(x) = -\frac{4}{\pi}\cos(\frac{\pi x}{2}) - \frac{6}{\pi}\sin(\frac{\pi x}{2}) + C.$$

where C is a constant.

Prob. 31. Given the function as,

$$f(x) = \sec^2(2x)$$

Integrating, we get

$$\int f(x) = F(x) = \frac{\tan(2x)}{2} + C.$$

where C is a constant.

Prob. 33. Given the function as,

$$f(x) = \sec^2\left(\frac{x}{3}\right)$$

Integrating, we get

$$\int f(x) = F(x) = 3\tan\left(\frac{x}{3}\right) + C.$$

where C is a constant.

Prob. 35. Given the function as,

$$f(x) = \frac{\sec x + \cos x}{\cos x}$$

We can rewrite it as,

$$f(x) = \sec^2 x + 1$$

Integrating, we get

$$\int f(x) = F(x) = \tan x + x + C$$

where C is a constant.

Prob. 37. $F(x) = \frac{x^{-6}}{-6} + \frac{3x^6}{6} - \frac{1}{2}\cos(2x) + C.$

Prob. 39. $F(x) = \frac{1}{3}\tan(2x - 1) + \frac{x^2}{2} - 3\ln|x| + C.$

Prob. 41. $F(x) = \frac{e^{(a+1)x}}{a(a+1)} + C$

Prob. 43. $F(x) = \frac{1}{a}\ln|ax + 3| + C.$

Prob. 45. $F(x) = \frac{x^{a+3}}{a+3} - \frac{a^{x+2}}{\ln a} + C.$

Prob. 47. Given the equation as,

$$\frac{dy}{dx} = \frac{2}{x} - x$$

Thus, we have the general solution by finding the antiderivative of the above function.

Hence,

$$y = 2\ln x - \frac{x^2}{2} + C$$

where C is a constant.

Prob. 49. Given the equation as,

$$\frac{dy}{dx} = x(1+x)$$

We can rewrite the above function as,

$$\frac{dy}{dx} = x^2 + x$$

Thus, we have the general solution by finding the antiderivative of the above function. Hence,

$$y = \frac{x^3}{3} + \frac{x^2}{2}$$

where C is a constant.

Prob. 51. Given the equation as,

$$\frac{dy}{dt} = t(1-t)$$

We can rewrite the above function as,

$$\frac{dy}{dt} = t - t^2$$

Thus, we have the general solution by finding the antiderivative of the above function. Hence,

$$y = \frac{t^2}{2} - \frac{t^3}{3} + C$$

where C is a constant.

Prob. 53. Given the equation as,

$$\frac{dy}{dt} = e^{-\frac{t}{2}}$$

Making the substitution, $x = -\frac{t}{2}$, we have $dx = -\frac{1}{2}dt$ Thus, the function becomes,

$$\frac{dy}{dx} = -2e^x$$

and we have the general solution by finding the antiderivative of the above function. Hence,

$$y = -2e^x$$

Substituting the value of x we have,

$$y = -2e^{-\frac{t}{2}} + C$$

where C is a constant.

Prob. 55. Given the equation as,

$$\frac{dy}{dx} = \sin(\pi x)$$

Thus, we have the general solution by finding the antiderivative of the above function. Hence,

$$y = \frac{1}{\pi}\cos(\pi x) + C$$

where C is a constant.

Prob. 57. Given the equation as,

$$\frac{dy}{dx} = \sec^2 \frac{x}{2}$$

Thus, we have the general solution by finding the antiderivative of the above function. Hence,

$$y = 2\tan \frac{x}{2} + C$$

where C is a constant.

Prob. 59. It should be $1 = 0 + C$, $c = 1$. Equation is $y = x^3 + 1$.

Prob. 61. Given the equation as,

$$\frac{dy}{dx} = 2\sqrt{x}$$

and given that when,

$$y = 2, x = 1$$

Thus, we have the general solution by finding the antiderivative of the above function. Hence,

$$y = \frac{4}{3}x^{\frac{3}{2}} + C$$

where C is a constant. Substituting the values of y and x we get,

$$2 = \frac{4}{3} + C$$

Hence, we get

$$C = \frac{2}{3}$$

Thus, we have final equation as

$$y = \frac{4}{3}x^{\frac{3}{2}} + \frac{2}{3}$$

Prob. 63. Given the equation as,

$$\frac{dN}{dt} = \frac{1}{t}$$

and given that when,

$$N(1) = 10$$

Thus, we have the general solution by finding the antiderivative of the above function. Hence,

$$N = \ln t + C$$

where C is a constant. Substituting the values of $N(1)$ we get,

$$10 = \ln 1 + C$$

Hence we get

$$C = 10$$

Thus, we have final equation as

$$N = \ln t + C$$

Prob. 65. Given the equation as,

$$\frac{dW}{dt} = e^t$$

and given that,

$$W(0) = 5$$

Thus, we have the general solution by finding the antiderivative of the above function. Hence,

$$W = e^t + C$$

where C is a constant. Substituting the values of $N(0)$ we get,

$$1 = 1 + C$$

Hence, we have

$$C = 0$$

Thus, we have final equation as

$$W = e^t$$

Prob. 67. Given the equation as, $\frac{dW}{dt} = e^{-3t}$ and given that, $W(0) = \frac{2}{3}$ Substituting $x = -3t$ we have $dx = -3dt$ Thus, we have the new function as,

$$\frac{dW}{dx} = \frac{-1}{3}e^x$$

and when $t = 0, x = 0$ Thus, we have the general solution by finding the antiderivative of the above function. Hence,

$$W = \frac{-1}{3}e^x + C$$

where C is a constant. Substituting the values of $N(0)$ we get,

$$\frac{2}{3} = \frac{-1}{3} + C$$

Hence, we have

$$C = 1$$

Thus, we have equation as

$$W = \frac{-1}{3}e^x + 1$$

Substituting we have the final equation, as

$$W = \frac{-1}{3}e^{-3t} + 1$$

Prob. 69. Given the equation as,

$$\frac{dT}{dt} = \sin(\pi t)$$

and given that,

$$W(0) = 3$$

Substituting $x = \pi t$ we have

$$dx = \pi dt$$

Thus, we have the new function as,

$$\frac{dT}{dx} = \frac{1}{\pi}\sin x$$

and when

$$t = 0, x = 0$$

Thus, we have the general solution by finding the antiderivative of the above function. Hence,

$$T = -\frac{1}{\pi}\cos x + C$$

where C is a constant. Substituting the values of $T(0)$ we get,

$$3 = -\frac{1}{\pi} + C$$

Hence, we have

$$C = 3 + \frac{1}{\pi}$$

Thus, we have equation as

$$T = -\frac{1}{\pi}\cos x + 3 + \frac{1}{\pi}$$

Substituting we have the final equation, as

$$T = -\frac{1}{\pi}\cos(\pi t) + 3 + \frac{1}{\pi}$$

Prob. 71. Given the equation as,

$$\frac{dy}{dx} = \frac{e^x + e^{-x}}{2}$$

and given that,

$$y = 0, x = 0$$

We can rewrite the above equation as

$$\frac{dy}{dx} = \frac{e^x}{2} + \frac{e^{-x}}{2}$$

Thus, we have the general solution by finding the antiderivative of the above function. Hence,

$$Y = \frac{e^x}{2} - \frac{e^{-x}}{2} + C$$

where C is a constant. Substituting the values of $y = 0$ and $x = 0$ we get,

$$0 = \frac{1}{2} - \frac{1}{2} + C$$

Hence, we have

$$C = 0$$

Thus, we have final equation as

$$y = \frac{e^x}{2} - \frac{e^{-x}}{2}$$

Prob. 73. Given the equation as,

$$\frac{dL}{dx} = e^{-.1x}$$

and $x \geq 0$. Also given that

$$L_\infty = \lim_{x \to \infty} L(x) = 25$$

First, we find, $L(x)$.

$$L(x) = -10e^{-.1x} + C$$

where C is a constant. Substituting this, in above equation, we have

$$L_\infty = \lim_{x \to \infty} (-10e^{-.1x} + C) = 25$$

$$C = 25$$

Thus we have,

$$L(x) = -10e^{-.1x} + 25$$

Hence, we have $L(0)$ as,

$$L(0) = -10 + 25$$

$$= 15$$

Prob. 75. We know that the equation of a freely falling object from rest is

$$v^2 = 2ah$$

where v is the final velocity and h is the height, and a is the acceleration. Thus putting the value,

$$v^2 = 6400$$

$$v = 80$$

Again, we know that,

$$v = at$$

where t is the time taken in falling. Thus putting the values, we have,

$$t = \frac{v}{a}$$

$$= \frac{5}{2} = 2.5$$

Prob. 77. We are given the volume change at time t is proportional to $t(24 - t)$, and that we water the plant at a constant rate of 4 per hour.

(a) Thus the net rate of loss of water is

$$\frac{dV}{dt} = -at(24 - t) + 4$$

This is because the first part of equation denotes the loss of water due to evaporation and the second part denote the water added per hour. Here a is a constant which has been introduced to convert the proportionality to some constant value dependent on t only.

(b) $V(t) = -12at^2 + \frac{at^3}{3} + 9t + C$, need to have

$$V(0) = V(24), -12a(24)^2 + \frac{a(24)^3}{3} + 4(24) = 0$$

$$a = \frac{1}{24}.$$

5.10 Review Problems

Prob. 1. Given the function as,

$$f(x) = xe^{-x}, x \geq 0$$

(a) For evaluation $f(0)$, we put $x = 0$ in above equation. Substituting, the value we have, $f(0) = 0$. For calculating the limit, we have

$$F = \lim_{x \to \infty} xe^{-x}$$

$$= \lim_{x \to \infty} \frac{x}{e^x}$$

Applying L'Hospital rule, we have,

$$F = \lim_{x \to \infty} \frac{1}{e^x}$$

Substituting the values we have

$$F = 0$$

(b) For local maxima, we differentiate the below function,

$$f(x) = xe^{-x}$$

So, differentiating,

$$f'(x) = e^{-x} - xe^{-x}$$

Substituting to zero,we have

$$f'(x) = 0$$

$$e^{-x} - xe^{-x} = 0$$

$$e^{-x}(1 - x) = 0$$

So we have

$$x = 1$$

Thus the function has a absolute maxima at $x = 1$

(c) To calculate the inflection point, we differentiate it twice and equate to zero. Given,

$$f(x) = xe^{-x}$$

So,

$$f'(x) = e^{-x} - xe^{-x}$$

Differentiating again,

$$f''(x) = -e^{-x} - e^{-x} + xe^{-x}$$

$$= e^{-x}(x - 2)$$

Equating to zero, we have

$$e^{-x}(x - 2) = 0$$

Hence,

$$x = 2$$

This is the point of inflection.

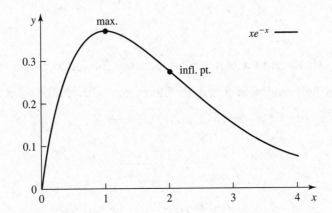

(d)

Prob. 3. Given the following definitions,

$$\sinh x = \frac{e^x - e^{-x}}{2}$$

$$\cosh x = \frac{e^x + e^{-x}}{2}$$

$$\tanh x = \frac{e^x - e^{-x}}{e^x + e^{-x}}$$

(a) From above we have,

$$\tanh x = \frac{e^x - e^{-x}}{e^x + e^{-x}}$$

So differentiating it, we have

$$f' = \frac{(e^x + e^{-x})(e^x + e^{-x}) - (e^x - e^{-x})(e^x - e^{-x})}{(e^x + e^{-x})^2}$$

$$= \frac{4}{(e^x + e^{-x})^2}$$

Thus we see that f' is always positive and hence is a strictly increasing function. Also,

$$\lim_{x \to \infty} \tanh x = \lim_{x \to \infty} \frac{e^x - e^{-x}}{e^x + e^{-x}}$$

$$= 1$$

Again,

$$\lim_{x \to -\infty} \tanh x = \lim_{x \to -\infty} \frac{e^x - e^{-x}}{e^x + e^{-x}}$$

$$= 1$$

(b) The function $f(x) = \tanh x$ is invertible because the function has the same values at both the limiting points and is strictly increasing with a minimum value at $x = 1$ Its inverse function can be easily calculated. We have,

$$y = \frac{e^x - e^{-x}}{e^x + e^{-x}}$$

$$e^{-x}(y + 1) = e^x(1 - y)$$

$$e^{-2x} = \frac{1 - y}{1 + y}$$

$$-2x = \ln\left(\frac{1 - y}{1 + y}\right)$$

$$x = \frac{1}{2}\ln\left(\frac{1 + y}{1 - y}\right)$$

Thus we have,

$$f^{-1}x = y = \frac{1}{2}\ln\left(\frac{1 + x}{1 - x}\right)$$

(c) To find the differential, we have

$$f^{-1}x = y = \frac{1}{2}\ln\left(\frac{1+x}{1-x}\right)$$

Differentiating,

$$f' = \frac{1}{2}\left(\frac{1-x}{1+x}\frac{-2}{(1-x)^2}\right)$$

Thus we have,

$$f' = \frac{1}{1-x^2}$$

Hence, we get the result.

(d) Given that,

$$\tanh x = \frac{\sinh x}{\cosh x}$$

Also given,

$$\cosh^2 x - \sinh^2 x = 1$$

Thus we have to find, $\frac{d(\tanh x)}{dx}$. Thus we have

$$\tanh x = \frac{e^x - e^{-x}}{e^x + e^{-x}}$$

$$\frac{d(\tanh x)}{dx} = \frac{4}{(e^x + e^{-x})^2}$$

Now we know that,

$$\cosh x = \frac{e^x + e^{-x}}{2}$$

So,

$$\cosh^2 x = \frac{(e^x + e^{-x})^2}{4}$$

Substituting in above, equation we have,

$$\frac{d(\tanh x)}{dx} = \frac{1}{\cosh^2 x}$$

Prob. 5. We are given a function, such that,

$$R(P) = \alpha P e^{-\beta P}, P \geq 0$$

(a) To, find the first derivative, we differentiate the above function,

$$R'(P) = \alpha e^{-\beta P} - \alpha \beta P e^{-\beta P}$$

$$= \alpha e^{-\beta P}(1 - \beta P)$$

To find the second differential, we differentiate the above function,

$$R''(P) = -\beta \alpha e^{-\beta P} - \beta \alpha e^{-\beta P}(1 - \beta P)$$

$$= -\beta \alpha e^{-\beta P}(2 - \beta P)$$

Hence, we get both the differentials.

(b) We have the differential of $R(P)$ as

$$R'(P) = \alpha e^{-\beta P}(1 - \beta P)$$

Equating it to zero, we have

$$\alpha e^{-\beta P}(1 - \beta P) = 0$$

$$1 - \beta P = 0$$

$$P = \frac{1}{\beta}$$

Thus $R'(P) = 0$ only when $P = \frac{1}{\beta}$ Also $R''(\frac{1}{\beta})$ equals,

$$R''(P) = -\beta \alpha e^{-\beta P}(2 - \beta P)$$

$$R''(\frac{1}{\beta}) = -\beta \alpha e^{-\beta \frac{1}{\beta}}(2 - \beta \frac{1}{\beta})$$

$$= -\beta \alpha e^{-1}$$

$$< 0$$

Thus at $P = \frac{1}{\beta}$, we have a local maxima.

Also, we see from above,

$$R(P) = \alpha P e^{-\beta P}, P \geq 0$$

Substituting $P = \frac{1}{\beta}$, we have

$$R(\frac{1}{\beta}) = \alpha \frac{1}{\beta} e^{-\beta \frac{1}{\beta}}$$

$$= \frac{\alpha}{\beta} e^{-1}$$

(c) To find the global maxima, we have to find $R(0)$ and

$$\lim_{x \to \infty} R(P)$$

So, finding the limit,

$$\lim_{x \to \infty} R(P) = \lim_{x \to \infty} \alpha P e^{-\beta P}$$

Rearranging the function, we get,

$$\lim_{x \to \infty} R(P) = \lim_{x \to \infty} \frac{\alpha P}{e^{\beta P}}$$

Applying L'Hospital Rule,

$$= \lim_{x \to \infty} \frac{\alpha}{\beta e^{\beta P}}$$

Substituting the value we get,

$$\lim_{x \to \infty} R(P) = 0$$

Thus, we have a global maxima at $P = \frac{1}{\beta}$.

(d) To find the inflection point we substitute the second derivative to zero. We have

$$R''(P) = -\beta \alpha e^{-\beta P}(2 - \beta P)$$

Equating it to zero, we have

$$-\beta \alpha e^{-\beta P}(2 - \beta P) = 0$$

$$2 - \beta P = 0$$

$$P = \frac{2}{\beta}$$

Also, we see that that the function changes sign at this value. Hence $P = \frac{2}{\beta}$ is a point of inflection.

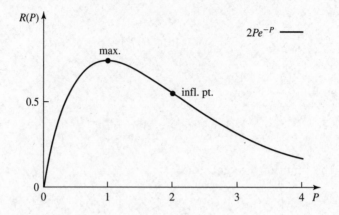

(e)

Prob. 7. The monod growth curve is given by

$$f(x) = \frac{cx}{k + x}$$

where c and k are constants.

(a) To find the horizontal asymptote, we find the limit tending to infinity,

$$\lim_{x \to \infty} f(x) = \lim_{x \to \infty} \frac{cx}{k + x}$$

Applying L'Hospital rule we have,

$$\lim_{x \to \infty} f(x) = \lim_{x \to \infty} c = c$$

This is called the saturation value.

(b) For $x \geq 0$ the function is strictly increasing . This is because when we find its derivative we have,

$$f'(x) = \frac{c(k + x) - cx}{(k + x)^2}$$

$$= \frac{ck}{(k + x)^2}$$

$$> 0$$

Hence the function is strictly increasing. Its concave down as its second derivative is less than zero for all $x > 0$.

(c) For $x = k$ we have

$$f(k) = \frac{ck}{k + k}$$

$$= \frac{c}{2}$$

Thus this value is half of the saturation value.

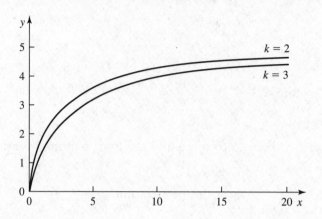

(d)

(e) For the three function given below,

$$g(x) = \frac{10x}{3+x}$$

$$h(x) = \frac{10x}{5+x}$$

$$J(x) = \frac{8x}{5+x}$$

We can find their saturation value and half saturation value directly. The higher the saturation value, the larger the function, and when the saturation value is same, we check the half saturation value, the greater it is the smaller the function. Thus, based on this analysis we can write,

$$\frac{10x}{3+x} > \frac{10x}{5+x} > \frac{8x}{5+x}$$

Prob. 9. Given that

$$L(\theta) = p_{AA}^{n_1} p_{Aa}^{n_2} p_{aa}^{n_3}$$

(a) Give that,

$$n_1 = 8, n_2 = 6, n_3 = 3$$

So our above equation becomes,

$$L(\theta) = p_{AA}^{8} p_{Aa}^{6} p_{aa}^{3}$$

Also given

$$p_{AA} = \theta^2, p_{Aa} = 2\theta(1 - \theta), p_{aa} = (1 - \theta)^2$$

Substituting above, we can get the value of $L(\theta)$.

(b) We have the equation,

$$L(\theta) = p_{AA}^8 p_{Aa}^6 p_{aa}^3$$

Taking ln, we have

$$\ln(L(\theta)) = n_1 \ln(p_{AA}) + n_2 \ln(p_{Aa}) + n_3 \ln(p_{aa})$$

Thus differentiating above and putting the values of the above three variables and equating to zero, we have the maximum value of $L(\theta)$ and that will be the same for the maximum value of $\ln(L(\theta))$.

(c) Put the given values and solve for x.

Prob. 11. Given the equation as,

$$\frac{dc}{dt} = -0.1e^{-0.3t}$$

(a) To find c, we integrate the above equation,

$$c = \int -0.1e^{-0.3t} \, dt \qquad = -\frac{1}{3}e^{-.3t} + K$$

where K is a constant. Given at $t = \infty$, $c = 0$ so $k = 0$. Hence, $C = -\frac{1}{3}e^{-.3t}$

(b) At $t = 0$, we have $c = \frac{1}{3}$. To become the value to half, we have,

$$\frac{2}{3} = -\frac{1}{3}e^{-.3t} \qquad -\ln 2 = -.3t \qquad t = \frac{\ln 2}{.3}$$

Thus at $t = \frac{\ln 2}{.3}$ the concentration becomes equal to half its initial amount.

Prob. 13. Given the equation of the height as,

$$h(t) = v_0 t - \frac{1}{2}gt^2$$

(a) To find the time at which, the object reaches the maximum height, we differentiate the below equation.

$$h(t) = v_0 t - \frac{1}{2} g t^2 h'(t) = v_0 - gt$$

Equating to zero, we have

$$v_0 - gt = 0 \qquad t = \frac{v_0}{g}$$

(b) The height reached at $t = \frac{v_0}{g}$ is,

$$h = v_0 \frac{v_0}{g} - \frac{1}{2} g (\frac{v_0}{g})^2 = \frac{(v_0)^2}{g}$$

This is the maximum height reached . So maximum height is,

$$\texttt{Max height} = \frac{(v_0)^2}{g}$$

(c) Velocity of the particle as it reaches the maximum height is calculated as follows. we know that $V = v_0 - gt$ Substituting the value of $t = \frac{v_0}{g}$, we have $V = 0$. Thus the velocity at the maximum point is zero.

(d) To reach the same height again, the particle will take the amount of time equal to two times the time taken to reach the maximum height. Thus we have from previous calculation, $t = \frac{v_0}{g}$. Thus the time taken to reach the same height again is, $T = 2\frac{v_0}{g}$

Chapter 6

Integration

6.1 The Definite Integral

Prob. 1. The left endpoints of the four intervals are $0, \frac{1}{4}, \frac{1}{2}, \frac{3}{4}$.

The approximation is then $\frac{1}{4} \cdot 0 + \frac{1}{4}\frac{1}{16} + \frac{1}{4}\frac{1}{4} + \frac{1}{4}\frac{9}{16} = \frac{7}{32}$.

Prob. 3. The right endpoints of the four intervals are $\frac{1}{4}, \frac{1}{2}, \frac{3}{4}, 1$.

The approximation is then $\frac{1}{4}\frac{1}{16} + \frac{1}{4}\frac{1}{4} + \frac{1}{4}\frac{9}{16} + \frac{1}{4} \cdot 1 = \frac{15}{32}$.

Prob. 5.
$$\sum_{k=1}^{4} \sqrt{k} = \sqrt{1} + \sqrt{2} + \sqrt{3} + \sqrt{4}.$$

Prob. 7.
$$\sum_{k=2}^{6} 3^k = 3^2 + 3^3 + 3^4 + 3^5 + 3^6.$$

Prob. 9.
$$\sum_{k=0}^{3} = (0+1)^0 + (1+1)^1 + (2+1)^2 + (3+1)^3.$$

Prob. 11.
$$\sum_{k=0}^{3} (-1)^{k+1} = (-1)^1 + (-1)^2 + (-1)^3 + (-1)^4.$$

Prob. 13.
$$\sum_{k=1}^{n} \left(\frac{k}{n}\right)^2 \frac{1}{n} = \left(\frac{1}{n}\right)^2 \frac{1}{n} + \left(\frac{2}{n}\right)^2 \frac{1}{n} + \cdots + \left(\frac{n}{n}\right)^2 \frac{1}{n}.$$

Prob. 15.
$$2 + 4 + 6 + \cdots + 2n = \sum_{k=1}^{n} 2k$$

291

Prob. 17.

$$\ln 2 + \ln 3 + \ln 4 + \ln 5 + \ln 6 = \sum_{k=2}^{5} \ln k.$$

Prob. 19.

$$-\frac{1}{4} + \frac{1}{6} + \frac{2}{7} + \frac{3}{8} = \sum_{k=2}^{6} \frac{k-3}{k+2}.$$

Prob. 21. $\displaystyle\sum_{k=0}^{n-1} q^k.$

Prob. 23.

$$\sum_{k=1}^{15}(2k+3) = \sum_{k=1}^{15} 2k + \sum_{k=1}^{15} 3 = 2 \cdot \left(\frac{15(15+1)}{2}\right) + 3 \cdot 15 = 15 \cdot 19 = 285.$$

Prob. 25.

$$\sum_{k=0}^{6} k(k+1) = \sum_{k=0}^{6}(k^2 + k) = \sum_{k=1}^{6} k + \sum_{k=1}^{6} k^2 = \frac{6 \cdot 7}{2} + \frac{6 \cdot 7 \cdot 13}{6} = 112.$$

Prob. 27.

$$\sum_{k=1}^{n} 4(k-1)^2 = 4 \sum_{k=1}^{n-1} k^2 = \frac{4(n-1)(n)(2n-1)}{6}.$$

Prob. 29. $\displaystyle\sum_{k=1}^{10}(-1)^k = -1 + 1 - 1 + 1 - 1 + 1 - 1 + 1 - 1 + 1 = 0.$

Prob. 31.

(a) All the terms collapse except the first and the last, so

$$\sum_{k=1}^{n}[(k+1)^3 - k^3] = -1^3 + (n+1)^3.$$

(b) We have

$$
\begin{aligned}
\sum_{k=1}^{n}[(k+1)^3 - k^3] &= \sum_{k=1}^{n}(3k^2 + 3k + 1) \\
&= 3\sum_{k=1}^{n} k^2 + 3\sum_{k=1}^{n} k + \sum_{k=1}^{n} 1 \\
&= 3\sum_{k=1}^{n} k^2 + 3\frac{n(n+1)}{2} + n
\end{aligned}
$$

(c) We have $3\displaystyle\sum_{k=1}^{n} k^2 = (n+1)^3 - 1 - 3\frac{n(n+1)}{2} - n,$

therefore, $\displaystyle\sum_{k=1}^{n} k^2 = \frac{n(n+1)(2n+1)}{6}.$

Prob. 33. $0.4[(1-(-0.8)^2)+(1-(-0.4)^2)+(1-0^2)+(1-(0.4)^2)+(1-(0.8)^2)]=1.36$.

Prob. 35. $(2+(-2)^2)+(2+(-1)^2)+(2+0^2)+(2+1^2)=14$.

Prob. 37. $\frac{\pi}{2}[\sin\frac{\pi}{2}+\sin\pi+\frac{3\pi}{2}]=0$.

Prob. 39. $\frac{b^2}{2}-\frac{a^2}{2}$ equals the area of $\triangle(0,0)(b,0)(b,b)$ minus the area of $\triangle(0,0)(a,0)(a,a)$ (see previous exercise.)

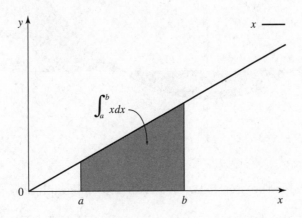

Prob. 41. $\int_1^2 2x^3 dx$.

Prob. 43. $\int_{-3}^2 (2x-1)\,dx$.

Prob. 45. $\int_2^3 \frac{x-1}{x+2}dx$.

Prob. 47. $\int_{-5}^2 e^x\,dx$.

Prob. 49. $\lim\limits_{\|p\|\to 0}\sum\limits_{k=1}^{n}(c_k+1)^{1/3}\,\triangle x_k,\quad p=[2,6]$.

Prob. 51. $\lim\limits_{\|P\|\to 0}\sum\limits_{k=1}^{n}\ln c_k\,\Delta x_k$, where P is a partition of $[1,e]$.

Prob. 53. $\lim\limits_{\|P\|\to 0}\sum\limits_{k=1}^{n}g(c_k)\,\Delta x_k$, where P is a partition of $[0,5]$.

Prob. 55. The integral is the area between x-axis and the graph of $y = x^2 - 1$ from $x = 1$ to $x = 2$ − area between x-axis and the graph of $y = x^2 - 1$ from $x = -1$ to $x = 1$.

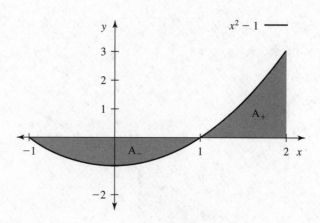

Prob. 57. The integral is the area between x-axis and the graph of $y = e^{-x}$ from $x = 0$ to $x = 5$.

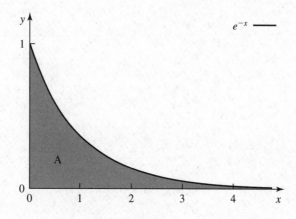

Prob. 59. The integral is the area between x-axis and the graph of $y = \ln x$ from $x = 1$ to $x = 4$ − area between x-axis and the graph of $y = \ln x$ from $x = 1/4$ to $x = 1$.

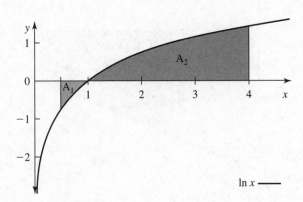

Prob. 61. The integral equals the sum of the areas of the triangles with vertices $(-2,0), (0,0), (-2,2)$ and $(0,0), (3,0), (3,3)$, that is $2 + 4.5 = 6.5$.

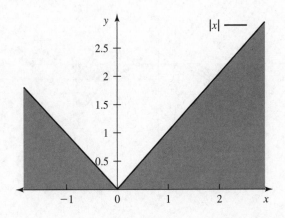

Prob. 63. The integral is negative. Its absolute value equals the area of the trapezoid with basis 3/2 and 3 and height 3, that is 27/4.

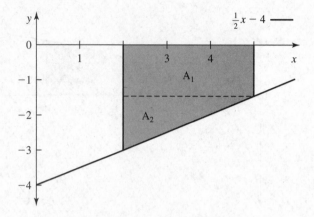

Prob. 65. The integral is negative. Its absolute value equals the area between the x-axis and the graph of the function $\sqrt{4 - x^2} - 2$. That area can be computed as the area of the rectangle of sides 2 and 4 minus half of the are of the circle of radius 2, that is, $2\pi - 8$.

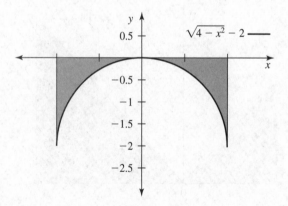

Prob. 67. The integral can be seen as the difference between two areas. One rectangle with vertices $(-3, 0), (-3, 4), (0, 4), (0, 0)$ and one fourth of the area of the disc centered at $(0, 4)$ with radius 3. Therefore the integral's value is $3 \times 4 - \frac{3}{4}\pi$.

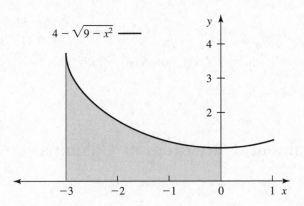

Prob. 69. 0.

Prob. 71. $\int_{-2}^{2} \frac{x^3}{3} dx = O$, look at the graph and use symmetry.

Prob. 73. 0.

Prob. 75. For $x \in [0, 1]$ we have $x^2 \le x$, so the inequality follows from property (7).

Prob. 77. For $x \in [0, 4]$ we have $0 \le \sqrt{x} \le 2$, so the inequality follows from property (8).

Prob. 79. The part of the disc centered at the origin with radius 1 contains the square with vertices $(0, 0), (1/\sqrt{2}, 0), (1/\sqrt{2}, 1/\sqrt{2}), (0, 1/\sqrt{2})$ (which has area 1/2), but is contained in the square with vertices $(0, 0), (1, 0), (1, 1), (0, 1)$ (which has area 1).

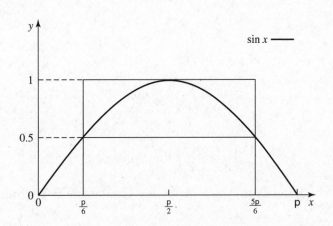

Prob. 81. $\cos x$ becomes negative for $x > \pi/2$, therefore the answer is $a = \pi/2$.

Prob. 83. For $f(x) = (x-2)^3$ we have $f(2+h) = -f(2-h)$, which gives the graph simmetry. We get then $a = 3$.

Prob. 85. We have

$$
\begin{aligned}
\int_t^{t+1} r(s)\,dt &= \int_0^{t+1} r(s)\,dt - \int_0^t r(s)\,dt = -\ln\frac{N(t+1)}{N(0)} + \ln\frac{N(t)}{N(0)} \\
&= -\ln N(t+1)) + \ln N(0) + \ln N(t) - \ln N(t) - \ln N(0) \\
&= -\ln\frac{N(t+1)}{N(t)}.
\end{aligned}
$$

6.2   The Fundamental Theorem of Calculus

Prob. 1. $\frac{dy}{dx} = 2x^3$

Prob. 3. $\frac{dy}{dx} = 9x^2 - 3$

Prob. 5. $\frac{dy}{dx} = \sqrt{1 + 2x}$

Prob. 7.

$$ y = \int_0^x \sqrt{1 + \sin^2 u}\,du, \, x > 0. $$

Hence

$$ \frac{dy}{dx} = \frac{d}{dx}\int_0^x \sqrt{1 + \sin^2 u}\,du. $$

So, from Fundamental Theorem of Calculus as the function $\sqrt{1 + \sin^2 u}$ is continuous between 0 and x

$$ \frac{dy}{dx} = \sqrt{1 + \sin^2 x}. $$

Prob. 9.

$$ y = \int_3^x u e^{4u}\,du. $$

Hence

$$ \frac{dy}{dx} = \frac{d}{dx}\int_3^x u e^{4u}\,du. $$

Since $f(x) = xe^{4x}$ is continuous for $x > 3$. Applying Fundamental Theorem of Calculus, we get

$$\frac{dy}{dx} = xe^{4x}.$$

Prob. 11.

$$y = \int_{-2}^{x} \frac{1}{u+3}\, du, x > -2.$$

Hence

$$\frac{dy}{dx} = \frac{d}{dx} \int_{-2}^{x} \frac{1}{u+3}\, du.$$

Since $f(x) = 1 \div x + 3$ is continuous for $x > -2$. Applying Fundamental Theorem of Calculus, we get

$$\frac{dy}{dx} = \frac{1}{x+3}.$$

Prob. 13.

$$y = \int_{\frac{\Pi}{2}}^{x} \sin(u^2 + 1)\, du.$$

Hence

$$\frac{dy}{dx} = \frac{d}{dx} \int_{\frac{\Pi}{2}}^{x} \sin(u^2 + 1)\, du.$$

Since $f(x) = \sin(x^2 + 1)$ is continuous for all x. Applying Fundamental Theorem of Calculus, we get

$$\frac{dy}{dx} = \sin(x^2 + 1).$$

Prob. 15. $h(x) = 3x$, $g(x) = 0$, $\frac{dy}{dx} = (1 + (3x)^2) \cdot 3$.

Prob. 17.

$$y = \int_{0}^{1-4x} (2t^2 + 1)\, dt.$$

Here the function $f(x) = 2x^2 + 1$ is continuous for all values of x, especially for $x > 0$. Hence we can apply *Leibniz's rule*.

$$h(x) = 1 - 4x, g(x) = 0$$

$$h'(x) = -4, g'(x) = 0$$

Hence applying equation (1)

$$\frac{dy}{dx} = (2(1-4x)^2 + 1)(-4);$$

Prob. 19. $h(x) = x^2 + 1$, $g(x) = 4$, $\frac{dy}{dx} = \sqrt{x^2 + 1} \cdot 2x$.

Prob. 21.

$$y = \int_0^{3x} (1 + e^t)\, dt.$$

Here the function $f(x) = 1 + e^{3x}$ is continuous for all values of x, especially for $x > 0$. Hence we can apply *Leibniz's rule*.

$$h(x) = 3x, g(x) = 0$$

$$h'(x) = 3, g'(x) = 0$$

Hence applying equation (1)

$$\frac{dy}{dx} = (1 + e^{3x})3.;$$

Prob. 23.

$$y = \int_1^{3x^2 + x} (1 + te^t)\, dt.$$

Here the function $f(x) = 1 + (3x^2 + x)e^{3x^2 + x}$ is continuous for all values of x, especially for $x > 0$. Hence we can apply *Leibniz's rule*.

$$h(x) = 3x^2 + x, g(x) = 1$$

$$h'(x) = 6x + 1, g'(x) = 0$$

Hence applying equation (1)

$$\frac{dy}{dx} = (1 + (3x^2 + x)e^{3x^2 + x})(6x + 1);$$

Prob. 25.

$$y = \int_x^3 (1 + t)\, dt.$$

Here the function $f(x) = 1 + x$ is continuous for all values x. Hence we can apply *Leibniz's rule*.

$$h(x) = 3, g(x) = x$$

$$h'(x) = 0, g'(x) = 1$$

Hence applying equation (1)

$$\frac{dy}{dx} = -(1+x);$$

Prob. 27. $h(x) = 3$, $g(x) = 2x$, $\frac{dy}{dx} = -(1 + \sin(2x)) \cdot 2$.

Prob. 29.

$$y = \int_x^5 (\frac{1}{u^2})\, du, x > 0.$$

Here the function $f(x) = \frac{1}{x^2}$ is continuous for all values $x > 0$. Hence we can apply *Leibniz's rule.*

$$h(x) = 5, g(x) = x$$

$$h'(x) = 0, g'(x) = 1$$

Hence applying equation (1)

$$\frac{dy}{dx} = -(\frac{1}{x^2});$$

Prob. 31.

$$y = \int_{x^2}^1 (\sec t)\, dt, -1 < x < 1.$$

Here the function $f(x) = \sec x^2$ is continuous for all values $-1 < x < 1$. Hence we can apply *Leibniz's rule.*

$$h(x) = 1, g(x) = x^2$$

$$h'(x) = 0, g'(x) = 2x$$

Hence applying equation (1)

$$\frac{dy}{dx} = -(\sec x^2)(2x);$$

Prob. 33.

$$y = \int_x^{2x} (1+t^2)\, dt.$$

Here the function $f(x) = 1 + x^2$ is continuous for all values x. Hence we can apply *Leibniz's rule.*

$$h(x) = 2x, g(x) = x$$

$$h'(x) = 1, g'(x) = 1$$

Hence applying equation (1)

$$\frac{dy}{dx} = (1 + 4x^2)(2) - (1 + x^2) = 1 + 7x^2;$$

Prob. 35. $h(x) = x^3$, $g(x) = x^3$, $\frac{dy}{dx} = \ln(x^3 - 3) \cdot 3x^2 - \ln(x^2 - 3) \cdot 2x$.

Prob. 37. $h(x) = x + x^3$, $g(x) = 2 - x^2$, $\frac{dy}{dx} = \sin(x + x^3) \cdot (1 + 3x^2) - \sin(2 - x^2) \cdot (-2x)$.

Prob. 39. $\int (1 + 3x^2) dx = x + x^3 + C$.

Prob. 41. $\int (\frac{1}{3}x^2 - \frac{1}{2}x) dx = \frac{x^3}{9} - \frac{1}{4}x^2 + C$.

Prob. 43.

$$y = \int \left(\frac{x^2}{2} + 3x - \frac{1}{3}\right) dx$$

We can write the above integral as

$$y = \int \left(\frac{x^2}{2}\right) dx + \int 3x \, dx - \int \frac{1}{3} dx$$

Thus computing the above integral, we get

$$y = \frac{x^3}{6} + \frac{3x^2}{2} - \frac{x}{3} + C$$

where C is a constant.

Prob. 45.

$$y = \int \left(\frac{2x^2 - x}{\sqrt{x}}\right) dx$$

We let $t = \sqrt{x}$ So

$$\frac{dy}{dx} = \frac{dy}{dt}\frac{dt}{dx}$$

So,

$$\frac{dt}{dx} = \frac{1}{2\sqrt{x}}$$

$$y = \int 4t^4 - 2t^2$$

Finding above integral we get

$$y = \frac{4t^5}{5} - \frac{2t^3}{3} + C$$

Substituting value of t in the above equation we get

$$y = \frac{4x^{\frac{5}{2}}}{5} - \frac{2x^{\frac{3}{2}}}{3} + C$$

where C is a constant.

Prob. 47.

$$y = \int (x^2 \sqrt{x}) \, dx$$

Hence we can rewrite the above integrand as $x^{\frac{5}{2}}$ Thus computing the integral we have

$$y = \frac{2}{7} x^{\frac{7}{2}} + C$$

where C is a constant.

Prob. 49.

$$y = \int (x^{\frac{7}{2}} + x^{\frac{2}{7}}) \, dx$$

We can split the integral as

$$y = \int (x^{\frac{7}{2}}) \, dx + \int (x^{\frac{2}{7}}) \, dx$$

Now applying standard integration methods to each of the integrals, we get

$$y = \frac{2}{9} x^{\frac{9}{2}} + \frac{7}{9} x^{\frac{9}{7}} + C$$

where C is a constant.

Prob. 51.

$$y = \int (\sqrt{x} + \frac{1}{\sqrt{x}}) \, dx$$

Thus we can split the above integral and rewrite as

$$y = \int x^{\frac{1}{2}} \, dx + \int x^{\frac{-1}{2}} \, dx$$

Now applying standard integration methods to each of the integrals, we get

$$y = \frac{2}{3} x^{\frac{3}{2}} + 2x^{\frac{1}{2}} + C$$

where C is a constant.

Prob. 53.

$$y = \int ((x - 1)(x + 1)) \, dx$$

Thus we can expand and split the above integral using algebraic formulae and rewrite as

$$y = \int x^2 \, dx - \int 1 \, dx$$

Now applying standard integration methods to each of the integrals, we get

$$y = \frac{x^3}{3} - x + C$$

where C is a constant.

Prob. 55.

$$y = \int \left((x-2)(3-x) \right) dx$$

Thus we can expand and split the above integral using algebraic formulae and rewrite as

$$y = -\int x^2 \, dx + \int 5x \, dx - \int 6 \, dx$$

Now applying standard integration methods to each of the integrals, we get

$$y = -\frac{x^3}{3} + \frac{5}{2}x^2 - 6x + C$$

where C is a constant.

Prob. 57.

$$y = \int e^{2x} \, dx$$

Let, $t = 2x$ Then $dt = 2dx$ Thus we have

$$y = \int \frac{e^t}{2} \, dt$$

Computing above integral yields

$$y = \frac{e^t}{2} + C$$

Substituting value of t we get the result as

$$y = \frac{e^{2x}}{2} + C$$

where C is a constant.

Prob. 59.

$$y = \int 3e^{-x} \, dx$$

Let, $t = -x$ Then $dt = -dx$ Thus we have

$$y = -\int 3e^t \, dt$$

Computing above integral yields

$$y = -3e^t + C$$

Substituting value of t we get the result as

$$y = -3e^{-x} + C$$

where C is a constant.

Prob. 61. $\int xe^{-\frac{x^2}{2}} dx = -e^{-x^2/2} + C.$

Prob. 63.

$$y = \int \sin 2x \, dx$$

Let, $t = 2x$ then we have $dt = 2dx$ Substituting in above integral, we get

$$y = \frac{1}{2} \int \sin t \, dt$$

Computing the above integral from standard Integration formulae, we get

$$y = -\frac{\cos t}{2} + C$$

Substituting, value of t we get

$$y = -\frac{\cos 2x}{2} + C$$

where C is a constant.

Prob. 65. $\int \cos(3x)dx = \frac{1}{3}\sin(3x) + C.$

Prob. 67.

$$y = \int \sec^2(3x) \, dx$$

Let, $t = 3x$ then we have $dt = 3dx$ Substituting in above integral we have

$$y = \int \frac{1}{3} \sec^2 t \, dt$$

Computing the above integral, we get

$$y = \frac{\tan t}{3} + C$$

Substituting the value of t we get the result as

$$y = \frac{\tan(3x)}{3} + C$$

where C is a constant.

Prob. 69.

$$y = \int \frac{\sin x}{1 - \sin^2 x}\, dx$$

Substituting $\cos x = t$ so that $-\sin x dx = dt$ And knowing that $1 - \sin^2 x = \cos^2 x$, we can transform the above integral as

$$y = -\int \frac{1}{t^2}\, dt$$

Computing the above integral we get,

$$y = \frac{1}{t} + C$$

Substituting value of t we get the result as,

$$y = \frac{1}{\cos x} + C$$

where C is a constant.

Prob. 71.

$$y = \int \tan(2x)\, dx$$

Let $t = \cos(2x)$, hence $dt = -2\sin(2x)dx$ and writing the integral as

$$y = \int \frac{\sin(2x)}{\cos(2x)}\, dx$$

We get after doing the transformation,

$$y = -\int \frac{1}{2t}\, dt$$

Computing the above integral we have

$$y = -\frac{\ln(t)}{2} + C$$

Substituting value of t we get the result as,

$$y = -\frac{\ln(\cos(2x))}{2} + C$$

where C is a constant.

Prob. 73. $\int (\sec^2 x + \tan x)dx = \int \sec^2 x dx + \int \tan x dx = \tan x + \ln|\sec x| + C.$

Prob. 75.

$$y = \int \frac{4}{1 + x^2} \, dx$$

Here, we take the constant out of the integration sign and what we are left is with standard integration of $\int \frac{1}{1+x^2}$ Thus we can rewrite the above integral as

$$y = 4 \int \frac{1}{1 + x^2} \, dx$$

Computing the integral we have

$$y = 4 \arctan x + C$$

where C is a constant.

Prob. 77.

$$y = \int \frac{1}{\sqrt{1 - x^2}} \, dx$$

The above integral is a standard integral of $\arcsin x$ and hence the result can be directly used. Thus using the standard result we get

$$y = \arcsin x + C$$

where C is constant.

Prob. 79. $\int \frac{1}{x+2} dx = \ln|x + 2| + C.$

Prob. 81. $\int \frac{2x-1}{3x} dx = \int (\frac{2}{3} - \frac{1}{3x}) dx = \frac{2}{3}x - \frac{1}{3} \ln|x| + C.$

Prob. 83.

$$y = \int \frac{x + 3}{x^2 - 9} \, dx$$

Now we know we can write

$$x^2 - 9 = (x + 3)(x - 3)$$

Thus writing the denominator of integral as above and carrying out proper algebraic transformation, we can rewrite the integral as

$$y = \int \frac{1}{x - 3} \, dx$$

Now let

$$t = x - 3$$

so we get

$$dt = dx$$

Now substituting in above integral, we get

$$y = \int \frac{1}{t} \, dt$$

which is a standard integral of $\ln t$. Hence after integration we have,

$$y = \ln t + C$$

Substituting value of t in terms of x we have

$$y = \ln(x - 3) + C$$

where C is a constant .

Prob. 85.

$$y = \int \frac{3 - x}{x^2 - 9} \, dx$$

Now we know we can write

$$x^2 - 9 = -(3 - x)(3 + x)$$

Thus writing the denominator of integral as above and carrying out proper algebraic transformation, we can rewrite the integral as

$$y = -\int \frac{1}{x + 3} \, dx$$

Now let

$$t = x + 3$$

so we get

$$dt = dx$$

Now substituting in above integral, we get

$$y = -\int \frac{1}{t} \, dt$$

which is a standard integral of $\ln t$. Hence after integration we have,

$$y = -\ln t + C$$

Substituting value of t in terms of x we have

$$y = -\ln(x+3) + C$$

where C is a constant .

Prob. 87.

$$y = \int \frac{5x^2}{x^2+1} \, dx$$

We can write the numerator of above integral as

$$5x^2 = 5(x^2+1) - 5$$

So the above integral takes the form as

$$y = 5 \int 1 \, dx - 5 \int \frac{1}{1+x^2} \, dx$$

Now both of them are standard integrals and hence we can compute them easily.

Computing both the integrals individually and combining them we get

$$y = 5x - 5 \arctan x + C$$

where C is a constant.

Prob. 89.

$$y = \int 3^x \, dx$$

The above integral is a standard one,and hence applying standard integration formulae to it we get the result as

$$y = \frac{3^x}{\ln 3} + C$$

where C is a constant.

Prob. 91. $\int 3^{-2x} dx = -\frac{3^{-2x}}{2 \cdot \ln(3)} + C.$

Prob. 93.

$$y = \int (x^2 + 2^x) \, dx$$

We can split the above integral and rewrite it as

$$y = \int x^2 \, dx + \int 2^x \, dx$$

Both of the above integrands are standard integrals and hence using standard integration formulae we have the result as

$$y = \frac{x^3}{3} + \frac{2^x}{\ln 2} + C$$

where C is a constant.

Prob. 95.

$$y = \int \left(\sqrt{x} + \sqrt{e^x} \right)$$

We can re write the above integral as

$$y = \int x^{\frac{1}{2}} \, dx + \int e^{\frac{x}{2}} \, dx$$

Now the first integral can be calculated from standard integration formulae, So

$$y_1 = \frac{2}{3} x^{\frac{3}{2}} + C_1$$

Now to compute the second integral we let,

$$t = \frac{x}{2}$$

so,

$$dt = \frac{1}{2} dx$$

Substituting we get

$$y_2 = 2 \int e^t \, dt$$

Computing the above integral from standard integration formulae we have,

$$y_2 = 2e^t + C_2$$

Substituting we get,

$$y_2 = 2e^{\frac{x}{2}} + C_2$$

Combining the above two, we get the result as

$$y = \frac{2}{3} x^{\frac{3}{2}} + 2e^{\frac{x}{2}} + C$$

where C is a constant.

Prob. 97.

$$y = \int_2^4 (3 - 2x)\, dx$$

The above integral can be rewritten as

$$y = \int_2^4 3\, dx - \int_2^4 2x\, dx$$

Hence computing the integral we get

$$y = [3x]_2^4 - [x^2]_2^4$$

Substituting the values we get

$$y = 6 - 12$$

$$y = -6$$

Prob. 99. $\int_0^1 (x^3 - x^{1/3})dx = \frac{x^4}{4} - \frac{x^{4/3}}{\frac{4}{3}}\Big|_0^1 = \frac{1}{4} - \frac{3}{4} = -\frac{1}{2}.$

Prob. 101.

$$y = \int_1^8 x^{\frac{-2}{3}}\, dx$$

Computing the above integral using standard integration formulae we have,

$$y = [3x^{\frac{1}{3}}]_1^8$$

Substituting the limits in the calculated integral, we get the result as

$$y = 3$$

Prob. 103.

$$y = \int_0^2 (2t - 1)(t + 3)\, dt$$

Exapnding the above integral using algebraic formulae, we have

$$y = \int_0^2 2t^2\, dt + \int_0^2 5t\, dt - \int_0^2 3\, dt$$

Computing the integral using standard integral formulae, we have

$$y = [\frac{2}{3}t^3 + \frac{5}{2}t^2 - 3t]_0^2$$

Substituting the values we get result as

$$y = \frac{16}{3} + 10 - 6$$

$$y = \frac{28}{3}$$

Prob. 105.

$$y = \int_0^{\frac{\pi}{4}} \sin(2x)\, dx$$

From the standard integration formulae for $\sin x$ and substituting $t = 2x$ we get after doing integration,

$$y = [-\frac{\cos(2x)}{2}]_0^{\frac{\pi}{4}}$$

Substituting the values we get result as

$$y = 0 + \frac{1}{2}$$

$$y = \frac{1}{2}$$

Prob. 107.

$$y = \int_0^{\frac{\pi}{8}} \sec^2(2x)\, dx$$

Using the standard integration formulae for $\sec x$ and using the substitution $t = 2x$ we get value as

$$y = [\frac{\tan(2x)}{2}]_0^{\frac{\pi}{8}}$$

Substituting the values, we get result as

$$y = \frac{1}{2}[1 - 0] = \frac{1}{2}$$

Prob. 109.

$$y = \int_0^1 \frac{1}{1 + x^2}\, dx$$

The above integral is a standard integral, and hence using standard integration formulae, we have following

$$y = [\arctan x]_0^1$$

Substituting the values we have,

$$y = \frac{\pi}{4}$$

Prob. 111.

$$y = \int_0^{\frac{1}{2}} \frac{1}{\sqrt{1-x^2}} \, dx$$

The above integral is a standard integral, and hence using standard integration formulae, we have following

$$y = [\arcsin x]_0^{\frac{1}{2}}$$

Substituting the values we have,

$$y = \frac{\pi}{6}$$

Prob. 113.

$$y = \int_0^{\frac{\pi}{6}} \tan(2x) \, dx$$

Let $t = \cos(2x)$, hence $dt = -2\sin(2x)dx$ and writing the integral as

$$y = \int_0^{\frac{\pi}{6}} \frac{\sin(2x)}{\cos(2x)} \, dx$$

Finding limits, when

$$x = 0, t = 1$$

$$x = \frac{\pi}{6}, t = \frac{1}{2}$$

We get after doing the transformation,

$$y = -\int_1^{\frac{1}{2}} \frac{1}{2t} \, dt$$

Computing the above integral we have

$$y = -[\frac{\ln(t)}{2}]_1^{\frac{1}{2}}$$

Substituting value of t we get the result as,

$$y = -[-\frac{\ln(2)}{2} - 0]$$

$$y = \frac{\ln(2)}{2}$$

Prob. 115.

$$y = \int_{-1}^{0} e^{3x} \, dx$$

First we convert it into standard form by substituting

$$t = 3x$$

then

$$dt = 3dx$$

Calculating the limits, when

$$x = -1, t = -3$$

$$x = 0, t = 0$$

Thus, our modified integral looks like,

$$y = \frac{1}{3} \int_{-3}^{0} e^t \, dt$$

Now using standard integration results, we get

$$y = \frac{1}{3} [e^t]_{-3}^{0}$$

Substituting the limits, we get the result as

$$y = \frac{1 - e^{-3}}{3}$$

Prob. 117.

$$y = \int_{-1}^{1} |x| \, dx$$

Since the above integrand is even, we can transform the above integral as,

$$y = 2 \int_{0}^{1} x \, dx$$

Now using standard integration formulae, ewe have

$$y = 2[\frac{x^2}{2}]_{0}^{1}$$

Substituting the limits, we get

$$y = 1$$

Prob. 119.

$$y = \int_{1}^{e} \frac{1}{x} \, dx$$

Above integral is a standard one and hence using standard integration formulae, we have

$$y = [\ln x]_1^e$$

Substituting the limits we have the result as,

$$y = 1 - 0$$

$$y = 1$$

Prob. 121.

$$y = \int_{-2}^{-1} \frac{1}{1-u} \, du$$

We do the following transformation. Let

$$t = 1 - u$$

then

$$dt = -du$$

Computing the limits, when

$$u = -2t = 3$$

$$u = -1, t = 2$$

Thus the above integral changes into,

$$y = -\int_3^2 \frac{1}{t} \, dt$$

Using standard integration formulae, we have

$$y = -[\ln t]_3^2$$

Substituting the limits we have the result as,

$$y = -\ln 2 + \ln 3$$

$$y = \ln \frac{3}{2}$$

Prob. 123.

$$\lim_{x \to 0} \frac{1}{x^2} \int_0^x \sin t \, dt$$

Using L'Hospital rule, we first differentiate the numerator. We have the value after differentiation as $\sin x$ Similarly doing the differentiation of numerator we have value as $2x$ Thus our limit becomes as

$$\lim_{x \to 0} \frac{\sin x}{2x}$$

Differentiating it once again, we get the value as

$$\lim_{x \to 0} \frac{\cos x}{2}$$

Thus, we can now substitute the limit, and hence the result is

$$\lim_{x \to 0} \frac{\cos x}{2} = \frac{1}{2}$$

Prob. 125. Given

$$\int_0^x f(t)\, dt = 2x^2$$

Thus differentiating the two sides with respect to x we get

$$f(x) = 4x$$

Thus, we have the result

$$f(x) = 4x$$

6.3   Applications of Integration

Prob. 1. From the graphs we see that $y = x + 2$ always sits on top, need to find intersections:

$$x + 2 = x^2 - 4 \quad x^2 - x - 6 = 0, \quad (x - 3)(x + 2) = 0, \quad x = -2, x = 3.$$

So, the area:
$$\int_{-2}^3 (x + 2 - x^2 + 9)dx = \frac{x^2}{2} - \frac{x^3}{3} + 6x \Big|_{-2}^3 = \frac{125}{6}.$$

Prob. 3. From the graphs we see that $y = e^{x/2}$ always sits on top. Hence, area:

$$\int_0^2 (e^{x/2} + x)dx = 2e^{x/2} + \frac{x^2}{2} \Big|_0^2 = 2e.$$

Prob. 5. Given

$$y = x^2 + 1, y = 4x - 2$$

First, we find the point of intersection by equating the two equation .

$$x^2 + 1 = 4x - 2$$

$$x^2 - 4x + 3 = 0$$

$$(x - 1)(x - 3) = 0$$

Hence we get values as,

$$4x = 1, x = 3$$

Thus the area under the curve in the 1st quadrant is,

$$Area = \int_1^3 (4x - 2) - (x^2 + 1)\, dx$$

$$= \int_1^3 -x^2 + 4x - 3\, dx$$

$$= [-\frac{x^3}{3} + 2x^2 - 3x]_1^3$$

$$= [-\frac{26}{3} + 16 - 6]$$

$$= 10 - \frac{26}{3}.$$

Prob. 7. From the graphs we see that the area is:

$$\int_{1/4}^1 (4 - \frac{1}{x})dx + \int_1^2 (4 - x^2)dx = 4x - \ln|x|\Big|_{1/4}^1 + 4x - \frac{x^3}{3}\Big|_1^2 = \frac{14}{3}.$$

Prob. 9. From the graphs we see that $\sin x \leq 1$ on $[0, \frac{\pi}{2}]$, so the area is:

$$\int_0^{\pi/2} (1 - \sin x)dx = x + \cos x\Big|_0^{\pi/2} = \frac{\pi}{2} - 1.$$

Prob. 11. Given

$$y = x^2, y = x^3$$

and

$$0 < x < 2.$$

The above two equations intersect at

$$x = 0 \, and \, x = 1.$$

Also for $0 < x < 1$ we have

$$x^2 > x^3$$

and for $1 < x < 2$ we have

$$x^3 > x^2.$$

Thus the area can be computed by splitting it into two parts and calculating area for each of them.

$$Area = \int_0^1 x^2 - x^3 \, dx + \int_1^2 x^3 - x^2 \, dx$$

$$= [\frac{x^3}{3} - \frac{x^4}{4}]_0^1 + [\frac{x^4}{4} - \frac{x^3}{3}]_1^2$$

$$= \frac{1}{3} - \frac{1}{4} + \frac{15}{4} - \frac{7}{3}$$

$$= \frac{7}{2} - 2$$

Prob. 13. Given

$$y = x^2, y = (x - 2)^2, y = 0$$

where,

$$0 < x < 2$$

Since, we have to integrate in terms of y, we first calculate y in terms of x.

$$x = \pm\sqrt{y}, x = 2 + \pm\sqrt{y}, y = 0$$

We calculate the point of intersection of the curves

$$x = \pm\sqrt{y}, x = 2 + \pm\sqrt{y}$$

$$\sqrt{y} = 2 - \sqrt{y}$$

$$2\sqrt{y} = 2$$

$$y = 1$$

Hence area under the curve can be calculated as,

$$Area = \int_0^1 (2 - \sqrt{y}) - \sqrt{y}\, dy$$

$$= 2 \int_0^1 1 - \sqrt{y}\, dy$$

$$= 2[y - \frac{2}{3} y^{\frac{3}{2}}]_0^1$$

$$= 2[1 - \frac{2}{3}]$$

$$= \frac{2}{3}$$

Prob. 15. Given

$$x = (y - 1)^2 + 3, x = 1 - (y - 1)^2$$

where

$$0 < y < 2$$

We have to find the area enclosed by the figure, only for 1st quadrant. First we find the point of intersection for the two curves

$$x = (y - 1)^2 + 3, x = 1 - (y - 1)^2$$

Equating the two values, we get

$$(y - 1)^2 + 3 = 1 - (y - 1)^2$$

$$2(y - 1)^2 + 2 = 0$$

$$(y - 1)^2 + 1 = 0$$

Thus the two curves never meet. Also,

$$(y - 1)^2 + 3 > 1 - (y - 1)^2$$

for

$$0 < y < 2$$

Hence the area can be directly computed as

$$Area = \int_0^2 (y - 1)^2 + 3 - (1 - (y - 1)^2)\, dy$$

$$= 2 \int_0^2 (y-1)^2 + 1 \, dy$$

$$= 2[\frac{(y-1)^3}{3} + y]_0^2$$

$$= 2[2\frac{1}{3} + 2]$$

$$= \frac{16}{3}$$

Prob. 17. Given

$$\frac{dN}{dt} = e^{-t}$$

(a) To find $N(t)$ we have to integrate the above equation with respect to t. Integrating

$$N(t) = \int e^{-t} \, dt$$

$$N(t) = -e^{-t} + C$$

where C is a constant. To evaluate this constant, we have to apply boundary condition. Given that at

$$t = 0, N(0) = 100$$

Hence,

$$N(0) = C - 1 = 100$$

$$C = 101$$

Hence we get the result as

$$N(t) = 101 - e^{-t}$$

(b) To find the cumulative value of $N(t)$ between $0 < t < 5$, we integrate $N(t)$.

Integrating

$$\int_0^5 N(t) \, dt = \int_0^5 101 - e^{-t} \, dt$$

$$= [101t + e^{-t}]_0^5$$

$$= 505 + e^{-5} - 1$$

$$= 504 + e^{-5}$$

which is the cumulative change in the population.

(c) For any time t, the cumulative change in population between that period can be written as

$$CumulativePopulation = \int_0^t 101 - e^{-x}\,dx$$

where x is a variable . The above integral means that at any time t, the total population is sum total of all the population increase in that period. Geometrically its the area bounded by the curve e^{-x} and the x axis.

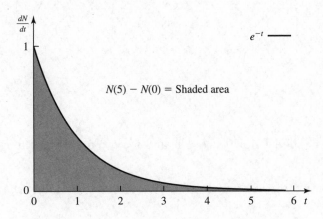

Prob. 19. Given

$$v(t) = -(t-2)^2 + 1$$

for

$$0 \le t \le 5$$

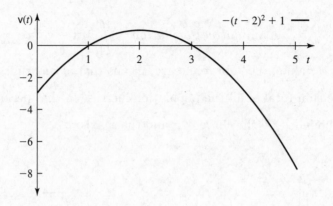

(a)

(b) Particle moves to the left for $0 < t < 1$ and $3 < t < 5$, and to the right for $1 < t < 3$.

(c) To find the position of the particle $s(t)$ at any time t, we know that,

$$v(t) = \frac{dx}{dt}$$

Hence

$$x = \int dx$$

Thus we get,

$$s(t) = \int dx$$

$$s(t) = \int_0^t v(x)\, dx$$

$$s(t) = \int_0^t -(x-2)^2 + 1\, dx$$

$$s(t) = [-\frac{(x-2)^3}{3} + x]_0^t$$

$$s(t) = -\frac{(t-2)^3}{3} + t - \frac{2}{3}$$

Thus $s(t)$ gives the distance traveled by the particle at time t.

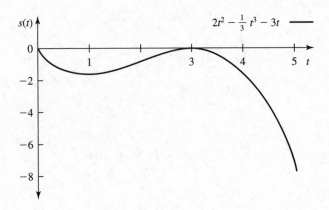

(d)

Prob. 21. Given that

$$\frac{dl}{dt}$$

represents the growth of an organism. Hence,

$$\frac{\int_2^7 dl}{dt\,dt}$$

represents the cumulative growth of the organism between the months $[2,7]$. Or in other words, it represents the cumulative growth in 5 months.

Prob. 23. Given that

$$\frac{db}{dt}$$

represents the rate of change of biomass at time t. Then

$$\int_1^6 \frac{dB}{dt}\,dt$$

represents the cumulative biomass between time 1 to 6 . That is, total biomass accumulated between time units $[1,6]$.

Prob. 25. Given that

$$f(x) = x^2 - 2$$

and

$$0 \leq x \leq 2.$$

To compute the average value of $f(x)$ we proceed as,

$$Avg(f(x)) = \frac{1}{2-0} \int_0^2 f(x)\, dx$$

$$= \frac{1}{2} \int_0^2 x^2 - 2\, dx$$

$$= \frac{1}{2}[\frac{x^3}{3} - 2x]_0^2$$

$$= \frac{1}{2}[\frac{8}{3} - 4]$$

$$= -\frac{2}{3}$$

Thus the average value of $f(x)$ is $-\frac{2}{3}$

Prob. 27. We are given that the temperature t in Fahrenheit varies as,

$$T(t) = 68 + \sin\frac{\pi t}{12}$$

and that

$$0 \leq t \leq 24.$$

To compute the average value of the temperature, we proceed as follows,

$$Avg(T) = \frac{1}{24-0} \int_0^2 4T(t)\, dt$$

$$= \frac{1}{24} \int_0^2 468 + \sin\frac{\pi t}{12}\, dt$$

$$= \frac{1}{24}[68t - \frac{12}{\pi}\cos\frac{\pi t}{12}]_0^24$$

$$= \frac{1}{24}[1632 - \frac{12}{\pi}(1-1)]$$

$$= 68$$

Hence the average temperature is $6d$.

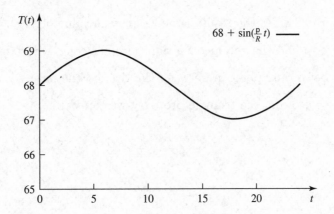

$$68 + \sin\!\left(\tfrac{p}{R}\,t\right) \text{——}$$

Prob. 29. The function $\tan x$ is odd, so for every positive value it assumes from $[0, 1]$, there is an equivalent negative value being assumed in the interval $[-1, 0]$, so the integral

$$\int_{-x}^{x} x^x \tan x \, dx = 0.$$

Prob. 31. Given

$$f(x) = 2x$$

and

$$0 \le x \le 2$$

Hence average value of $f(x)$ can be computed as,

$$Avg(f) = \frac{1}{2 - 0} \int_0^2 f(x) \, dx$$

$$= \frac{1}{2} \int_0^2 2x \, dx$$

$$= \frac{1}{2} [x^2]_0^2$$

$$= 2$$

Hence the average value of $f(x)$ is 4. Now to find x such that $f(x)$ is equal to avg(x). We, do the following,

$$f(x) = 2$$

$$2x = 2$$

$$x = 1$$

Hence the value of $x = 2$ gives the result same as average value of $f(x)$.

Prob. 33. We have to find the volume of a right circular cone, whose base radius is r and whose height is h. Let at any point in between the tip and the base, $height = x, radius = y$. From the triangle property we can write,

$$\frac{h - x}{h} = \frac{y}{r}$$

$$x = h - \frac{hy}{r}$$

Hence,

$$dx = \frac{h}{r} dy$$

Thus the volume of that small circular section can be written as

$$dV = \pi y^2 dx$$

Writing dx in terms of dy, we have

$$V = \int_0^r \pi y^2 \frac{h}{r} dy$$

$$= \frac{h}{r} [\pi \frac{y^3}{3}]_0^r$$

$$= \frac{1}{3} \pi r^2 h$$

Thus, we get the volume of cone in terms of $r and x$ as

$$V = \frac{1}{3} \pi r^2 h.$$

Prob. 35. Given

$$y = 4 - x^2, y = 0, x = 0$$

When we plot the graph, we find that that the graph intersects the x-axis at $x = 2$. Thus when we rotate the figure we find that the area at any point x can be written as

$$area = \pi y^2$$

Thus the volume can be calculated as,

$$V = \int_0^2 \pi(4 - x^2)^2 \, dx$$

$$= \pi \int_0^2 16 + x^4 - 8x^2 \, dx$$

$$= \pi[16x + \frac{x^5}{5} - \frac{8x^3}{3}]_0^2$$

$$= \pi[32 + \frac{32}{5} - \frac{64}{3}]$$

$$= \frac{256}{15}\pi$$

Above gives the volume bounded by the curve, which is

$$V = \frac{256}{15}\pi.$$

Prob. 37. Given $y = \sqrt{\sin x}, y = 0$ and $0 \le x \le \pi$. When we draw the curve and rotate it about the x-axis we get sphere whose diameter is π. Area at any point x can be written as

$$area = \pi y^2$$

We can then easily compute the volume, as follows,

$$V = \int y^2 \, dx$$

$$= \int_0^\pi \sin x \, dx$$

$$= \pi[-\cos x]_0^\pi$$

$$= 2\pi$$

Thus the volume of the figure, when rotated is $V = 2$.

Prob. 39. Given

$$y = \sec x, y = 0$$

and

$$\frac{-\pi}{3} \le x \le \frac{\pi}{3}$$

When we draw the above curve, we find that the area is symmetric about the y-axis. Thus we need to compute the volume of only one part and double it. Area at any point x can be written as

$$area = \pi y^2$$

We can then easily compute the volume, as follows,

$$V = \pi \int y^2 \, dx$$

$$= \pi \int_0^{\frac{\pi}{3}} \sec^2 x \, dx$$

$$= \pi [\tan x]_0^{\frac{\pi}{3}}$$

$$= \pi [\sqrt{3} - 0]$$

$$= \pi \sqrt{3}$$

Hence the volume of the enclosed figure is twice this volume. Thus the volume is

$$V = \pi 2 \sqrt{3}.$$

Prob. 41. Given $y = x^2, y = x$ and $0 \leq x \leq 1$ When we draw the curve we see that for $0 \leq x \leq 1$, $x > x^2$. Also to compute the volume we can compute the volume of each of them and then subtract them to get the enclosed volume . Volume of a small disc can be written as

$$dV = \pi x^2 dx - \pi x^4 dx$$

$$V = \pi \int_0^1 x^2 - x^4 \, dx$$

$$= \pi [\frac{x^3}{3} - \frac{x^5}{5}]_0^1$$

$$= \pi [\frac{1}{3} - \frac{1}{5}]$$

$$= \frac{2}{15} \pi$$

Thus the volume of the enclosed figure, when rotated around the x-axis is $V = \frac{2}{15} \pi$.

Prob. 43. $y = e^x, y = e^{-x}$ and $0 \leq x \leq 2$ When we draw the curve we see that for $0 \leq x \leq 2$, $e^x > e^{-x}$. Also to compute the volume we can compute the volume of each of them and then subtract them to get the enclosed volume . Volume of a figure can be written as

$$V = \pi \int_0^2 e^{2x} - e^{-2x} \, dx$$

$$= \pi \left[\frac{1}{2} e^{2x} + \frac{1}{2} e^{-2x} \right]_0^2$$

$$= \pi \frac{1}{2} [e^4 - 1 + e^{-4} - 1]$$

$$= \frac{\pi}{2} (e^4 + e^{-4} - 2)$$

Thus the volume of the enclosed figure, when rotated around the x-axis is

$$V = \frac{\pi}{2} (e^4 + e^{-4} - 2).$$

Prob. 45. Given

$$y = \sqrt{\cos x}, y = 1$$

and

$$0 \le x \le \frac{\pi}{2}$$

When we draw the curve we see that for $0 \le x \le \frac{\pi}{2}$,

$$1 > \sqrt{\cos x}.$$

Also to compute the volume we can compute the volume of each of them and then subtract them to get the enclosed volume . Volume of a small disc can be written as

$$V = \pi \int_0^{\frac{\pi}{2}} 1 - (\sqrt{\cos x})^2 \, dx$$

$$= \pi \int_0^{\frac{\pi}{2}} 1 - \cos x \, dx$$

$$= \pi [x - \sin x]_0^{\frac{\pi}{2}}$$

$$= \pi \left[\frac{\pi}{2} - 1 \right]$$

$$= (\frac{\pi}{2} - 1)\pi$$

Thus the volume of the enclosed figure, when rotated around the x-axis is

$$V = (\frac{\pi}{2} - 1)\pi.$$

Prob. 47. Given

$$y = \sqrt{x}, y = 2, x = 0$$

We have to find the volume by rotating around the y-axis. Hence we find y in terms of x.

$$x = y^2$$

Hence the volume can be calculated as

$$V = \pi \int_0^2 (y^2)^2 \, dy$$

$$= \pi \int_0^2 y^4 \, dy$$

$$= \pi [\frac{y^5}{5}]_0^2$$

$$= \pi [\frac{32}{5} - 0]$$

$$= \frac{32\pi}{5}$$

Hence the volume of the enclosed figure, when rotated around the y-axis is

$$V = \frac{32\pi}{5}.$$

Prob. 49. Given

$$y = \ln(x + 1), y = \ln 3, x = 0$$

Since, the enclosed figure has to be rotated around y-axis, we get x as a function of y.
Hence,

$$x = e^y - 1$$

and

$$0 \le y \le \ln 3$$

Hence the volume can be calculated as

$$V = \pi \int_0^{\ln(3)} (e^y - 1)^2 \, dy$$

$$= \pi \int_0^{\ln(3)} e^{2y} - 2e^y + 1 \, dy$$

$$= \pi[\frac{e^{2y}}{2} - 2e^y + y]_0^{\ln 3}$$

$$= \pi[4 - 4 + \ln 3]$$

$$= \pi \ln 3$$

Hence the volume of the enclosed figure, when rotated around the y-axis is

$$V = \pi \ln 3.$$

Prob. 51. Given

$$y = x^2, y = \sqrt{x}$$

and

$$0 \le x \le 1$$

Since, the enclosed figure has to be rotated around y-axis, we get x as a function of y. Hence,

$$x = \sqrt{y}, x = y^2$$

and

$$0 \le y \le 1$$

Also

$$\sqrt{y} > y^2$$

in the given interval Hence the volume can be calculated as,

$$V = \pi \int_0^1 (\sqrt{y})^2 - (y^2)^2 \, dy$$

$$= \pi \int_0^1 y - y^4 \, dy$$

$$= \pi[\frac{y^2}{2} - \frac{y^5}{5}]_0^1$$

$$= \pi[\frac{1}{2} - \frac{1}{5}]$$

$$= \frac{3\pi}{10}$$

Hence the volume of the enclosed figure, when rotated around the y-axis is

$$V = \frac{3\pi}{10}.$$

Prob. 53. Given, the equation of line as

$$y = 2x$$

and

$$0 \le x \le 2$$

(a) Using planar geometry first we calculate both the end points.

$$x = 0, y = 0$$

$$x = 2, y = 4$$

Hence, the length is

$$\sqrt{(2-0)^2 + (4-0)^2} = \sqrt{20}$$

(b) Using integral formulae, we have

$$y' = 2$$

Hence

$$length = \int_0^2 \sqrt{1 + 2^2}\, dx$$

$$= \int_0^2 \sqrt{5}\, dx$$

$$= [\sqrt{5}x]_0^2$$

$$= \sqrt{20}$$

Hence from both the ways, we get the same length.

Prob. 55. Given, the equation of line as

$$y^2 = x^3$$

and

$$1 \le x \le 4$$

Hence we calculate y',

$$y' = \frac{3}{2}\sqrt{x}$$

Applying the standard formulae for determining the length,

$$L = \int_a^b \sqrt{1 + [f'(x)]^2}\, dx$$

Substituting in above equation,

$$L = \int_1^4 \sqrt{1 + [\frac{3}{2}\sqrt{x}]^2}\, dx$$

$$= \int_1^4 \sqrt{1 + \frac{9x}{4}}\, dx$$

$$= [\frac{4}{9}\frac{3}{2}[1 + \frac{9x}{4}]^{\frac{3}{2}}]_1^4$$

$$= \frac{2}{3}[10^{\frac{3}{2}} - (\frac{13}{4})^{\frac{3}{2}}]$$

Above result gives us the length of the line.

Prob. 57. Given, the equation of line as

$$y = \frac{x^3}{6} + \frac{1}{2x}$$

and

$$1 \le x \le 3$$

Hence we calculate y',

$$y' = \frac{x^2}{2} - \frac{1}{2x^2}$$

Applying the standard formulae for determining the length,

$$L = \int_a^b \sqrt{1 + [f'(x)]^2}\, dx$$

Substituting in above equation,

$$L = \int_1^3 \sqrt{1 + [\frac{x^2}{2} - \frac{1}{2x^2}]^2}\, dx$$

$$= \int_1^3 \frac{1}{2}\sqrt{(x^4 + 2 + \frac{1}{x^4})}\, dx$$

$$= \frac{1}{2}\int_1^3 x^2 + \frac{1}{x^2}$$

$$= \frac{1}{2}[\frac{x^3}{3} - \frac{1}{x}]_1^3$$

$$= \frac{14}{3}$$

Above result gives us the length of the line.

Prob. 59. Given, the equation of line as

$$y = x^2$$

and

$$-1 \le x \le 1$$

Hence we calculate y',

$$y' = 2x$$

Applying the standard formulae for determining the length,

$$L = \int_a^b \sqrt{1 + [f'(x)]^2}\, dx$$

Substituting in above equation,

$$L = \int_{-1}^1 \sqrt{1 + [2x]^2}\, dx$$

Thus, above equation can be used for computing the length of the equation .

Prob. 61. Given, the equation of line as

$$y = e^{-x}$$

and

$$0 \le x \le 1$$

Hence we calculate y',

$$y' = -e^{-x}$$

Applying the standard formulae for determining the length,

$$L = \int_a^b \sqrt{1 + [f'(x)]^2}\, dx$$

Substituting in above equation,

$$L = \int_0^1 \sqrt{1 + [-e^{-x}]^2}\, dx$$

Thus, above equation can be used for computing the length of the equation .

Prob. 63. Given the equation of a quarter circle

$$y = \sqrt{1 - x^2}$$

and

$$0 \le x \le 1$$

We have to find the length.

(a) First we will find the length by using geometrical formulae. We can easily see that the radius of the circle is 1. Hence, semi perimeter of the circle is π.

(b) Now, we calculate using integral formulae. For that first we find f'.

$$y' = -\frac{x}{\sqrt{1 - x^2}}$$

Thus the length is equal to

$$Length = \int_0^1 \sqrt{1 + [-\frac{x}{\sqrt{1 - x^2}}]^2} \, dx$$

$$= \int_0^1 \frac{1}{\sqrt{1 - x^2}} \, dx$$

$$= [\arcsin x]_0^1$$

$$= \pi$$

Thus from, both ways we get the length as π.

Prob. 65. Given equation is

$$y = \frac{e^x + e^{-x}}{2}$$

and

$$0 \le x \le a.$$

We first calculate y',

$$y' = \frac{e^x - e^{-x}}{2}$$

Thus the length can be computed as

$$Length = \int_0^a \sqrt{1 + [\frac{e^x - e^{-x}}{2}]^2}\, dx$$

$$= \int_0^a \sqrt{(\frac{e^x + e^{-x}}{2})^2}\, dx$$

$$= \int_0^a \frac{e^x + e^{-x}}{2}$$

$$= [\frac{e^x - e^{-x}}{2}]_0^a$$

$$= \frac{e^a - e^{-a}}{2}$$

Thus above gives us the length. Also if we substitute a in the derivative of y, then

$$f'(a) = \frac{e^a - e^{-a}}{2}$$

Thus we see that the results are same. Hence, proved.

6.5   Review Problems

Prob. 1. We construct new column:

$h \cdot \bar{V}$
0.04816
0.16188
0.3082
0.40142
0.34222
0.24926
0.15521
0.048972

Total: 1.71512, since the step is $2m$, then

$$Q \approx 2 \cdot 1.71582 = 3.43\frac{m^3}{S}.$$

Prob. 3. We are given the theoretical velocity is given by

$$v(d) = \left(\frac{D-d}{a}\right)^{\frac{1}{c}})$$

where v(d) is the velocity at depth d below the water surface, c is a constant varying from 5 for coarse beds to 7 for smooth beds, D is the total depth of the channel, and a is a constant that is equal to the distance above the bottom of the channel at which velocity has unit value. Now depending upon the two methods, the second method is more efficient since it takes two different values and computes the average of them. We see that, more the number of samples take , better will be the result, as the error will get reduced. Hence, method 2 is a better method.

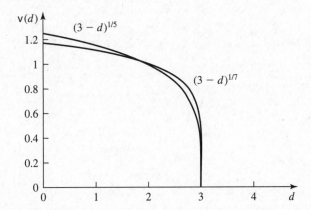

(a)

(b) substituting the values in the above equation we see that at $d = D$,

$$V(D) = \left(\frac{D-D}{a}\right)^{\frac{1}{c}})$$

$$= 0$$

and at $d = 0$, we have

$$v(0) = \left(\frac{D-0}{a}\right)^{\frac{1}{c}}) = \frac{D^{\frac{1}{c}}}{a}$$

The derivative of the function at $d = 0$ is zero, and hence has the maximal value.

Also, the second derivative is negative.

Prob. 5. Substituting d_1 in the average velocity equation at any depth, we find that

$$v(d_1) = (\frac{D - d_1}{a})^{\frac{1}{c}})$$

Since at this depth $v(d_1) = \overline{v}$, we get

$$\frac{c}{c+1}(\frac{D}{a})^{\frac{1}{c}} = (\frac{D - d_1}{a})^{\frac{1}{c}})$$

Raising each side by the power of c, we have

$$(\frac{c}{c+1})^c \frac{D}{a} = \frac{D - d_1}{a}$$

Thus we have the result,

$$\frac{d_1}{D} = 1 - (\frac{c}{c+1})^c$$

Chapter 7

Integration Techniques and Computational Methods

7.1 The Substitution Rule

Prob. 1. $\frac{2}{3}(x^2+3)^{3/2}+C$.

Prob. 3. Let $u = 10x^2$, $du = -2x\,dx$, we have:

$$-\frac{3}{2}\int u^{1/4}du = -\frac{3}{2}\cdot\frac{4}{5}\cdot u^{5/4}+C = -\frac{6}{5}(1-x^2)^{5/4}+C.$$

Prob. 5. $\frac{5}{3}\sin(3x)+C$.

Prob. 7. $-\frac{7}{12}\cos(4x^3)+C$.

Prob. 9. $\frac{1}{2}e^{2x+3}+C$.

Prob. 11. Let $u = -x^2/2$, $du = -x\,dx$, we have:

$$\int -e^u du = -e^u+C = -e^{-x^2/2}+C.$$

Prob. 13. $\frac{1}{2}\ln|x^2+4x|+C$.

Prob. 15. Let $u = x+u$, $du = dx$, $x = u-4$, we have:

$$\int \frac{3(u-4)}{u}du = 3\int(1-\frac{4}{u})du = 3u-4\cdot\ln|u|+C = 3(x+4)-4\ln|x+4|+C.$$

Prob. 17. $\frac{2}{3}(x+3)^{3/2}+C$.

Prob. 19. $\frac{2}{3}(2x^2-3x+2)^{3/2}+C$.

Prob. 21. Let $u = 1 + 4x - 2x^2$, $du = (4 - 4x)dx$, we have:

$$-\frac{1}{4}\int \frac{1}{u}du = -\frac{1}{4}\cdot \ln|u| + C = \frac{1}{4}\ln(1 + 4x - 2x^2) + C.$$

Prob. 23. $\frac{1}{2}\ln|1 + 2x^2| + C.$

Prob. 25. $\frac{3}{2}e^{x^2} + C.$

Prob. 27. $-\cot(\ln x) + C.$

Prob. 29. $-\frac{2}{3\pi}\cos\left(\frac{3\pi}{2} + \frac{\pi}{4}\right) + C.$

Prob. 31. $\frac{1}{2}\tan^2 x + C.$

Prob. 33. $\frac{(\ln x)^3}{3} + C.$

Prob. 35. $\frac{1}{15}(5 + x^2)^{3/2}(3x^2 - 10) + C$

Prob. 37. $\ln|ax^2 + bx + c| + C.$

Prob. 39. $\frac{1}{n+1}g^{n+1}(x) + C.$

Prob. 41. $-e^{-g(x)} + C.$

Prob. 43. Let $u = x^2 + 1$, $du = 2xdx$, we have:

$$\frac{1}{2}\int u^{1/2}du = \frac{1}{2}\cdot\frac{2}{3}u^{3/2} + C = \frac{1}{3}\cdot(x^2 + 1) + C.$$

Now,

$$F(3) - F(0) = \frac{1}{3}(10^{3/2} - 1).$$

Prob. 45. $\frac{7}{2025}$

Prob. 47. $1 - e^{-9/2}.$

Prob. 49. $\frac{3}{8}.$

Prob. 51. $\frac{1}{2}.$

Prob. 53. $4 + \ln 27.$

Prob. 55. Let $u = \ln x$, $du = \frac{1}{x}dx$, we have:

$$\int \frac{1}{u^2} = -u^{-1}C = -\frac{1}{\ln(x)} + C.$$

Now, $F(e^2) - F(e) = \frac{1}{2}.$

Prob. 57. $2e^{-1} - 2e^{-3}.$

Prob. 59. $\int \cot x\,dx = \int \cos x \sin^{-1} x\,dx = \ln|\sin x| + C.$

7.2 Integration by Parts

Prob. 1. $x \sin x + \cos x$, $(u = x,\ v' = \cos x)$.

Prob. 3. $\frac{2}{3}x \sin(3x - 1) + \frac{2}{9}\cos(3x - 1) + C$, $(u = 2x, C' = \cos(3x - 1))$.

Prob. 5. $-2x \cos(x - 1) + 2 \sin(x - 1) + C$, $(u = 2x, v' = \sin(x - 1))$.

Prob. 7. $(x - 1)e^x$, $(u = x, v' = e^x)$.

Prob. 9. $(x^2 - 2x + 2)e^x$, $(u = x^2, v' = e^x)$.

Prob. 11. $\frac{1}{2}x^2 \ln x - \frac{x^2}{4}$, $(u = \ln x, v' = x)$.

Prob. 13. $\frac{1}{2}x^2 \ln(3x) - \frac{x^2}{4}$, $(u = \ln(3x), v' = x)$.

Prob. 15. $\ln|\cos x| + x \tan x$, $(u = x, v' = \sec^2 x)$.

Prob. 17. $\frac{\sqrt{3}}{2} - \frac{\pi}{6}$, $(u = x, v' = \sin x)$.

Prob. 19. $2 \ln 2 - 1$, $(u = \ln x, v' = 1)$.

Prob. 21. $\frac{1}{2}(4 \ln 4 - 3)$, $(u = \ln x, v' = \frac{1}{2})$.

Prob. 23. $1 - \frac{2}{e}$, $(u = x, v' = e^{-x})$.

Prob. 25. $\frac{1}{4}(2 + (\sqrt{3} - 1)e^{\pi/3}$, $(u = \sin x, v' = e^x), u = -\cos x, v' = e^x$.

Prob. 27. $\frac{2e^{-3x}\left(-6\cos\left(\frac{\pi x}{2}\right) + \pi \sin\left(\frac{\pi x}{2}\right)\right)}{36 + \pi^2}$, $\left(u = \cos\left(\frac{\pi}{2}x\right), v' = e^{-3x}\right)$,
$\left(u = \frac{\pi}{2}\sin\left(\frac{\pi}{2}x\right), v' = \frac{-e^{-3x}}{3}\right)$.

Prob. 29. $\frac{1}{2}(\sin(\ln x) - \cos(\ln x))$, $(u = \sin(\ln x), v' = 1)$, $(u = \cos(\ln x), v' = 1)$.

Prob. 31. We have

$$
\begin{aligned}
\int \cos^2 x\,dx &= \sin x \cos x + \int \sin^2 x\,dx \\
&= \sin x \cos x + \int (1 - \cos^2 x)\,dx \\
&= \sin x \cos x + \int 1\,dx - \int \cos^2 x\,dx
\end{aligned}
$$

therefore $2 \int \cos^2 x\,dx = \sin x \cos x + x$ and we get $\int \cos^2 x\,dx = \frac{1}{2}(\sin x \cos x + x)$.

Prob. 33. We have

$$
\begin{aligned}
\int \arcsin x\,dx &= x \arcsin x - \int \frac{x}{\sqrt{1 - x^2}}\,dx \\
&= x \arcsin x - \int \frac{-1}{2\sqrt{u}}\,du \quad (1 - x^2 = u) \\
&= x \arcsin x + \sqrt{u} \\
&= x \arcsin x + \sqrt{1 - x^2}.
\end{aligned}
$$

Prob. 35. Taking $u = \ln x$, $v' = \frac{1}{x}$, we get

$$\int \frac{1}{x} \ln x \, dx = (\ln x)^2 - \int \frac{1}{x} \ln x \, dx.$$

Collecting similar terms we conclude that $\int \frac{1}{x} \ln x \, dx = \frac{1}{2}(\ln x)^2$.

Prob. 37.

(a) Taking $u = x^n$, $v' = e^{ax}$, we get

$$\int x^n e^{ax} \, dx = \frac{1}{a} x^n e^{ax} - \frac{n}{a} \int x^{n-1} e^{ax} \, dx.$$

(b) We use the previous part with $n = 2$, $a = -3$. We have

$$
\begin{aligned}
\int x^2 e^{-3x} \, dx &= -\frac{1}{3} x^2 e^{-3x} + \frac{2}{3} \int x e^{-3x} \, dx \\
&= -\frac{1}{3} x^2 e^{-3x} + \frac{2}{3}\left(-\frac{1}{3} x e^{-3x} + \frac{1}{3} \int e^x \, dx\right) \\
&= -\frac{1}{3} x^2 e^{-3x} + \frac{2}{3}\left(-\frac{1}{3} x e^{-3x} + \frac{1}{3} e^x\right) \\
&= \left(-\frac{x^2}{3} - \frac{-2x}{9} - \frac{2}{27}\right) e^{-3x}.
\end{aligned}
$$

Prob. 39. $2\cos\sqrt{x} + 2\sqrt{x}\sin\sqrt{x}$, $(\sqrt{x} = y)$, $(u = \cos y, v' = 2u)$.

Prob. 41. $-e^{-x^2/2}(2 + x^2)$, $(\frac{x^2}{2} = y)$ and see problem 8.

Prob. 43. $e^{\sin x}(\sin x - 1)$, $(\sin x = y)$, $(u = y, v' = e^y)$.

Prob. 45. $2e^{\sqrt{x}}(\sqrt{x} - 1)$, $(\sqrt{x} = y)$, $(u = y, v' = e^y)$.

Prob. 47. $\sqrt{x} - \frac{x}{2} + (x - 1)\ln(1 + \sqrt{x})$, $(\sqrt{x} + 1 = y)$ and see problem 11.

Prob. 49. $(-\frac{1}{4} - \frac{x}{2})e^{-2x}$, $(u = x, v' = e^{-2x})$.

Prob. 51. $\ln|\sin x|$, $(\sin x = y)$.

Prob. 53. $-\cos(x^2)$, $(x^2 = y)$.

Prob. 55. $\frac{1}{4}\arctan\frac{x}{4}$, $(\frac{x}{4} = y)$.

Prob. 57. $x - 3\ln|x + 3|$.

Prob. 59. $\frac{1}{2}\ln(x^2 + 3)$, $x^2 + 3 = y$.

Prob. 61.

(a) Taking $u = \ln x, v' = 1$, we get $\int \ln x \, dx = x \ln x - \int 1 \, dx = x \ln x - x$.

(b) Using $\ln x = y$, we have $x = e^y$, $dx = e^y dy$, therefore, $\int \ln x \, dx = \int y e^y dy$. Using integration by parts, $(u = y, v' = e^y)$, we see that this integral is $y e^y - e^y$. As $y = \ln x$, we get $x \ln x - x$.

Prob. 63. $(\frac{4x^2}{9} - \frac{8x}{45} - \frac{64}{45})(x-2)^{1/4}$, $((x-2)^{1/4} = y)$.

Prob. 65. $2e^2$.

Prob. 67. $\frac{\pi}{2}$.

Prob. 69. $\frac{1}{2}$.

7.3 Practicing Integration and Partial Fractions

Prob. 1. $2x + 1 - \frac{3}{x+2}$

Prob. 3. $3x - 2 + \frac{2x}{x^2+1}$

Prob. 5. $\frac{5}{x+1} - \frac{3}{x}$

Prob. 7. $\frac{2}{x} - \frac{1}{x-3} + \frac{3}{x+1}$

Prob. 9. $\frac{2}{x-1} + \frac{3}{x+1}$

Prob. 11. $\frac{3}{x-5} + \frac{1}{x+2}$

Prob. 13. (a) $a = -\frac{1}{2}, b = \frac{1}{2}$. (b) $\frac{1}{2} \ln |\frac{x-2}{x}|$.

Prob. 15. Partial fractions give: $\frac{1}{4}\frac{1}{x-3} - \frac{1}{4}\frac{1}{x+1}$, hence the integral is:

$\frac{1}{4} \ln |x - 3| - \frac{1}{4} \ln |x + 1| + C$.

Prob. 17. Partial fractions give: $-\frac{1}{x^2} - \frac{1}{2x} + \frac{3}{2(x+2)}$, hence the integral is:

$\frac{3}{2} \ln |x + 2| + \frac{1}{x} - \frac{1}{2} \ln |x| + C$.

Prob. 19. $-\tan^{-1} x + \frac{1}{2} \ln |x^2 + 4| + C$.

Prob. 21. $2 \tan^{-1} x + \frac{3}{2(x^2+1)} + C$

Prob. 23. $\int \frac{1}{x^2 - 2x + 2} dx = \int \frac{1}{(x-1)^2 + 1} dx = \tan^{-1}(x - 1) + C$

Prob. 25. $\frac{1}{3} \arctan(\frac{1}{3}(x - 2))$.

Prob. 27. $\frac{1}{5} \ln \frac{x-3}{x+2}|$.

Prob. 29. $\frac{1}{6} \ln \frac{x-3}{x+3}|$.

Prob. 31. $\frac{1}{3}\ln\frac{x-2}{x+1}|$.

Prob. 33. $2\ln(x+1) - 5\ln(x+2) + x$.

Prob. 35. $x + 2\ln\frac{x-2}{x+2}|$.

Prob. 37. $2 + \ln\frac{3}{5}$.

Prob. 39. $\frac{\ln 2}{2}$.

Prob. 41. $-\ln 2$.

Prob. 43. $\frac{\pi}{2} - \frac{1}{2}\ln 2$.

Prob. 45. $\frac{1}{1+x} + \ln\left|\frac{x}{x+1}\right|$.

Prob. 47. $-\frac{2}{x+1} + \ln\left|\frac{x+1}{x-1}\right|$.

Prob. 49. $-\frac{x}{18(x^2-9)} + \frac{1}{108}\ln\left|\frac{x+3}{x-3}\right|$.

Prob. 51. $-\frac{1}{x} - \arctan x$.

Prob. 53.

(a) Long division gives:

$$\frac{x^4(1-x)^4}{1+x^2} = x^6 - 4x^5 + 5x^4 - 4x^2 + 4 - \frac{4}{1+x^2}.$$

(b) Since

$$\frac{x^4(1-x)^4}{2} \leq \frac{x^4(1-x)^4}{1+x^2} \leq x^4(1-x)^4$$

for $0 \leq x \leq 1$, we have:

$$\int_0^1 \frac{x^4(1-x)^4}{2}dx \leq \int_0^1 \frac{x^4(1-x)^4}{1+x^2} \leq \int_0^1 \frac{x^4(1-x)^4}{1}dx.$$

But then,

$$\int_0^1 \frac{x^4(1-x)^4}{2} = \frac{1}{1260},$$

$$\int_0^1 \frac{x^4(1-x)^4}{1+x^2} = \frac{22}{7} - 17, \quad \int_0^1 x^4(1-x)^4 dx = \frac{1}{630}.$$

Therefore,

$$\frac{1}{1260} \leq \frac{22}{7} - \pi \leq \frac{1}{630},$$

meaning

$$3.140 \leq \pi \leq 3.142.$$

7.4 Improper Integrals

Prob. 1. The integral is improper because it has infinite upper limit. Its value is 1.

Prob. 3. The integral is improper because it has infinite upper limit. Its value is π.

Prob. 5. The integral is improper because it has infinite upper limit. Its value is 2.

Prob. 7. The integral is improper because it has infinite limits. Its value is 2.

Prob. 9. The integral is improper because it has infinite limits. Its value is 0.

Prob. 11. The integral is improper because the integrand has infinite limit at upper endpoint. Its value is 6.

Prob. 13. The integral is improper because the integrand has infinite limit at upper endpoint. Its value is 2.

Prob. 15. The integral is improper because the integrand is discontinuous in the interval. Its value is -2.

Prob. 17. The integral is convergent. Its value is $\frac{1}{2}$.

Prob. 19. The integral is divergent.

Prob. 21. $\int_0^2 \frac{1}{(x-1)^{1/3}}dx$, discontinuous at $x = 1$,

$$\lim_{t\to 1^-}\int_0^t \frac{1}{(x-1)^{1/3}}dx = \lim_{t\to 1^-}\frac{3}{2}(x-1)^{2/3}\Big|_0^t$$

$$= \lim_{t\to 1^-}\frac{3}{2}(t-1)^{2/3} - \frac{3}{2}(-1)^{2/3} = -\frac{3}{2}$$

$$\lim_{t\to 1^+}\int_t^2 \frac{1}{(x-1)^{1/3}}dx = \lim_{t\to 1^-}\frac{3}{2}(x-1)^{2/3}\Big|_t^2 = \frac{3}{2},$$

hence we get 0, convergent.

Prob. 23. The integral is divergent.

Prob. 25. The integral is divergent.

Prob. 27. The integral is convergent. Its value is 0.

Prob. 29. The integral is divergent.

Prob. 31. $c = 3$.

Prob. 33.

(a) We have, for $p > 1$, $\int \frac{1}{x^p} dx = \frac{x^{1-p}}{1-p}$. Also $\int \frac{1}{x} dx = \ln x$, and the expressions for $A(z)$ follow.

(b) We have, for $0 < p \leq 1$, $\lim_{z\to\infty} A(z) = \infty$ because $\lim_{z\to\infty} z^{1-p} = \infty$.

(c) We have, for $p > 1$, $\lim_{z\to\infty} A(z) = \frac{1}{p-1}$ because $\lim_{z\to\infty} z^{1-p} = 0$.

Prob. 35.

(a) The exponential function is increasing in all its domain, and, for $x \geq 1$, we have
$-x^2 \leq -x$.

(b) We know that $\int_1^\infty e^{-x} dx$ converges, therefore, by part a), so does $\int_1^\infty e^{-x^2} dx$.

Prob. 37.

(a) For $x \geq 1$ we have $4x^2 \geq 1 + x^2$, then, exctracting square roots, we get
$2x \geq \sqrt{1+x^2}$, therefore $\frac{1}{\sqrt{1+x^2}} \geq \frac{1}{2x}$.

(b) We know that $\int_1^\infty \frac{1}{2x} dx$ diverges, therefore, by part a), so does $\int_1^\infty \frac{1}{\sqrt{1+x^2}}$.

Prob. 39. Convergent, compare with $\int_{-\infty}^\infty e^{-|x|} dx$.

Prob. 41. Divergent, compare with $\int_1^\infty \frac{1}{\sqrt{2x}} dx$.

Prob. 43.

(a) We have $\lim_{x\to\infty} \frac{\ln x}{\sqrt{x}} = \lim_{x\to\infty} \frac{\sqrt{x}}{2x} = 0$.

(b) For $x \geq 75$ we have $2\ln x \leq \sqrt{x}$.

(c) It is enough to study the integral over the interval $[75, \infty]$, since $\int_0^{75} e^{-\sqrt{x}} dx$ offers no problems. We have, from part b), $\ln x^2 \leq \sqrt{x}$, so $\ln \frac{1}{x^2} \geq -\sqrt{x}$, and $\frac{1}{x^2} \geq e^{-\sqrt{x}}$. As $\int_0^\infty \frac{1}{x^2} dx$ converges, so does $\int_0^\infty e^{-\sqrt{x}}$.

The graph of $e^{-\sqrt{x}}$ stays, from some point on, below the graph of $\frac{1}{x^2}$, therefore it defines a finite area with the x-axis.

7.5 Numerical Integration

Prob. 1. 2.32813.

Prob. 3. 0.629204.

Prob. 5. 0.69122. The exact value is 0.6931...

Prob. 7. 5.38382. The exact value is 5.3333...

Prob. 9. 2.3438

Prob. 11. 0.637963.

Prob. 13. 20.32. The exact value is 20.

Prob. 15. 1.81948. The exact value is 1.88562...

Prob. 17. We use the error formula to get: $n = 82$.

Prob. 19. We use the error formula to get: $n = 58$.

Prob. 21. We use the error formula to get: $n = 92$.

Prob. 23. We use the error formula to get: $n = 50$.

Prob. 25.

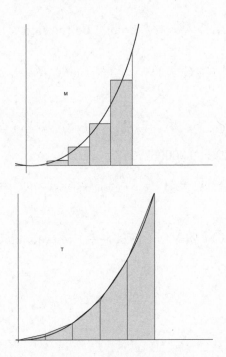

(a) We have $M_5 = 0.245$, $T_5 = 0.268$, the exact value being 0.25.

(b) The same reason as in a) applies.

(c) Reversing the concavity reverses the inequalities above.

7.6   The Taylor Approximation'

Prob. 1. $1 + 2x$.

Prob. 3. $1 + x$.

Prob. 5. $\ln 2$.

Prob. 7. $1 - \frac{x^2}{2} + \frac{x^4}{24}$.

Prob. 9. x^5.

Prob. 11. $P_3(x) = \sqrt{2} + \frac{x}{2\sqrt{2}} - \frac{x^2}{16\sqrt{2}} + \frac{x^3}{64\sqrt{2}}$. $\sqrt{2 + 0.1} = 1.4491376...$,
$P_3(0.1) = 1.44913800...$

Prob. 13. $P_5(x) = x - \frac{x^3}{6} + \frac{x^5}{120}$. $\sin 1 = 0.841471$, $P_5(1) = 0.841667...$.

Prob. 15. $P_2(x) = x$. $\tan 0.1 = 0.100335...$, $P_2(0.1) = 0.1$.

Prob. 17.

(a) $P_3(x) = x - \frac{x^3}{6}$.

(b) $\lim_{x \to 0} \frac{\sin x}{x}$ and $\lim_{x \to 0} \frac{x - \frac{x^3}{6}}{x} = 1$ "should" be the same.

Prob. 19. $P_3(x) = 1 + \frac{x-1}{2} - \frac{1}{8}(x-1)^2 + \frac{1}{16}(x-1)^3$. $\sqrt{3} = 1.4142135623...$,
$P_3(2) = 1.4375$.

Prob. 21. $P_3(x) = \frac{\sqrt{3}}{2} - \frac{1}{2}(x - \pi/6) - \frac{1}{4}\sqrt{3}(x - \pi/6)^2 + \frac{1}{12}(x - \pi/6)^3$.
$\cos \frac{\pi}{7} = 0.900968867...$, $P_3(2) = 0.9009677287...$

Prob. 23. $P_3(x) = e + e(x - 1) + \frac{e}{2}(x - 1)^2 + \frac{e}{6}(x - 1)^3$. $e^{2.1} = 8.1661...$,
$P_3(-0.9) = 7.9559...$

Prob. 25. Let g be defined by $g(N) = rN(1 - \frac{N}{K})$ where r and K are constants. The linearization near 0 of g is rN.

Prob. 27. 10.

Prob. 29. 2.

Prob. 31. $P_2(x) = 0$, $\frac{x^3}{6}\left(\frac{e^{-1/c}}{x^6} - \frac{6e^{-1/x}}{x^5} + \frac{6e^{-1/x}}{x^4}\right)$.

Prob. 33.

(a) Let $g(x) = \tan^{-1} x$. We have $g(0) = 0$, $g'(0) = 0$, $g'''(0) = -2$, and so on ($g^{(ek)(0)} = 0$, $g^{(2k+1)}(0) = (-1)^k(2k)!$. The expression is just the Taylor polynomial for g.

(b) Noting that $\tan^{-1} 1 = \frac{\pi}{4}$ and plugging in $x = 1$ in the expression in part a), we get the given equality.

7.7 Table of Integrals

Prob. 1. $\frac{x}{2} + \frac{3}{4} \ln|2x - 3| + C$.

Prob. 3. $\frac{1}{2}\sqrt{x^2 - 16} - 8\ln|x + \sqrt{x^2 - 16}| + C$.

Prob. 5. $6 - \frac{16}{e}$.

Prob. 7. $\frac{1}{9}(1 + 2e^3)$.

Prob. 9. $\frac{e^{\pi/6}}{2} + \frac{1}{4} - \frac{\sqrt{3}}{4}$.

Prob. 11. $-e^{-x/2}(2x^2 + 8x + 14) + C$.

Prob. 13. $\frac{1}{10}(5x - 3) - \frac{1}{20}\sin(10x - 6) + C$.

Prob. 15. $\frac{x}{2}\sqrt{9 + 4x^2} + \frac{9}{4}\ln|x + \frac{1}{2}\sqrt{9 + 4x^2}| + C$.

Prob. 17. $\frac{e^{2x+1}}{\pi^2 + 16}(4\sin\frac{\pi x}{2} - \pi\cos\frac{\pi x}{2}) + C$.

Prob. 19. 1.38629.

Prob. 21. $\frac{x}{2}(\cos(\ln|3x|) + \sin(\ln|3x|)) + C$.

7.9 Review Problems

Prob. 1. $-\frac{1}{9}(1 - x^3)^3 + C$.

Prob. 3. $-2e^{-x^2} + C$.

Prob. 5. $\frac{6}{7}(1 + \sqrt{x})^{7/3} - \frac{3}{2}(1 + \sqrt{x})^{4/3} + C$.

Prob. 7. $\frac{1}{6}\tan(3x^2) + C$.

Prob. 9. $\frac{x^2}{4}(2\ln x - 1) + C$.

Prob. 11. $\tan x(\ln|\tan x| - 1) + C$.

Prob. 13. $\frac{1}{2}\arctan\frac{x}{2} + C$.

Prob. 15. $-\ln|\cos x| + C$.

Prob. 17. $\frac{e^{2x}}{5}(2\sin x - \cos x) + C$.

Prob. 19. $2\sqrt{e^x} + C$.

Prob. 21. $\frac{1}{2}(x - \cos x \sin x) + C$.

Prob. 23. $\ln\left|\frac{x-1}{x}\right| + C$.

Prob. 25. $x - 5\ln|x+5| + C$.

Prob. 27. $\ln|x+5| + C$.

Prob. 29. $\frac{1}{2}x(x+6) + 4\ln|x-1|$.

Prob. 31. $4 + \ln 3$.

Prob. 33. $1 - \frac{1}{\sqrt{e}}$.

Prob. 35. $\frac{\pi}{8}$.

Prob. 37. 4.

Prob. 39. $\frac{\pi}{6}$.

Prob. 41. Diverges.

Prob. 43. Diverges.

Prob. 45. 2.

Prob. 47. $-\frac{1}{4}$.

Prob. 49. $e - e^{1/\sqrt{2}}$.

Prob. 51. $M_4 = 0.625, T_4 = 0.75$.

Prob. 53. $M_5 = 0.631068, T_5 = 0.634226$.

Prob. 55. $2x - \frac{4x^3}{3}$.

Prob. 57. $(x-1) - \frac{1}{2}(x-1)^2 + \frac{1}{3}(x-1)^3$.

Prob. 59. The area to be shaded is the one below the horizontal line $y = K$, above the graph of $y = f_{avg}(x)$.

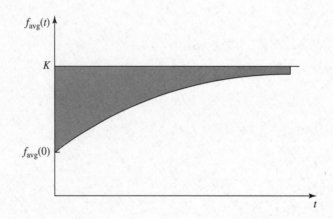

Chapter 8

Differential Equations

8.1 Solving Differential Equations

Prob. 1.

$$y(x) \;=\; y(0) + \int_0^x u + \sin U \, du$$

$$=\; \frac{U^2}{2} - \cos U\big]_0^x$$

$$=\; \frac{x^2}{2} - \cos x + \cos 0$$

$$=\; 1 + \frac{x^2}{2} - \cos x$$

Prob. 3.

$$y(x) = y(1) + \int_1^x \frac{1}{U} du = \ln U\Big|_1^x = \ln x - \ln 1 = \ln x$$

Prob. 5.

$$x(t) = x(0) + \int_0^t \frac{1}{1-U} du = 2 - \ln(1-U)\,|_0^t = 2 - \ln(1-t)$$

Prob. 7.

$$s(t) = s(0) + \int_0^t \sqrt{3U+1}\, du \;\; = 1 + \tfrac{2}{3}(3U+1)^{3/2} \cdot \tfrac{1}{3}\Big|_0^t$$

$$= 1 + \tfrac{2}{9}(3t+1)^{3/2} - \tfrac{2}{9}$$

$$= \tfrac{7}{9} + \tfrac{2}{9}(3t+1)^{3/2}$$

Prob. 9.

$$v(t) = v(0) + \int_0^t \cos U \, du = 5 + \sin U\Big|_0^t = 5 + \sin t$$

Prob. 11.

$$\frac{dy}{3y} = dx$$

so $\int \frac{dy}{3y} = \int dx$, then $\ln|3y| = x + c_1$ so $|3y| = e^{x+c_1}$ and taking out the absolute value signs

$$3y = \pm e^{x+c_1}$$

Now using the condition that the point $(0, 2)$ belongs to the function

$$3y = ce^x$$

we find $c = 6$, so the solution is

$$y = 2e^x$$

Prob. 13. $\int \frac{dx}{-2x} = \int dt$, so $-\frac{1}{2}\int \frac{dx}{x} = \int dt$ and integrating both sides we have

$$
\begin{aligned}
-\frac{1}{2}\ln|x| &= t + c_1 \\
\ln|x| &= -2t + c_2 \\
|x| &= e^{-2t+c_2} \\
x &= \pm e^{-2t+c_2} \\
x &= ce^{-2t}.
\end{aligned}
$$

Now using the condition that $x(1) = 5$, we get $5 = ce^{-2}$, that is, $c = 5e^2$, this the solution is

$$x(t) = 5e^2 e^{-2t}$$

or

$$x(t) = 5e^{2-2t}$$

Prob. 15. $\int \frac{dh}{2h+1} = \int ds$, so integrating both sides we have

$$
\begin{aligned}
\frac{\ln|2h+1|}{2} &= s + c_1 \\
\ln|2h+1| &= 2s + 2c_1 \\
|2h+1| &= e^{25+c_2} \\
2h+1 &= \pm e^{2s+c_2} \\
2h+2 &= c_3 e^{25}
\end{aligned}
$$

$$2h = -1 + c_3 e^{25}$$

$$h = -\tfrac{1}{2} + ce^{2s}$$

with the initial condition $h(0) = 4$, we have:

$$4 = -\frac{1}{2} + c \quad \textbf{or} \quad c = \frac{9}{2}$$

and the final solution is:

$$h = -\frac{1}{2} + \frac{9}{2}ce^{25}$$

Prob. 17. $\int \frac{dN}{0.3N} = \int dt$, so integrating both sides

$$\frac{\ln|0.3N|}{0.3} = t + c_1$$

$$|0.3N| = e^{0.3t + c_2}$$

$$0.3N = \pm e^{0.3t + c_2}$$

$$N(t) = ce^{0.3t}$$

with $N(0) = 20$, the final solution is $c = 20$ and

$$N(t) = 20e^{0.3t}$$

so the population at time $= 5$ is

$$N(5) = 20 \cdot e^{0.3 \cdot 5}$$

$$= 20 \cdot e^{1.5}$$

$$\simeq 89 \text{ individuals.}$$

Prob. 19.

(a) $\int \frac{dN}{N} = rdt$, so integrating both sides we have

$$\ln|N| = rt + c_1$$

$$|N| = e^{rt + c_1}$$

$$N = \pm e^{rt + c_2}$$

$$N(t) = ce^{rt}$$

(b) Transforming the coordinates into a semilog, the equation changes to

$$\ln N(t) = \ln c + rt$$

so the constant r will be the slope of the live in the $\ln N(t) x t$ graph.

(c) Find the best possible line that fits the data on the semilog plot described above, and r will be the slope of the line.

Prob. 21. (a) $N^{-2}dN = \frac{1}{100}dt$ so integrating both sides

$$-\frac{1}{N} = \frac{t}{100} + c_1$$

so

$$N(t) = \frac{1}{C - \frac{t}{100}}$$

with $N(0) = 10$ we get $10 = \frac{1}{c}$, or $c = \frac{1}{10}$, and the final equation is

$$N(t) = \frac{1}{\frac{1}{10} - \frac{t}{100}}$$

or

$$N(t) = \frac{100}{10 - t}$$

(b)

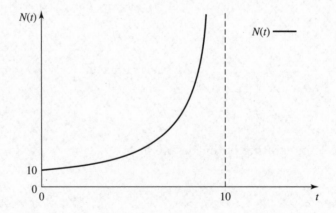

When $t \to 10$, $N(t)$ will approach $+\infty$, meanig that the population will grow without bounds.

Prob. 23.

(a) In this case $L_\infty = 123$, and the equation becomes

$$\frac{dh}{dt} = k(123 - L(t)) \quad \text{with} \quad L(0) = 1$$

and the solution according to Section 8.1.2 is

$$L(t) = 123 \left[1 - \left(1 - \frac{1}{123}\right)e^{-kt}\right]$$

now substituting $L(27) = \frac{123}{2} = 61.5$ we get

$$\frac{123}{2} = 123\left[1 - \left(1 - \frac{1}{123}\right)e^{-27k}\right]$$

$$\frac{1}{2} = \left(1 - \frac{1}{123}\right)e^{-27k}$$

$$\frac{123}{2 \times 122} = e^{-27k}$$

and taking ln on both sides we have

$$\ln\frac{2 \times 122}{123} = 27k$$

$$k \simeq 0.025$$

(b)

$$L(10) = 123\left[1\left(1 - \frac{1}{123}\right)e^{-0.025 \times 10}\right]$$

$$= 123\left[1 - \frac{122}{123} \cdot e^{-0.25}\right] \qquad \simeq 28 \text{ inches}$$

(c) 90% of the asymptotic length will be

$$\frac{9}{10}123 = 123\left[1\left(1 - \frac{1}{123}\right)e^{-0.025t}\right]$$

$$\frac{1}{10} = \left(1 - \frac{1}{123}\right)e^{-.025t}$$

$$\frac{123}{1220} = e^{-0.25t}$$

$$\ln\frac{1220}{123} = 0.025t$$

$$t \simeq 92 \text{ months.}$$

Prob. 25. We will use below the fact that

$$\frac{1}{(y-a)(y-b)} = \frac{1}{a-b}\left[\frac{1}{y-a} - \frac{1}{y-b}\right], \quad a \neq b$$

so the equation becomes

$$\int \frac{dy}{y(1+y)} = \int dx$$

on the equation $yy + 1 = ce^x$ we can easily see that $c = \frac{2}{3}$, so the final solution is

$$y = \frac{(2.3)e^x}{1-(2/3)e^x} = \frac{2e^x}{3-2e^x}$$

Prob. 27. $\frac{1}{y(y-5)} = \frac{1}{5}\left[\frac{1}{y} - \frac{1}{y-5}\right]$ so

$$\frac{1}{5}\int\left(\frac{1}{y} - \frac{1}{y-5}\right)dy = \int dx$$

$$\ln|y| - \ln|y-5| = 5x + c_1$$

$$\ln\left|\frac{y}{y-5}\right| - 5x + c_1$$

$$\frac{y}{y-5} = ce^{5x}$$

with the initial condition we can see that $c = -\frac{1}{4}$ and the solution can be written as

$$\frac{y}{y-5} = -\frac{1}{4}e^{5x}$$

cross multiplying and solving for y in terms of x, we have

$$y = (y-5)(-\frac{1}{4})e^{5x}$$

$$4y = (5-y)e^{5x}$$

$$4y + ye^{5x} = 5e^{5x}$$

$$y(4 + e^{5x}) = 5e^{5x}$$

and finally the solution is

$$y = \frac{5e^{5x}}{4 + e^{5x}}$$

Prob. 29.

$$\frac{1}{2y(3-y)} = \frac{1}{2}\left(\frac{1}{3}\left(\frac{1}{y-3} - \frac{1}{y}\right)\right) = \frac{1}{6}\left(\frac{1}{y-3} - \frac{1}{y}\right)$$

Integrating both sides:

$$\frac{1}{6}\int\left(\frac{1}{y-3} - \frac{1}{y}\right)dy = \int dx$$

$$\ln|y-3| - \ln|y| = 6x + C_1$$

$$\frac{y-3}{y} = e^{6x+C_1}$$

Using initial conditions:

$$\frac{5-3}{5} = Ce^6, \quad C = \frac{2}{5}e^{-6},$$

hence, $y(x) = \frac{3}{1-\frac{2}{5}e^{-6(x-1)}}$.

Prob. 31. $\frac{dy}{y(1+y)} = dx$, so using the fact that

$$\frac{1}{y(y+1)} = \frac{1}{y} - \frac{1}{y+1}$$

we get

$$\int\left(\frac{1}{y} - \frac{1}{y+1}\right) = \int dx,$$

so integrating both sides

$$\ln|y| - \ln|y+1| = x + c_1$$

$$\ln\left|\frac{y}{y+1}\right| = x + c_1$$

$$\frac{y}{y+1} = ce^x$$

solving for y in terms of x, we get

$$y = \frac{ce^x}{1-ce^x}$$

Prob. 33. $\frac{dy}{(1+y)^3} = dx$, is integrating both sides we get

$$\int\frac{dy}{(1+y)^3} = \int dx$$

$$\frac{1}{-2(1+y)^2} = x + c_1$$

$$(1+y)^2 = -\frac{1}{x+c_1}$$

$$1+y = \pm\sqrt{1\frac{1}{x+c}} \quad \text{so}$$

$$y = -1 \pm \sqrt{-\frac{1}{x+c}}$$

Prob. 35.

(a)

$$\int \frac{du}{u^2 - a^2} = \int \frac{du}{(u+a)(u-a)}$$

$$= \frac{1}{2a} \int \left(\frac{1}{u-a} - \frac{1}{u+a} \right) du$$

$$= \frac{1}{2a} (\ln |u - a| - \ln |u + a|)$$

$$= \frac{1}{2a} \ln \left| \frac{u-a}{u+a} \right| + c$$

(b) The differential equation can be changed to $\frac{dy}{y^2-4} = dx$ and integrating both sides we get $\int \frac{dy}{y^2-4} = \int dx$ and using the integral we have

$$\frac{1}{4} \ln \left| \frac{y-2}{y+2} \right| = x + c_1$$

$$\ln \left| \frac{y-2}{y+2} \right| = 4x + c_2$$

$$\left| \frac{y-2}{y+2} \right| = ce^{4x}.$$

So for the first point $(0,0)$, we can calculate $c = -1$; for $(0,2)$ we get $c = 0$ and for $(0,4)$ we get $c = \frac{1}{3}$.

Now cross multiplying and solving for y in terms of x we obtain

$$y - 2 = ce^{4x}(y + 2)$$

$$y = \frac{2(ce^{4x}+1)}{1-ce^{4x}}$$

so the different solutions are

(i) $(0,0) \rightarrow c = -1 \rightarrow y = \frac{2(1-e^{4x})}{1+e^{4x}}$

(ii) $(0,2) \rightarrow c = 0 \rightarrow y = 2$

(iii) $(0,4) \rightarrow c = \frac{1}{3} \rightarrow y = 2\frac{3+e^{4x}}{3-e^{4x}}$

Prob. 37. This is the Logistic Equation (8.27) with $r = 0.34$ and $k = 200$, so the solution is

$$N(t) = \frac{200}{1 + \left(\frac{200}{50} - 1 \right) e^{-0.34t}} = \frac{200}{1 + 3e^{-0.34t}}$$

and

$$\lim_{t \to \infty} N(t) = k = 200.$$

Prob. 39.

(a) The solution is given by

$$N(t) = \frac{k}{1 + \left(\frac{k}{N_0} - 1\right) e^{-rt}}$$

so for $k = 50$, $r = 1.5$ as given we have

$$N(t) = \frac{50}{1 + \left(\frac{50}{10} - 1\right) e^{-1.5t}} = \frac{50}{1 + 4e^{-1.5t}}$$

(b)

$$N(t) = \frac{50}{1 + \left(\frac{50}{90} - 1\right) e^{-1.5t}} = \frac{50}{1 - \frac{4}{9}e^{-1.5t}}$$

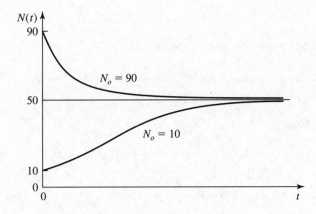

(c)

Prob. 41.

(a) If $r = 5$ and $k = 30$, the equation will be given by

$$\frac{dN}{dt} = 5N \left(1 - \frac{N}{30}\right)$$

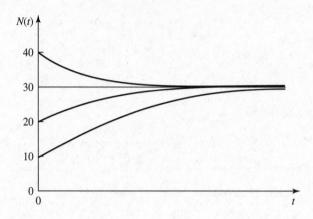

(b)

Prob. 43.

(a) From the partial fraction expansion we can see that

$$\frac{1}{p(1-p)} = \frac{1}{p} - \frac{1}{p-1}$$

and using it to separate the variables we have

$$\frac{dp}{p(1-p)} = \frac{s}{2}dt$$

integrating both sides

$$\int \left(\frac{1}{p} - \frac{1}{p-1} \right) dp = \int \frac{s}{2}\, dt$$

$$\ln|p| - \ln|p-1| = \frac{s}{2}t + c_1$$

$$\ln \left| \frac{p}{p-1} \right| = \frac{s}{2}t + c_1$$

$$\frac{p}{p-1} = ce^{\frac{s}{2}t}$$

and substituting the initial condition we can find the value of $c = \frac{p_0}{p_0-1}$, so the solution, can now be computed as

$$p = (p-1)ce^{\frac{s}{2}t}$$

$$p(1 - ce^{-\frac{s}{2}t}) = -ce^{\frac{s}{2}t}$$

so

$$p = \frac{ce^{\frac{st}{2}}}{ce^{\frac{st}{2}} - 1} = \frac{p_0 e^{\frac{st}{2}}}{p_o e^{\frac{st}{2}} + 1 - p_0}$$

(b) For these values the solution becomes

$$p = \frac{0.1e^{0.005t}}{0.1e^{0.005t} + 0.9} = \frac{e^{0.005t}}{e^{0.005t} + 9}$$

and the time can the be solved by

$$0.5 = \frac{e^{0.005t}}{e^{0.005t} + 9}$$

$$0.5e^{0.005t} = 4.5$$

$$e^{0.005t} = 9$$

and applying ln to both sides we have

$$0.005t \simeq 2.1972$$

so $t = 439$ units of time

(c)

$$\lim_{t \to \infty} p(t) = \lim_{t \to \infty} \frac{p_0 e^{\frac{st}{2}}}{p_0 e^{\frac{st}{2}} + 1 - p_0} = 1$$

This means that the frequency of the A_1 will become closer to 1 as times goes by, that is, eventually the population will consist of only $A_1 A_1$ types.

Prob. 45. Separating the variable we have $ydy = (x+1)dx$, so integrating both sides:

$$\frac{y^2}{2} = \frac{(x+1)^2}{2} + c$$

or

$$y^2 = (x+1)^2 + c$$

for our initial conditions ($y_0 = 2$ ir $x_0 = 0$) we find $c = 4 - 1 = 3$, so the final solution is

$$y = \pm\sqrt{x^2 + 2x + 4},$$

and then the initial condition again we see that the branch in case is positive one, that is,

$$y = \sqrt{x^2 + 2x + 4}$$

Prob. 47. Separating the variables we have $\frac{dy}{y+1} = e^{-x}dx$, and integrating both sides

$$\ln|y+1| = -c^{-x} + c_1$$

now substituting the initial condition we find

$$c_1 = 1 + \ln 3$$

and substituting back into the equation

$$\ln|y+1| = 1 + \ln 3 - e^{-x}$$

so

$$|y+1| = \exp[1 + \ln 3 - e^{-x}]$$

giving us two branches $\pm$, but because of the initial condition we are interested in the positive only, that is

$$y \qquad = -1 + \exp[1 + \ln 3 - e^{-x}]$$

$$= -1 + \exp(1) \cdot \exp(\ln 3) \cdot \exp(-e^{-x})$$

$$= -1 + 3e \cdot \exp(-e^{-x})$$

$$= -1 + 3\exp[1 - e^{-x}]$$

Prob. 49. Separating the variables we have $\frac{dy}{y+1} = \frac{dx}{x-1}$, and integrating both sides

$$\ln|y+1| = \ln|x-1| + c_1$$

$$\ln\left|\frac{y+1}{x-1}\right| = c_1$$

$$\frac{y+1}{x-1} = c$$

using the initial condition we see that $c = 6$ and the final solution is $y = 6x - 7$.

Prob. 51. Separating the variables we have $\frac{dr}{r} = e^{-t}dt$ and integrating both sides

$$\ln|r| = -c^{-t} + c_1$$

$$|r| = \exp(-e^{-t} + c_1)$$

so

$$r = c\exp(-e^{-t})$$

from the initial conditions we see that $c = 1$, and

$$r = \exp(1 - e^{-t})$$

Prob. 53. The line that relate the two quantities is given by

$$\ln O_2 = 0.8\ln m + c$$

and derivating both sides

$$\frac{dO_2}{2} = 0.8\frac{dm}{m}$$

or the final equation

$$\frac{dO_2}{dm} = 0.8\frac{O_2}{m}$$

Prob. 55. The relation among the logarithm of the two quantities is

$$\ln P = \frac{1}{7.7}\ln p + k$$

where P is the amount of phosphorous in the Daphusia, and p in the algal food

$$\frac{dP/dp}{P} = \frac{1}{7.7} \cdot \frac{1}{p},$$

so

$$\frac{dP}{P} = \frac{1}{7.7}\frac{dp}{p}$$

Prob. 57. Separation of variables and integration gives:

$$\ln|N| = 2t - \frac{1}{\pi}\cos(2\pi t) + C_1$$

$$N = \pm Ce^{2t - 1/\pi \cos(2\pi t)}$$

with initial conditions: $t = \pm Ce^{-1/\pi}$, so

$$N(t) = 5\exp(\frac{1}{\pi} + 2t - \frac{1}{\pi}\cos(2\pi t)).$$

8.2   Equilibria and Their Stability

Prob. 1.

(a) $y = 0, 2$

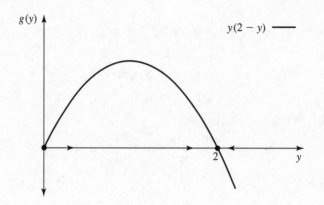

(b) $y = 0$ is unstable; $y = 2$ is locally stable.

(c) Eigenvalue associated with $y = 0$ is $2 > 0$, hence $y = 0$ is unstable; eigenvalue associated with $y = 2$ is $-2 < 0$, hence $y = 2$ is locally stable.

Prob. 3. (a) $y = 0, 1, 2$

(b)

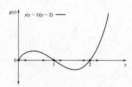

$y = 0$ and $y = 2$ are unstable; $y = 1$ is locally stable.

(c) Eigenvalue associated with $y = 0$ is $2 > 0$, hence $y = 0$ is unstable; eigenvalue associated with $y = 1$ is $-1 < 0$, hence $y = 1$ is locally stable; eigenvalue associated with $y = 2$ is $2 > 0$, hence $y = 2$ is unstable.

Prob. 5. (a) $\frac{dN}{dt} = 1.5N\left(1 - \frac{N}{100}\right)$

(b)

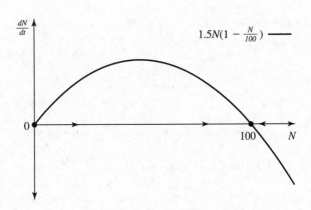

$N = 0$ is unstable; $N = 100$ is stable

(c) Eigenvalue associated with $N = 0$ is $1.5 > 0$, hence $N = 0$ is unstable; eigenvalue associated with $N = 100$ is $-1.5 < 0$, hence $N = 100$ is locally stable.

Prob. 7. (a) $K = 2000$ (b) $t = \frac{1}{2}\ln 199 \approx 2.65$ (c) 2000

Prob. 9. (a) $N \approx 52.79$ is unstable; $N \approx 947.21$ is locally stable

(b) The maximal harvesting rate is $rK/4$.

Prob. 11.

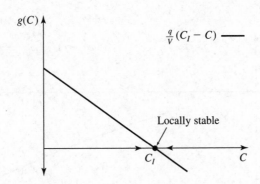

The equilibrium C_I is locally stable.

Prob. 13. (a) $\frac{dC}{dt} = \frac{0.2}{400}(3 - C)$

(b) $C(t) = 3 - 3e^{-t/2000}$, $t \geq 0$; $\lim_{t \to \infty} C(t) = 3$

(c) $C = 3$ is locally stable.

Prob. 15. (a) Equilibrium concentration: $C_I = 254$

(b) $T_R = \frac{1}{0.37} \approx 2.703$

(c) $T_R = \frac{1}{0.37} \approx 2.703$

(d) They are the same.

Prob. 17. We know from the solution of the genaral equantion that

$$C(t) = C_I \left[1 - \left(1 - \frac{C_0}{C_I} \right) e^{-(q/V)t} \right]$$

so multiplying it out, we have:

$$C(t) - C_I = -(C_I - C_0)\, e^{-(q/V)t}$$

which reduces to

$$\frac{C(t) - C_I}{C_0 - C_I} = e^{-(q/V)t}$$

now integrating both sides from 0 to ∞,

$$\int_0^\infty \frac{C(t) - C_I}{C_0 - C_I} = \int_0^\infty e^{-(q/V)t} = -\frac{V}{q} e^{-(q/v)t} \Big|_0^\infty = -\frac{V}{q}(0 - 1) = \frac{V}{q} = T_R$$

Prob. 19. $T_R = \frac{12.3 \times 10^9}{220}$ seconds ≈ 647.1 days; $C(T_R) \approx 0.806 \frac{\text{mg}}{\text{l}}$

Prob. 21. (a)

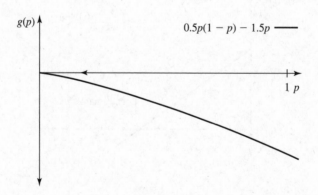

(b) $p = 0$ is locally stable

(c) $g'(0) = -1 < 0$, which implies that 0 is locally stable.

Prob. 23. (a) $\frac{dp}{dt}$ describes the rate of change of $p(t)$; $cp(1 - p - D)$ describes colonization of vacant undestroyed patches; $-mp$ describes extinction.

(b)

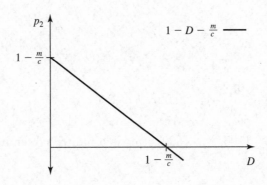

(c) $D < 1 - \frac{m}{c}$; $p_1 = 0$ is unstable; $p_2 = 1 - D - \frac{m}{c}$ is locally stable

Prob. 25. (a) $N = 0$, $N = 17$, and $N = 200$

(b) $N = 0$ is locally stable; $N = 17$ is unstable; $N = 200$ is locally stable

(c)

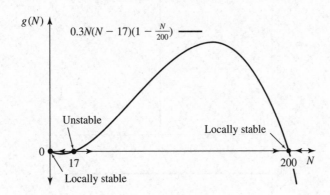

8.3   Systems of Autonomous Equations

Prob. 1. (a) $R_0 = 1.5 > 1$, the disease will spread

(b) $R_0 = \frac{1}{2} < 1$, the disease will not spread

Prob. 3. $R_0 = 0.9999 < 1$, the disease will not spread

Prob. 5. (a)

$$\frac{dN}{dt} = N_I - 5N - 0.02NX + X$$

$$\frac{dX}{dt} = 0.02NX - 2X$$

(b) equilibrium: $(\hat{N}, \hat{X}) = (100, N_I - 500)$, this is a nontrivial equilibrium provided $N_I > 500$

Prob. 7. (a)

$$\frac{dN}{dt} = 200 - N - 0.01NX + 2X$$

$$\frac{dX}{dt} = 0.01NX - 3X$$

(b)

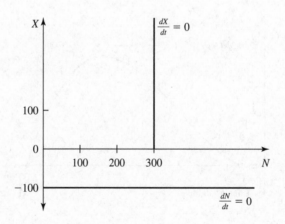

(c) No nontrivial equilibria

Prob. 9. (a) Equilibria: $(0,0)$, $(0,2/3)$, $(1/2,0)$

(b) Since $\frac{dp_2}{dt} < 0$ when $p_1 = 1/2$ and p_2 is small, species 2 cannot invade.

Prob. 11. (a)

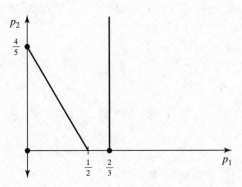

(b) equilibria: $(0,0)$, $(2/3,0)$, $(0,4/5)$

8.5 Review Problems

Prob. 1. (a) $\frac{dT}{dt}$ is proportional to the difference between the temperature of the object and the temperature of the surrounding medium.

(b) $t = \frac{1}{0.013} \ln \frac{9}{4} \approx 62.38$ minutes

Prob. 3. (a) $N(t) = N(0)e^{r_e t}$

(b) $N(t) = \dfrac{K}{1 + \left(\frac{K}{N_0} - 1\right)e^{-r_l t}}$

(c) $r_e \approx 0.691$; $K = 1001$: $r_l \approx 1.382$; $K = 10,000$: $r_l \approx 0.701$

Prob. 5. (a)

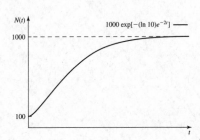

(b) We have the following:

$$\lim_{N \to 0} N \ln N = \lim_{N \to 0} \frac{N}{\frac{1}{\ln N}},$$

by L'Hospital rule we see that it is indeed zero. Hence,

$$\lim_{N\to 0} \frac{dN}{dt} = \lim_{N\to 0} (KNlnK - KNlnN) = 0.$$

(c)

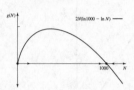

$N = 0$ is unstable; $N = 1000$ is locally stable

K is the carrying capacity

Prob. 7. (a) For 8.94, the rate of change of the number of bacteria present is calculated by taking the difference in rate of growth and the input/output flow rate.

(b) For 8.95, the rate of nutrients supply is calculated by subtracting current nutrient level rate and the growth rate from the initial nutrient level rate.

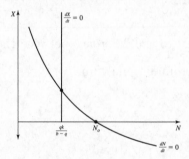

(c) $\hat{N} = \frac{qk}{b-q} > 0$ if $b > q$, $\hat{X} = \left(k + \frac{qk}{b-q}\right)\frac{q}{b}\left(\frac{N_0(b-q)}{qk} - 1\right)$

Chapter 9

Linear Algebra and Analytic Geometry

9.1 Linear Systems

Prob. 1.

$$x - y = 1 \qquad (9.1.1)$$

$$x - 2y = -2 \qquad (9.1.2)$$

Subtracting (2) from (1) i.e., (1) - (2)

$$y = 3 \qquad (9.1.3)$$

so,

$$x = 4 \qquad (9.1.4)$$

for the eqn (1)   for the eqn (2)

at y = 1, x = 2   at y = 1, x = 0

at y = 2, x = 3   at y = 2, x = 2

at y = 3, x = 4   at y = 3, x = 4

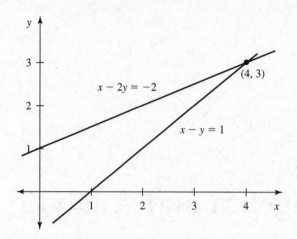

If we draw the graph for the above points, we can get the intersection point of these two lines.

Prob. 3.

$$x - 3y \;=\; 6 \tag{9.1.5}$$

$$y \;=\; 3 + \frac{1}{3} \tag{9.1.6}$$

Multiplying the eqn (9.1.6) by 3 & rearranging

$$-x + 3y \;=\; 9$$

$$\text{\textit{That is,}} \; x - 3y \;=\; -9 \tag{9.1.7}$$

So eqn (9.1.5) and eqn (9.1.7) are contradictory to themselves. So there will not be any solution for this system.

If we give values for both the equations

for the eqn (9.1.5)   for the eqn (9.1.7)

at x = 3, y = -1   at x = 3, y = 4

at x = 6, y = 0   at x = 6, y = 5

at x = 9, y = 1   at y = 9, x = 6

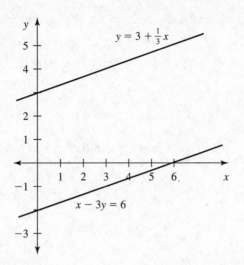

If we draw the graph for the above points, we can see that the lines will not intersect at any point.

Prob. 5.

$$2x - 3y = 5 \qquad (9.1.8)$$

$$4x - 6y = c \qquad (9.1.9)$$

(a) There will be an infinite number of solutions when the lines are identical so, divide the eqn (9.1.9) by 2

$$2x - 6y = \frac{c}{2} \Rightarrow \frac{c}{2} = 5 \Rightarrow c = 10$$

(b) No solution is only when both lines are parallel and disjoint i.e., $c \neq 10$

(c) There is no possibility for that as the slopes are equal.

Prob. 7. With the equation given by:

$$a_{11}x_1 + a_{12}x_2 = b_1$$

$$a_{21}x_1 + a_{22}x_2 = b_2$$

we can find the equation of the lines defined by it in the form $y = mx + b$, and we obtain:

$$x_2 = \frac{1}{a_{12}}(b_1 - a_{11}x_1) \qquad \text{and} \qquad x_2 = \frac{1}{a_{22}}(b_2 - a_{21}x_1)$$

So equating the x_2 in order to find the intersection of the line we get:

$$\frac{1}{a_{12}}(b_1 - a_{11}x_1) = \frac{1}{a_{22}}(b_2 - a_{21}x_1)$$

and solving for x_1 we get:

$$a_{22}(b_1 - a_{11}x_1) = a_{12}(b_2 - a_{21}x_1)$$

$$a_{22}b_1 - a_{22}a_{11}x_1 = a_{12}b_2 - a_{12}a_{21}x_1$$

$$(a_{22}a_{11} - a_{12}a_{21})x_1 = a_{22}b_1 - a_{12}b_2$$

from where the first coordinate follows:

$$x_1 = \frac{a_{22}b_1 - a_{12}b_2}{(a_{22}a_{11} - a_{12}a_{21})}.$$

the coordinate x_2 follows from straight substitution of this result in the equation of either lines.

Prob. 9. Subtract $\frac{1}{2}$ Row 1 to get:

$$2x - y = 3$$

$$-2.5y = 5.5$$

Hence, $y = -\frac{11}{5}$, $x = \frac{2}{5}$.

Prob. 11. Subtract $\frac{3}{7}$ Row 1 to get:

$$7x - y = 4$$

$$\frac{17}{7}y = -\frac{5}{7}$$

Hence $y = -\frac{5}{17}$, $x = \frac{9}{17}$.

Prob. 13. Add Row 1 to Row 2 to get:

$$3x - y = 1$$

$$0 = 5$$

No solution.

Prob. 15. Re-write the system as:

$$x + 2y = 3$$

$$2x + 4y = 6$$

Subtract 2 Row 1 from Row 2 to get:

$$x + 2y = 3$$

$$0 = 0$$

Intinitely many solutions. $x = t$, $y = \frac{3}{2} - \frac{1}{2}t$, $t \in \mathbb{R}$.

Prob. 17. Solving by augmented matrix method

$$Aug.\ Matrix = \begin{bmatrix} 1 & 1 & 0 & | & 3 \\ 0 & -1 & 1 & | & -1 \\ 1 & 0 & 1 & | & 2 \end{bmatrix}$$

$$R_3 = R_3 - R_1 = \begin{bmatrix} 1 & 1 & 0 & | & 3 \\ 0 & -1 & 1 & | & -1 \\ 1 & 1 & 0 & | & 3 \end{bmatrix}$$

The first and last row have the same elements. So it has infinite number of solutions.

Prob. 19. This system is underdetermined because it has 3 unknowns and 2 equations.

Solving by augmented matrix method

$$Aug.\ Matrix = \begin{bmatrix} 1 & -2 & 1 & | & 3 \\ 2 & -3 & 1 & | & 8 \end{bmatrix}$$

$$R_2 = R_2 - 2R_1 = \begin{bmatrix} 1 & -2 & 1 & | & 3 \\ 0 & 1 & -1 & | & 2 \end{bmatrix}$$

$$y - z = 2 \tag{9.1.10}$$

$$x - 2y + z = 3 \tag{9.1.11}$$

So,

$$y = 2 + z \tag{9.1.12}$$

Substituting (9.1.25) in the eqn. (9.1.26),

$$x - 2(2 + z) + z = 3$$

$$x - 4 - z = 3$$

$$x = 7 + z \tag{9.1.13}$$

$$If \; z = t$$

$$x = 7 + t$$

$$y = 2 + t$$

$$z = t$$

Prob. 21. Solving by augmented matrix method

$$Aug. \; Matrix = \begin{bmatrix} 2 & -1 & | & 3 \\ 1 & -1 & | & 4 \\ 3 & -1 & | & 1 \end{bmatrix}$$

$$R_3 = R_3 + R_2 = \begin{bmatrix} 2 & -1 & | & 3 \\ 1 & -1 & | & 4 \\ 2 & 0 & | & -3 \end{bmatrix}$$

$$R_2 = R_1 - R_2 = \begin{bmatrix} 2 & -1 & | & 3 \\ 1 & 0 & | & -1 \\ 2 & 0 & | & -3 \end{bmatrix}$$

From the above matrix we are getting the equations like

$$2x = -3 \tag{9.1.14}$$

$$x = -1 \tag{9.1.15}$$

Equations (9.1.27) & (9.1.28) are giving contradictory solutions

So this is an inconsistent system.

Prob. 23. The composition of three different types of fertilizers are written in the equation form

$$SL1 : 24N + 4P + 8K$$

$$SL2 : 21N + 7P + 12K$$

$$SL1 : 17N + 0P + 0K$$

N for Nitrogen, P for Phosphate & K for Potassium

As given,

$$x(24N + 4P + 8K) + y(21N + 7P + 12K)$$

$$+z(17N + 0P + 0K) = 500N + 100P + 180K$$

So separating N, P & K

$$N(24x + 21y + 17z) = 500N$$
$$P(4x + 7y + 0z) = 100P$$
$$K(8x + 12y + 0z) = 180K$$

If we write the above equations in the matrix form

$$\begin{pmatrix} 24 & 21 & 17 \\ 4 & 7 & 0 \\ 8 & 12 & 0 \end{pmatrix} \begin{pmatrix} x \\ y \\ z \end{pmatrix} = \begin{pmatrix} 500 \\ 100 \\ 180 \end{pmatrix}$$

Solving by augmented matrix method

$$Aug.\ Matrix = \begin{pmatrix} 24 & 21 & 17 & | & 500 \\ 4 & 7 & 0 & | & 100 \\ 8 & 12 & 0 & | & 180 \end{pmatrix}$$

$$R_3 = 2R_2 - R_3 \quad = \quad \begin{pmatrix} 24 & 21 & 17 & | & 500 \\ 4 & 7 & 0 & | & 100 \\ 0 & 2 & 0 & | & 20 \end{pmatrix}$$

So,

$$2y \;=\; 20$$

$$y \;=\; 10 \tag{9.1.16}$$

$$4x + 7y \;=\; 100 \tag{9.1.17}$$

Substituting (9.4.2) in the above eqn,

$$4x + 70 \;=\; 100$$

$$4x \;=\; 30$$

$$x \;=\; \frac{15}{2} \tag{9.1.18}$$

Substituting (9.4.2) & (9.4.2) in the eqn,

$$24x + 21y + 17Z \;=\; 500$$

$$180 + 210 + 17z \;=\; 500$$

$$So, \tag{9.1.19}$$

$$z = \frac{110}{17} \tag{9.1.20}$$

So, $\frac{15}{2}$ amount of SL1, 10 amount of SL2 and $\frac{110}{17}$ amount of SL3 should be applied.

Prob. 25. Solving by augmented matrix method

$$Aug.\ Matrix \quad = \quad \begin{bmatrix} -1 & -2 & 3 & | & -9 \\ 2 & 1 & -1 & | & 5 \\ 4 & -3 & 5 & | & -9 \end{bmatrix}$$

$$R_3 = R_3 - 2R_2 \quad = \quad \begin{bmatrix} -1 & -2 & 3 & | & -9 \\ 2 & 1 & -1 & | & 5 \\ 0 & -5 & 7 & | & -19 \end{bmatrix}$$

$$R_2 = R_2 + 2R_1 \quad = \quad \begin{bmatrix} -1 & -2 & 3 & | & -9 \\ 0 & -3 & 5 & | & -13 \\ 0 & -5 & 7 & | & -19 \end{bmatrix}$$

$$R_3 = 5R_2 - 3R_3 \ = \ \begin{bmatrix} 1 & -2 & 3 & | & -9 \\ 0 & -3 & 5 & | & -13 \\ 0 & 0 & 4 & | & -8 \end{bmatrix}$$

$$-4z \ = \ -8$$

So,

$$z \ = \ -2 \qquad\qquad (9.1.21)$$

$$-3y + 5z \ = \ -13$$

Substituting (9.1.21),

$$y \ = \ 1 \qquad\qquad (9.1.22)$$

$$-x - 2y + 3z \ = \ -9$$

Substituting (9.1.21) & (9.1.22) change ever occurrence to $x = 1$

$$x \ = \ 2$$

So,

$$x \ = \ 2$$

$$y \ = \ 1$$

$$z \ = \ -2$$

Prob. 27. Solving by augmented matrix method

$$Aug. \ Matrix \ = \ \begin{bmatrix} 1 & 1 & 0 & | & 3 \\ 0 & -1 & 1 & | & -1 \\ 1 & 0 & 1 & | & 2 \end{bmatrix}$$

$$R_3 = R_3 - R_1 \ = \ \begin{bmatrix} 1 & 1 & 0 & | & 3 \\ 0 & -1 & 1 & | & -1 \\ 1 & 1 & 0 & | & 3 \end{bmatrix}$$

The first and last row have the same elements. So it has infinite number of solutions.

Prob. 29. This system is under determined because it has 3 unknowns and 2 equations.

Solving by augmented matrix method

$$Aug.\ Matrix \ = \ \begin{bmatrix} 1 & -2 & 1 & | & 3 \\ 2 & -3 & 1 & | & 8 \end{bmatrix}$$

$$R_2 = R_2 - 2R_1 \ = \ \begin{bmatrix} 1 & -2 & 1 & | & 3 \\ 0 & 1 & -1 & | & 2 \end{bmatrix}$$

$$y - z \ = \ 2 \tag{9.1.23}$$

$$x - 2y + z \ = \ 3 \tag{9.1.24}$$

$$So,$$

$$y \ = \ 2 + z \tag{9.1.25}$$

Substituting (9.1.25) in the eqn. (9.1.26),

$$x - 2(2 + z) + z \ = \ 3$$

$$x - 4 - z \ = \ 3$$

$$x \ = \ 7 + z \tag{9.1.26}$$

$$If\ z = t$$

$$x \ = \ 7 + t$$

$$y \ = \ 2 + t$$

$$z \ = \ t$$

Prob. 31. Solving by augmented matrix method

$$Aug.\ Matrix \ = \ \begin{bmatrix} 2 & -1 & | & 3 \\ 1 & -1 & | & 4 \\ 3 & -1 & | & 1 \end{bmatrix}$$

$$R_3 = R_3 + R_2 \ = \ \begin{bmatrix} 2 & -1 & | & 3 \\ 1 & -1 & | & 4 \\ 2 & 0 & | & -3 \end{bmatrix}$$

$$R_2 = R_1 - R_2 = \begin{bmatrix} 2 & -1 & | & 3 \\ 1 & 0 & | & -1 \\ 2 & 0 & | & -3 \end{bmatrix}$$

From the above matrix we are getting the equations like

$$2x = -3 \tag{9.1.27}$$

$$x = -1 \tag{9.1.28}$$

Equations (9.1.27) & (9.1.28) are giving contradictory solutions

So this is an inconsistent system.

Prob. 33. Under determined: 2 equations, 3 variables Row 1 - 2 Row 2:

$$x + y - 2z = 4$$

$$-9y - 4z = -6$$

Let $z = t$, then $y = \frac{4}{9}t - \frac{2}{3}$, $x = \frac{14t}{9} + \frac{14}{3}$, $t \in \mathbb{R}$.

Prob. 35. The composition of three different types of fertilizers are written in the equation form

$$SL1 : 24N + 4P + 8K$$

$$SL2 : 21N + 7P + 12K$$

$$SL1 : 17N + 0P + 0K$$

N for Nitrogen, P for Phosphate & K for Potassium

As given,

$$x(24N + 4P + 8K) + y(21N + 7P + 12K)$$

$$+z(17N + 0P + 0K) = 500N + 100P + 180K$$

So separating N, P & K

$$N(24x + 21y + 17z) = 500N$$

$$P(4x + 7y + 0z) = 100P$$

$$K(8x + 12y + 0z) = 180K$$

If we write the above equations in the matrix form

$$
\begin{pmatrix}
24 & 21 & 17 \\
4 & 7 & 0 \\
8 & 12 & 0
\end{pmatrix}
\begin{pmatrix}
x \\
y \\
z
\end{pmatrix}
=
\begin{pmatrix}
500 \\
100 \\
180
\end{pmatrix}
$$

Solving by augmented matrix method

$$
Aug.\ Matrix =
\left(
\begin{array}{ccc|c}
24 & 21 & 17 & 500 \\
4 & 7 & 0 & 100 \\
8 & 12 & 0 & 180
\end{array}
\right)
$$

$$
R_3 = 2R_2 - R_3 =
\left(
\begin{array}{ccc|c}
24 & 21 & 17 & 500 \\
4 & 7 & 0 & 100 \\
0 & 2 & 0 & 20
\end{array}
\right)
$$

So,

$$2y = 20$$

$$y = 10 \qquad (9.1.29)$$

$$4x + 7y = 100 \qquad (9.1.30)$$

Substituting (9.4.2) in the above eqn,

$$4x + 70 = 100$$

$$4x = 30$$

$$x = \frac{15}{2} \qquad (9.1.31)$$

Substituting (9.4.2) & (9.4.2) in the eqn,

$$24x + 21y + 17Z = 500$$

$$180 + 210 + 17z = 500$$

$$So,$$ (9.1.32)

$$z = \frac{110}{17}$$ (9.1.33)

So, $\frac{15}{2}$ amount of SL1, 10 amount of SL2 and $\frac{110}{17}$ amount of SL3 should be applied.

9.2 Matrices

Prob. 1.

$$2C = \begin{bmatrix} 2 & -4 \\ 2 & -2 \end{bmatrix}$$

$$A - B + 2C = \begin{bmatrix} -1-0+2 & 2-1-4 \\ 0-2+2 & -3-4-2 \end{bmatrix}$$

$$= \begin{bmatrix} 1 & -3 \\ 0 & -9 \end{bmatrix}$$

Prob. 3.

$$A + B = 2A - B + D$$

$$D = A + B - 2A + B$$

$$= -A + 2B$$

To find D

$$2B = \begin{bmatrix} 0 & 2 \\ 4 & 8 \end{bmatrix}$$

$$-A + 2B = \begin{bmatrix} 1+0 & -2+2 \\ 0+4 & 3+8 \end{bmatrix}$$

$$= \begin{bmatrix} 1 & 0 \\ 4 & 11 \end{bmatrix}$$

Prob. 5. To Prove $(A + B) + C = A + (B + C)$

$$A + B = \begin{bmatrix} -1 & 3 \\ 2 & 1 \end{bmatrix}$$

$$(A + B) + C = \begin{bmatrix} -1 + 1 & 3 - 2 \\ 2 + 1 & 1 - 1 \end{bmatrix}$$

$$= \begin{bmatrix} 0 & 1 \\ 3 & 0 \end{bmatrix}$$

$$B + C = \begin{bmatrix} 0 + 1 & 1 - 2 \\ 2 + 1 & 4 - 1 \end{bmatrix}$$

$$= \begin{bmatrix} 1 & -1 \\ 3 & 3 \end{bmatrix}$$

$$A + (B + C) = \begin{bmatrix} -1 + 1 & 2 - 1 \\ 0 + 3 & -3 + 3 \end{bmatrix}$$

$$= \begin{bmatrix} -1 + 1 & 3 - 2 \\ 2 + 1 & 1 - 1 \end{bmatrix}$$

$$So, \ (A + B) + C = A + (B + C)$$

Prob. 7.

$$2A + 3B - C = \begin{bmatrix} 2 + 15 + 2 & 6 - 3 - 0 & -2 + 12 - 4 \\ 4 + 6 - 1 & 8 + 0 + 3 & 2 + 3 - 1 \\ 0 - 3 - 0 & -4 - 9 - 0 & 4 - 9 - 2 \end{bmatrix} = \begin{bmatrix} 19 & 3 & 6 \\ 9 & 11 & 4 \\ -3 & -13 & -7 \end{bmatrix}$$

Prob. 9.

$$D = -A - B - C$$

$$D = -[A + B + C]$$

$$A + B + C =$$

$$\begin{bmatrix} 1 + 5 - 2 & 3 - 1 + 0 & -1 + 4 + 4 \\ 2 + 2 + 1 & 4 + 0 - 3 & 1 + 1 + 1 \\ 0 + 1 + 0 & -2 - 3 + 0 & 2 - 3 + 2 \end{bmatrix}$$

$$A + B + C = \begin{bmatrix} 4 & 2 & 7 \\ 5 & 1 & 3 \\ 1 & -5 & 1 \end{bmatrix}$$

$$-(A + B + C) = \begin{bmatrix} -4 & -2 & -7 \\ -5 & -1 & -3 \\ -1 & 5 & -1 \end{bmatrix}$$

$$D = \begin{bmatrix} -4 & -2 & -7 \\ -5 & -1 & -3 \\ -1 & 5 & -1 \end{bmatrix}$$

Prob. 11.

$$A + B = \begin{bmatrix} 1+5 & 3-1 & -1+4 \\ 2+2 & 4+0 & 1+1 \\ 0+1 & -2-3 & 2-3 \end{bmatrix} = \begin{bmatrix} 6 & 2 & 3 \\ 4 & 4 & 2 \\ 1 & -5 & -1 \end{bmatrix}$$

$$B + A = \begin{bmatrix} 5+1 & -1+3 & 4-1 \\ 2+2 & 0+4 & 1+1 \\ 1+0 & -3-2 & -3+2 \end{bmatrix} = \begin{bmatrix} 6 & 2 & 3 \\ 4 & 4 & 2 \\ 1 & -5 & -1 \end{bmatrix}$$

So $A + B = B + A$

Prob. 13. Write A, B, C as generic $m \times n$ matrices, and use the rules of matrix addition and the fact that $a_{ij} + b_{ij} = c_{ij}$ implies $a_{ij} = c_{ij} - b_{ij}$, meaning $A + B = C$ implies $A = C - B$.

Prob. 15. Transpose of A is $[\mathbf{A}]^t = \begin{bmatrix} -1 & 2 \\ 0 & 1 \\ 3 & -4 \end{bmatrix}$

Prob. 17. Write A and B as generic $m \times n$ matrices with entries a_{ij} and b_{ij}, use the rules of transposition to show $(A + B)^T = A^T + B^T$.

Prob. 19. Write kA as a generic $m \times n$ matrix with entries kA_{ij}, use the rule of transposition to show $(kA)^T = kA^T$.

Prob. 21.

$$AB = \begin{bmatrix} -2+0 & -3+0 \\ 2-2 & 3+2 \end{bmatrix}$$

$$= \begin{bmatrix} -2 & -3 \\ 0 & 5 \end{bmatrix}$$

$$BA = \begin{bmatrix} -2+3 & 0+6 \\ 1+1 & 0+2 \end{bmatrix}$$

$$= \begin{bmatrix} 1 & 6 \\ 2 & 2 \end{bmatrix}$$

So, $AB \neq BA$

Prob. 23.

$$AC = \begin{bmatrix} -1 & 0 \\ 1 & 2 \end{bmatrix} \begin{bmatrix} 1 & 2 \\ 0 & -1 \end{bmatrix}$$

$$= \begin{bmatrix} -1+0 & -2+0 \\ 1+0 & 2-2 \end{bmatrix}$$

$$= \begin{bmatrix} -1 & -2 \\ 1 & 0 \end{bmatrix}$$

$$CA = \begin{bmatrix} 1 & 2 \\ 0 & -1 \end{bmatrix} \begin{bmatrix} -1 & 0 \\ 1 & 2 \end{bmatrix}$$

$$= \begin{bmatrix} -1+2 & 0+4 \\ 0-1 & 0-2 \end{bmatrix}$$

$$= \begin{bmatrix} 1 & 4 \\ -1 & -2 \end{bmatrix}$$

So, $AC \neq CA$

Prob. 25. To prove $(A+B)C = AC + BC$

From the previous problems

$$AC = \begin{bmatrix} -1 & -2 \\ 1 & 0 \end{bmatrix}$$

$$BC = \begin{bmatrix} 2 & 1 \\ -1 & -3 \end{bmatrix}$$

So $AC + BC$ is,

$$AC + BC = \begin{bmatrix} 1 & -1 \\ 0 & -3 \end{bmatrix}$$

To find (A+B)C

$$A + B = \begin{bmatrix} -1 & 0 \\ 1 & 2 \end{bmatrix} + \begin{bmatrix} 2 & 3 \\ -1 & 1 \end{bmatrix}$$

$$= \begin{bmatrix} 1 & 3 \\ 0 & 3 \end{bmatrix}$$

$$[A + B] + C = \begin{bmatrix} 1 & 3 \\ 0 & 3 \end{bmatrix} \begin{bmatrix} 1 & 2 \\ 0 & -1 \end{bmatrix}$$

$$= \begin{bmatrix} 1+0 & 2-3 \\ 0+0 & 0-3 \end{bmatrix}$$

$$= \begin{bmatrix} 1 & -1 \\ 0 & -3 \end{bmatrix}$$

So, (A+B)C = AC+BC

Prob. 27. A = 3 x 4 and B = 4 x 2 so AB = 3 x <u>4 4</u> x 2 = 3 x 2

Prob. 29.

(a) D'= 3 x 4 and BD': = 1 x <u>3 3</u> x 4, so BD' = 1 x 4

(b) D'A : = 3 x <u>4 4</u> X 3 and so, D'A = 3 x 3

(c) ACB: = 4 x <u>3 3</u> x 1 1 x 3 = 4 x <u>1 1</u> x 3 = 4 x 3

Prob. 31.

$$AB = \begin{bmatrix} 1+6 & 2+3 & 0+9 & -1+0 \\ 0-4 & 0-2 & 0-6 & 0+0 \end{bmatrix}$$

$$= \begin{bmatrix} 7 & 5 & 9 & -1 \\ -4 & -2 & -6 & 0 \end{bmatrix}$$

$$B' = \begin{bmatrix} 1 & 2 \\ 2 & 1 \\ 0 & 3 \\ -1 & 0 \end{bmatrix}$$

$$B'A = \begin{bmatrix} 1+0 & 3-4 \\ 2+0 & 6-2 \\ 0+0 & 0-6 \\ -1+0 & -3+0 \end{bmatrix}$$

$$= \begin{bmatrix} 1 & -1 \\ 2 & 4 \\ 0 & -6 \\ -1 & -3 \end{bmatrix}$$

Prob. 33.

$$A^2 = \begin{bmatrix} 2 & 1 \\ -1 & -3 \end{bmatrix} \begin{bmatrix} 2 & 1 \\ -1 & -3 \end{bmatrix} = \begin{bmatrix} 4-1 & 2-3 \\ -2+3 & -1+9 \end{bmatrix} = \begin{bmatrix} 3 & -1 \\ 1 & 8 \end{bmatrix}$$

$$A^3 = AA^2 = \begin{bmatrix} 2 & 1 \\ -1 & -3 \end{bmatrix} \begin{bmatrix} 2 & 1 \\ -1 & -3 \end{bmatrix} \begin{bmatrix} 2 & 1 \\ -1 & -3 \end{bmatrix} = \begin{bmatrix} 6+1 & -2+8 \\ -3+3 & 1-24 \end{bmatrix} = \begin{bmatrix} 7 & 6 \\ -6 & -23 \end{bmatrix}$$

$$A^4 = AA^3 = \begin{bmatrix} 2 & 1 \\ -1 & -3 \end{bmatrix} \begin{bmatrix} 7 & 6 \\ -6 & -23 \end{bmatrix} = \begin{bmatrix} 14-6 & 12-23 \\ -7+18 & -6+69 \end{bmatrix} = \begin{bmatrix} 7 & -11 \\ 11 & 63 \end{bmatrix}$$

Prob. 35. (a) $B^2 = \begin{bmatrix} 1 & 0 \\ 0 & 1 \end{bmatrix}$, $B^3 = \begin{bmatrix} 0 & 1 \\ 1 & 0 \end{bmatrix}$, $B^4 = \begin{bmatrix} 1 & 0 \\ 0 & 1 \end{bmatrix}$, $B^5 = \begin{bmatrix} 0 & 1 \\ 1 & 0 \end{bmatrix}$.

(b) When k is even, $B^k = \begin{bmatrix} 1 & 0 \\ 0 & 1 \end{bmatrix}$

When k is odd, $B^k = \begin{bmatrix} 0 & 1 \\ 1 & 0 \end{bmatrix}$

Prob. 37.

$$A = \begin{bmatrix} 1 & 3 \\ 0 & -2 \end{bmatrix}$$

$$I_2 = \begin{bmatrix} 1 & 0 \\ 0 & 1 \end{bmatrix}$$

$$AI_2 = \begin{bmatrix} 1+0 & 0+3 \\ 0+0 & 0-2 \end{bmatrix}$$

$$= \begin{bmatrix} 1 & 3 \\ 0 & -2 \end{bmatrix}$$

$$I_2 A = \begin{bmatrix} 1+0 & 3+0 \\ 0+0 & 0-2 \end{bmatrix}$$

So $AI_2 = I_2 A$

Prob. 39. Writing the equations in the matrix form

$$\begin{bmatrix} 2 & 3 & -1 \\ 0 & 2 & 1 \\ 1 & 0 & -2 \end{bmatrix} \begin{bmatrix} x_1 \\ x_2 \\ x_3 \end{bmatrix} = \begin{bmatrix} 0 \\ 1 \\ 2 \end{bmatrix}$$

Prob. 41. Writing the equations in the matrix form

$$\begin{bmatrix} 2 & -3 \\ -1 & 1 \\ 3 & 0 \end{bmatrix} \begin{bmatrix} x_1 \\ x_2 \end{bmatrix} = \begin{bmatrix} 4 \\ 3 \\ 4 \end{bmatrix}$$

Prob. 43. Inverse of matrix A is B :

$$A = \begin{bmatrix} 2 & 1 \\ 1 & 1 \end{bmatrix}$$

$$B = \begin{bmatrix} 1 & -1 \\ -1 & 2 \end{bmatrix}$$

i.e.,

$$AA^{-1} = I_2$$

Let us take $A^{-1} = B$

$$B = \begin{bmatrix} b_{11} & b_{12} \\ b_{21} & b_{22} \end{bmatrix}$$

So,

$$\begin{bmatrix} 2 & 1 \\ 1 & 1 \end{bmatrix} \begin{bmatrix} b_{11} & b_{12} \\ b_{21} & b_{22} \end{bmatrix} = \begin{bmatrix} 1 & 0 \\ 0 & 1 \end{bmatrix}$$

If we write the above matrix in the linear equations form

$$2b_{11} + b_{21} = 1 \qquad (9.2.1)$$

$$1b_{11} + b_{21} = 0 \qquad (9.2.2)$$

$$2b_{12} + b_{22} = 0 \qquad (9.2.3)$$

$$1b_{12} + b_{22} = 1 \qquad (9.2.4)$$

Solving eqn. (9.2.1) and eqn. (9.2.2)

eqn.9.2.1 - 2 x eqn. (9.2.2)

$$-b_{21} = 1$$

$$b_{21} = -1 \qquad (9.2.5)$$

$$(9.2.6)$$

Substituting the eqn (9.2.5) in eqn (9.2.1)

$$2b_{11} - 1 = 1$$

$$2b_{11} = 2 \; () \qquad (9.2.7)$$

Solving eqn. (9.2.3) and eqn. (9.2.4)

eqn. (9.2.3) - 2 x eqn. (9.2.4)

$$-b_{22} = -2$$

$$b_{22} = 2 \qquad (9.2.8)$$

Substituting the eqn (9.2.8) in eqn (9.2.3)

$$2b_{12} = -2$$

$$b_{12} = -1 \tag{9.2.9}$$

$$\begin{bmatrix} b_{11} & b_{12} \\ b_{21} & b_{22} \end{bmatrix} = \begin{bmatrix} 2 & -1 \\ -1 & 2 \end{bmatrix}$$

So, $B = (A)^{-1}$

$$= \begin{bmatrix} 2 & -1 \\ -1 & 2 \end{bmatrix}$$

Prob. 45.

$$A = \begin{bmatrix} -1 & 1 \\ 2 & 3 \end{bmatrix}$$

$$AA^{-1} = I_2$$

Let us take $A^{-1} = B$

$$B = \begin{bmatrix} b_{11} & b_{12} \\ b_{21} & b_{22} \end{bmatrix}$$

So,

$$\begin{bmatrix} -1 & 1 \\ 2 & 3 \end{bmatrix} \begin{bmatrix} b_{11} & b_{12} \\ b_{21} & b_{22} \end{bmatrix} = \begin{bmatrix} 1 & 0 \\ 0 & 1 \end{bmatrix}$$

If we write the above matrix in the linear equations form

$$-b_{11} + b_{21} = 1 \tag{9.2.10}$$

$$2b_{11} + 3b_{21} = 0 \tag{9.2.11}$$

$$-b_{12} + b_{22} = 0 \tag{9.2.12}$$

$$2b_{12} + 3b_{22} = 1 \tag{9.2.13}$$

Solving eqn. (9.2.10) and eqn. (9.2.11)

2 x eqn.9.2.10 - eqn. (9.2.11)

$$5b_{21} = 2$$

$$b_{21} = \frac{2}{5} \tag{9.2.14}$$

$$\tag{9.2.15}$$

From the equation (9.2.10)

$$
\begin{aligned}
b_{11} &= b_{21} - 1 \\
&= -\frac{3}{5}
\end{aligned}
\tag{9.2.16}
$$

Solving eqn. (9.2.12) and eqn. (9.2.13)

2 x eqn. (9.2.12) - eqn. (9.2.13)

$$
\begin{aligned}
5b_{22} &= 1 \\
b_{22} &= \frac{1}{5}
\end{aligned}
\tag{9.2.17}
$$

From the equation (9.2.12)

$$
b_{12} = b_{22}
$$

$$
b_{22} = b_{12} = \frac{1}{5}
\begin{bmatrix}
b_{11} & b_{12} \\
b_{21} & b_{22}
\end{bmatrix}
$$

$$
=
\begin{bmatrix}
-\frac{3}{5} & \frac{1}{5} \\
\frac{2}{5} & \frac{1}{5}
\end{bmatrix}
$$

So, $B = (A)^{-1}$

$$
=
\begin{bmatrix}
-\frac{3}{5} & \frac{1}{5} \\
\frac{2}{5} & \frac{1}{5}
\end{bmatrix}
$$

Prob. 47.

$$
A =
\begin{bmatrix}
-1 & 1 \\
2 & 3
\end{bmatrix}
\tag{9.2.18}
$$

From the problem no 37, we know that,

$$
A^{-1} =
\begin{bmatrix}
-\frac{3}{5} & \frac{1}{5} \\
\frac{2}{5} & \frac{1}{5}
\end{bmatrix}
So,
$$

$$
A^{-1}[A^{-1}]^{-1} = I_2
$$

Let us take $[A^{-1}]^{-1} = B$

$$B = \begin{bmatrix} b_{11} & b_{12} \\ b_{21} & b_{22} \end{bmatrix}$$

So,

$$\begin{bmatrix} -\frac{3}{5} & \frac{1}{5} \\ \frac{2}{5} & \frac{1}{5} \end{bmatrix} \begin{bmatrix} b_{11} & b_{12} \\ b_{21} & b_{22} \end{bmatrix} = \begin{bmatrix} 1 & 0 \\ 0 & 1 \end{bmatrix}$$

If we write the above matrix in the linear equations form

$$-\frac{3}{5}b_{11} + \frac{1}{5}b_{21} = 1 \tag{9.2.19}$$

$$\frac{2}{5}b_{11} + \frac{1}{5}b_{21} = 0 \tag{9.2.20}$$

$$-\frac{3}{5}b_{12} + \frac{1}{5}b_{22} = 0 \tag{9.2.21}$$

$$\frac{2}{5}b_{12} + \frac{1}{5}b_{22} = 1 \tag{9.2.22}$$

Solving eqn. (9.2.19) and eqn. (9.2.20)

eqn.9.2.19 - eqn. (9.2.20)

$$-\frac{5}{5}b_{11} = 1$$

$$b_{11} = -1 \tag{9.2.23}$$

$$\tag{9.2.24}$$

Substituting in the equation (9.2.19)

$$\frac{3}{5} + \frac{1}{5}b_{21} = 1$$

$$b_{21} = 2 \tag{9.2.25}$$

Solving eqn. (9.2.21) and eqn. (9.2.22)

eqn. (9.2.21) - eqn. (9.2.22)

$$-\frac{5}{5}b_{12} = -1$$

$$-b_{12} = -1$$

$$b_{12} = 1 \tag{9.2.26}$$

Substituting in the equation (9.2.21)

$$-\frac{3}{5} + \frac{1}{5}b_{22} = 0$$

$$\frac{1}{5}b_{22} = \frac{3}{5}$$

$$b_{22} = 3$$

$$\begin{bmatrix} b_{11} & b_{12} \\ b_{21} & b_{22} \end{bmatrix} = \begin{bmatrix} -\frac{3}{5} & \frac{1}{5} \\ \frac{2}{5} & \frac{1}{5} \end{bmatrix}$$

So, $B = [A^{-1}]^{-1}$

$$= \begin{bmatrix} -1 & 1 \\ 2 & 3 \end{bmatrix}$$

Prob. 49.

$$C = \begin{bmatrix} 2 & 4 \\ 3 & 6 \end{bmatrix}$$

$$CC^{-1} = I_2$$

Let us take $A^{-1} = B$

$$B = \begin{bmatrix} b_{11} & b_{12} \\ b_{21} & b_{22} \end{bmatrix}$$

So,

$$\begin{bmatrix} 2 & 4 \\ 3 & 6 \end{bmatrix} \begin{bmatrix} b_{11} & b_{12} \\ b_{21} & b_{22} \end{bmatrix} = \begin{bmatrix} 1 & 0 \\ 0 & 1 \end{bmatrix}$$

If we write the above matrix in the linear equations form

$$2b_{11} + 4b_{21} = 1 \tag{9.2.27}$$

$$3b_{11} + 6b_{21} = 0 \tag{9.2.28}$$

$$2b_{12} + 4b_{22} = 0 \tag{9.2.29}$$

$$3b_{12} + 6b_{22} = 1 \tag{9.2.30}$$

Solving eqn. (9.2.27) and eqn. (9.2.28)

3 x eqn.9.2.27 - 2 x eqn. (9.2.28)

$$6b_{11} + 12b_{21} = 3 \qquad (9.2.31)$$

$$6b_{11} + 12b_{21} = 0 \qquad (9.2.32)$$

Eqn (9.2.31) & eqn (9.2.32) are contradictory.

So there will not be any solution.

So this matrix is not invertible.

Prob. 51.

(a) Writing the above matrix in the equation form

$$-x + 0 = -2 \qquad (9.2.33)$$

$$2x - 3y = -5 \qquad (9.2.34)$$

From the eqn. (9.4.27),

$x = 2$ Substituting the value of x in eqn. (9.4.28)

$$2(2) - 3y = -5$$

$$-3y = -9$$

$$y = 3$$

$$So, \; detX = \begin{bmatrix} 2 \\ 3 \end{bmatrix} \qquad (9.2.35)$$

(b) Using inverse matrix method

$$A = \begin{bmatrix} -1 & 0 \\ 2 & -3 \end{bmatrix}$$

$$D = \begin{bmatrix} -2 \\ -5 \end{bmatrix}$$

$$AX = D$$

$$X = (A^{-1})D$$

$$A = \begin{bmatrix} -1 & 0 \\ 2 & -3 \end{bmatrix}$$

$$AA^{-1} = I_2$$

Let us take $A^{-1} = B$

$$B = \begin{bmatrix} b_{11} & b_{12} \\ b_{21} & b_{22} \end{bmatrix}$$

So,

$$\begin{bmatrix} -1 & 0 \\ 2 & -3 \end{bmatrix} \begin{bmatrix} b_{11} & b_{12} \\ b_{21} & b_{22} \end{bmatrix} = \begin{bmatrix} 1 & 0 \\ 0 & 1 \end{bmatrix}$$

If we write the above matrix in the linear equations form

$$-b_{11} + 0 = 1 \tag{9.2.36}$$

$$2b_{11} - 3b_{21} = 0 \tag{9.2.37}$$

$$-b_{12} + 0 = 0 \tag{9.2.38}$$

$$2b_{12} - 3b_{22} = 1 \tag{9.2.39}$$

From the eqn. (9.2.36) & 43.b.3

we get,

$$b_{11} = -1 \tag{9.2.40}$$

$$b_{12} = 0 \tag{9.2.41}$$

$$\tag{9.2.42}$$

Substituting b_{11} in eqn. (9.2.37)

$$-2 - 3b_{21} = 0$$

$$b_{21} = -\frac{2}{3} \tag{9.2.43}$$

Substituting b_{12} in eqn. (9.2.39)

$$0 - 3b_{22} = 1$$

$$b_{22} = -\frac{1}{3} \tag{9.2.44}$$

So, $B = (A)^{-1}$

$$= \begin{bmatrix} -1 & 0 \\ -\frac{2}{3} & -\frac{1}{3} \end{bmatrix}$$

$$X = (A^{-1})D$$

$$X = \begin{bmatrix} -1 & 0 \\ -\frac{2}{3} & -\frac{1}{3} \end{bmatrix} \begin{bmatrix} -2 \\ -5 \end{bmatrix}$$

$$= \begin{bmatrix} 2+0 \\ \frac{4}{3}+\frac{5}{3} \end{bmatrix}$$

$$= \begin{bmatrix} 2 \\ \frac{9}{3} \end{bmatrix}$$

$$X = \begin{bmatrix} 2 \\ 3 \end{bmatrix} \qquad (9.2.45)$$

Prob. 53.

$$A = \begin{bmatrix} 2 & -1 \\ 1 & 3 \end{bmatrix}$$

To check whether A is invertible, we compute the determinant of A.

$$[A] = (2)(3) - (1)(-1)$$

$$[A] = 6+1$$

$$[A] = 7$$

As $[A] \neq 0$, A is invertible.

Prob. 55.

$$A = \begin{bmatrix} 4 & -1 \\ 8 & -2 \end{bmatrix}$$

To check whether A is invertible, we compute the determinant of A.

$$[A] = (4)(-2) - (8)(-1)$$

$$[A] = -8+8$$

$$[A] = 0$$

As $[A] = 0$, A is not an invertible matrix.

Prob. 57.

(a)

$$A = \begin{bmatrix} 2 & 4 \\ 3 & 6 \end{bmatrix}$$

To check whether A is invertible, we compute the determinant of A.

$$[A] = (2)(6) - (3)(4)$$
$$[A] = 12 - 12$$
$$[A] = 0$$

As $[A] = 0$, A is not an invertible matrix.

(b)

$$x = \begin{bmatrix} x \\ y \end{bmatrix}$$

$$B = \begin{bmatrix} b_1 \\ b_2 \end{bmatrix}$$

$$AX = B$$

$$\begin{bmatrix} 2 & 4 \\ 3 & 6 \end{bmatrix} \begin{bmatrix} x \\ y \end{bmatrix} = \begin{bmatrix} b_1 \\ b_2 \end{bmatrix}$$

Writing the above matrix in the form of linear equations,

$$2x + 4y = b_1$$
$$3x + 6y = b_2$$

(c)

$$A = \begin{bmatrix} 2 & 4 \\ 3 & 6 \end{bmatrix}$$

$$x = \begin{bmatrix} x \\ y \end{bmatrix}$$

$$B = \begin{bmatrix} 3 \\ \frac{9}{2} \end{bmatrix}$$

$$AX = B$$

$$\begin{bmatrix} 2 & 4 \\ 3 & 6 \end{bmatrix} \begin{bmatrix} x \\ y \end{bmatrix} = \begin{bmatrix} 3 \\ \frac{9}{2} \end{bmatrix}$$

Writing the above matrix in the form of linear equations,

$$2x + 4y = 3 \tag{9.2.46}$$

$$3x + 6y = \frac{9}{2} \tag{9.2.47}$$

multiplying the eqn. (9.2.47) by 2

$$6x + 12y = 9 \tag{9.2.48}$$

$$\tag{9.2.49}$$

multiplying the eqn. (9.2.46) by 3

$$6x + 12y = 9 \tag{9.2.50}$$

$$\tag{9.2.51}$$

So, eqn. (9.2.48 & eqn. (9.2.50) are the same.

So they are identical and parallel.

So it will have infinite number of solutions.

Graphical Solution:

for the eqn (9.2.46)   for the eqn (9.2.47)

at x = 0, y = $\frac{3}{4}$ at x = 0, y = $\frac{3}{4}$

at x = 1, y = $\frac{1}{4}$ at x = 1, y = $\frac{1}{4}$

at x = 2, y = $-\frac{3}{4}$ at x = 2, y = $-\frac{3}{4}$

at x = 3, y = $-\frac{3}{4}$ at x = 3, y = $-\frac{3}{4}$

(d) To find AX=B has no solution.

$$2x + 4y = b_1 \tag{9.2.52}$$

$$3x + 6y = b_2 \tag{9.2.53}$$

multiplying the eqn. (9.2.52) by 2 & multiplying the eqn. (9.2.53) by 2

$$6x + 12y = 3b_1 \tag{9.2.54}$$

$$6x + 12y = 2b_2 \tag{9.2.55}$$

$$\tag{9.2.56}$$

Equating the above two equations by subtracting,

i.e., eqn. (9.2.54) - eqn. (9.2.55) gives

$$3b_1 - 2b_2 = 0$$

$$3b_1 = 2b_2$$

$$b_1 = \frac{2}{3} \tag{9.2.57}$$

when the eqn. (9.2.57) is satisfied, then there will be infinite number of solutions.

So, $b_1 \neq \frac{2}{3}$ is the condition to have no solution.

Prob. 59. $A^{-1} = -\frac{1}{5} \begin{bmatrix} -1 & -1 \\ -3 & 2 \end{bmatrix}$.

Prob. 61. $A^{-1} = -\frac{1}{21} \begin{bmatrix} 1 & -4 \\ -5 & -1 \end{bmatrix}$.

Prob. 63.

$$A = \begin{bmatrix} 1 & -1 \\ 0 & 2 \end{bmatrix}$$

$$AA^{-1} = I_2$$

To check whether A is invertible, we compute the determinant of A.

$$[A] = (1)(2) - (0)(-1)$$

$$[A] = 2 + 0$$

$$[A] = 2$$

As $[A] \neq 0$, A is invertible.

Let us take $A^{-1} = B$

$$B = \begin{bmatrix} b_{11} & b_{12} \\ b_{21} & b_{22} \end{bmatrix}$$

$$So,$$

$$\begin{bmatrix} 1 & -1 \\ 0 & 2 \end{bmatrix} \begin{bmatrix} b_{11} & b_{12} \\ b_{21} & b_{22} \end{bmatrix} = \begin{bmatrix} 1 & 0 \\ 0 & 1 \end{bmatrix}$$

If we write the above matrix in the linear equations form

$$b_{11} - b_{21} = 1 \tag{9.2.58}$$

$$2b_{21} = 0 \tag{9.2.59}$$

$$b_{12} - b_{22} = 0 \tag{9.2.60}$$

$$2b_{22} = 1 \tag{9.2.61}$$

From the eqn. (9.2.59) & eqn. (9.2.61) we get

$$b_{21} = 0 \tag{9.2.62}$$

$$b_{22} = \frac{1}{2} \tag{9.2.63}$$

Substituting eqn (9.2.62) in eqn. (9.2.58)

$$b_{11} = 1 + b_{21}$$

$$= 1 \tag{9.2.64}$$

Substituting eqn (9.2.63) in eqn. (9.2.59)

$$b_{12} = b_{22}$$

$$= \frac{1}{2} \tag{9.2.65}$$

$$So, B = (A)^{-1} = \begin{bmatrix} 1 & 1/2 \\ 0 & 1/2 \end{bmatrix}$$

$$AX = 0$$

$$So, X = 0xA^{-1}$$

$$= 0$$

$$So, X = \begin{bmatrix} 0 \\ 0 \end{bmatrix}$$

So this system has a trivial solution.

Prob. 65.

$$C = \begin{bmatrix} 1 & 3 \\ 1 & 3 \end{bmatrix}$$

To check whether C is invertible, we compute the determinant of C.

$$[C] = (1)(3) - (1)(3)$$

$$[C] = 3 - 3$$

$$[C] = 0$$

As $[C] = 0$, C is non invertible matrix.

Let us take $C^{-1} = B$

$$B = \begin{bmatrix} b_{11} & b_{12} \\ b_{21} & b_{22} \end{bmatrix}$$

$$So,$$

$$\begin{bmatrix} 1 & 3 \\ 1 & 3 \end{bmatrix} \begin{bmatrix} b_{11} & b_{12} \\ b_{21} & b_{22} \end{bmatrix} = \begin{bmatrix} 1 & 0 \\ 0 & 1 \end{bmatrix}$$

If we write the above matrix in the linear equations form

$$b_{11} + 3b_{21} = 1 \tag{9.2.66}$$

$$b_{11} + 3b_{21} = 0 \tag{9.2.67}$$

$$b_{12} + 3b_{22} = 0 \tag{9.2.68}$$

$$b_{12} + 3b_{22} = 1 \tag{9.2.69}$$

The eqns. (9.2.66) & (9.2.67) and also the eqns. (9.2.68)& (9.2.69) are contradictory

As this system is not having any solution we can't find the inverse So we can't find $CX = 0$

Prob. 67.

$$A = \begin{bmatrix} 2 & -1 & -1 \\ 2 & 1 & 1 \\ -1 & 1 & -1 \end{bmatrix}$$

(9.2.70)

To check whether A is invertible, we compute the determinant of A.

$$[A] = 2(-1-1) + (-2+1) - (2+1)$$
$$[A] = 2(-2) + (-1) - (3)$$
$$[A] = -4 - 1 - 3$$
$$[A] = -8$$

As $[A] \neq 0$, A is invertible.

Let us take $A^{-1} = B$

$$AA^{-1} = I_3$$

Let us take $A^{-1} = B$

$$B = \begin{bmatrix} b_{11} & b_{12} & b_{13} \\ b_{21} & b_{22} & b_{23} \\ b_{31} & b_{32} & b_{33} \end{bmatrix}$$

So,

$$\begin{bmatrix} 2 & -1 & -1 \\ 2 & 1 & 1 \\ -1 & 1 & -1 \end{bmatrix} \begin{bmatrix} b_{11} & b_{12} & b_{13} \\ b_{21} & b_{22} & b_{23} \\ b_{31} & b_{32} & b_{33} \end{bmatrix}$$

$$= \begin{bmatrix} 1 & 0 & 0 \\ 0 & 1 & 0 \\ 0 & 0 & 1 \end{bmatrix}$$

If we write the above matrix in the linear equations form

$$2b_{11} - b_{21} - b_{31} = 1 \tag{9.2.71}$$

$$2b_{11} + b_{21} + b_{31} = 0 \tag{9.2.72}$$

$$-b_{11} + b_{21} - b_{31} = 0 \tag{9.2.73}$$

$$2b_{12} - b_{22} - b_{32} = 0 \tag{9.2.74}$$

$$2b_{12} + b_{22} + b_{32} = 1 \tag{9.2.75}$$

$$-b_{12} + b_{22} - b_{32} = 0 \tag{9.2.76}$$

$$2b_{13} - b_{23} - b_{33} = 0 \tag{9.2.77}$$

$$2b_{13} + b_{23} + b_{33} = 0 \tag{9.2.78}$$

$$-b_{13} + b_{23} - b_{33} = 1 \tag{9.2.79}$$

Solving eqn. (9.4.40) and eqn. (9.4.41)

by adding those two eqns

$$4b_{11} = 1$$

$$b_{11} = \frac{1}{4} \tag{9.2.80}$$

Solving eqn. (9.4.41) and eqn. (9.4.42)

by adding those two eqns. and substituting eqn (9.2.80)gives

$$b_{11} + 2b_{21} = 0$$

$$2b_{21} = -\frac{1}{4}$$

$$b_{21} = -\frac{1}{8} \tag{9.2.81}$$

substituting eqn b_{11} & b_{21} in eqn (9.4.40) gives

$$b_{31} = -\frac{3}{8} \tag{9.2.82}$$

Solving eqn. (9.4.43) and eqn. (9.4.44)

by adding those two eqns

$$4b_{12} = 1$$

$$b_{12} = \frac{1}{4} \tag{9.2.83}$$

Solving eqn. (9.4.44) and eqn. (9.2.76)

by adding those two eqns. and substituting eqn (9.2.83)gives

$$b_{12} + 2b_{22} = 1$$
$$2b_{22} = 1 - \frac{1}{4}$$
$$b_{22} = \frac{3}{8} \tag{9.2.84}$$

substituting eqn b_{12} & b_{22} in eqn (9.2.76) gives

$$-\frac{1}{4} + \frac{3}{8} - b_{32} = 0$$
$$b_{32} = \frac{-2+3}{8}$$
$$b_{32} = \frac{1}{8} \tag{9.2.85}$$

Solving eqn. (9.2.77) and eqn. (9.2.78)

by adding those two eqns

$$4b_{13} = 0$$
$$b_{13} = 0 \tag{9.2.86}$$

Solving eqn. (9.2.78) and eqn. (9.2.79)

by adding those two eqns. and substituting eqn (9.2.86)gives

$$b_{13} + 2b_{23} = 1$$
$$2b_{23} = 1$$
$$b_{23} = \frac{1}{2} \tag{9.2.87}$$

substituting eqn b_{13} & b_{23} in eqn (9.2.79) gives

$$\frac{1}{2} - b_{33} = 1$$
$$b_{33} = -\frac{1}{2} \tag{9.2.88}$$

So, $B = (A)^{-1}$

$$= \begin{bmatrix} /4 & 1/4 & 0 \\ -1/8 & -3/8 & 1/2 \\ -3/8 & 1/8 & -1/2 \end{bmatrix}$$

Prob. 69.

$$A = \begin{bmatrix} -1 & 0 & -1 \\ 0 & -2 & 0 \\ -1 & 1 & 2 \end{bmatrix}$$

(9.2.89)

To check whether A is invertible, we compute the determinant of A.

$$[A] = -1(-4) - 1(0 - 2)$$
$$[A] = 4 + 2$$
$$[A] = 6$$

As $[A] \neq 0$, A is invertible.

Let us take $A^{-1} = B$

$$AA^{-1} = I_3$$

Let us take $A^{-1} = B$

$$B = \begin{bmatrix} b_{11} & b_{12} & b_{13} \\ b_{21} & b_{22} & b_{23} \\ b_{31} & b_{32} & b_{33} \end{bmatrix}$$

So,

$$\begin{bmatrix} -1 & 3 & -1 \\ 2 & -2 & 3 \\ -1 & 1 & 2 \end{bmatrix} \begin{bmatrix} b_{11} & b_{12} & b_{13} \\ b_{21} & b_{22} & b_{23} \\ b_{31} & b_{32} & b_{33} \end{bmatrix}$$

$$= \begin{bmatrix} 1 & 0 & 0 \\ 0 & 1 & 0 \\ 0 & 0 & 1 \end{bmatrix}$$

If we write the above matrix in the linear equations form

$$-b_{11} - b_{31} = 1$$

(9.2.90)

$$-2b_{21} = 0 \qquad (9.2.91)$$

$$-b_{11} + b_{21} + 2b_{31} = 0 \qquad (9.2.92)$$

$$-b_{12} - b_{32} = 0 \qquad (9.2.93)$$

$$-2b_{22} = 1 \qquad (9.2.94)$$

$$-b_{12} + b_{22} + 2b_{32} = 0 \qquad (9.2.95)$$

$$-b_{13} - b_{33} = 0 \qquad (9.2.96)$$

$$-2b_{23} = 0 \qquad (9.2.97)$$

$$-b_{13} + b_{23} + 2b_{33} = 1 \qquad (9.2.98)$$

From the equation (9.4.46)

$$b_{21} = 0 \qquad (9.2.99)$$

Solving eqn. (9.4.45) and eqn. (9.4.46)

by adding

$$b_{21} + b_{31} = 1$$

$$b_{31} = 1 \qquad (9.2.100)$$

substituting the value of b_{31} in eqn. (9.4.45)

$$-b_{11} = 2$$

$$b_{11} = -2 \qquad (9.2.101)$$

From the equation (9.2.94)

$$-2b_{22} = 1$$

$$b_{22} = -\frac{1}{2} \qquad (9.2.102)$$

Solving eqn. (9.2.95) and eqn. (9.2.93)

by subtracting i.e., eqn. (9.2.95) - eqn. (9.2.93)

$$b_{22} + 3b_{32} = 0$$

$$b_{32} = \frac{1}{6} \qquad (9.2.103)$$

substituting the value of b_{32} and b_{22} in eqn. (9.4.47)

$$-b_{12} = b_{32}$$

$$b_{12} = \frac{1}{6} \qquad (9.2.104)$$

$$(9.2.105)$$

From the equation (9.2.97)

$$b_{23} = 0 \qquad (9.2.106)$$

Solving eqn. (9.2.98) and eqn. (9.2.96)

by subtracting i.e., eqn. (9.2.98) - eqn. (9.2.96)

$$b_{23} + 3b_{33} = 1$$

$$b_{33} = \frac{1}{3} \qquad (9.2.107)$$

substituting the value of b_{23} in eqn. (9.2.96)

$$-b_{13} = b_{33}$$

$$b_{13} = -\frac{1}{3} \qquad (9.2.108)$$

$$(9.2.109)$$

So, $B = (A)^{-1}$

$$= \begin{bmatrix} 2 & -\frac{1}{6} & -\frac{1}{3} \\ 0 & -\frac{1}{2} & 0 \\ 1 & \frac{1}{6} & \frac{1}{3} \end{bmatrix}$$

Prob. 71.

$$N_1(t+1) = 0.2N_0(t)$$

$$N_2(t+1) = 0.7N_1(t)$$

For finding the values at t = 1

$$N(1) = \begin{bmatrix} 0 & 3.2 & 1.7 \\ 0.2 & 0 & 0 \\ 0 & 0.7 & 0 \end{bmatrix} \begin{bmatrix} 2000 \\ 800 \\ 200 \end{bmatrix}$$

$$= \begin{bmatrix} (3.2)(800) + (1.7)(200) \\ (0.2)(2000) \\ (0.7)(800) \end{bmatrix}$$

$$= \begin{bmatrix} 2560 + 340 \\ 400 \\ 560 \end{bmatrix}$$

$$= \begin{bmatrix} 2900 \\ 400 \\ 560 \end{bmatrix}$$

For finding the values at t = 2

$$N(2) = \begin{bmatrix} 0 & 3.2 & 1.7 \\ 0.2 & 0 & 0 \\ 0 & 0.7 & 0 \end{bmatrix} \begin{bmatrix} 2900 \\ 400 \\ 560 \end{bmatrix}$$

$$= \begin{bmatrix} (3.2)(400) + (1.7)(560) \\ (0.2)(2900) \\ (0.7)(400) \end{bmatrix}$$

$$= \begin{bmatrix} 1280 + 952 \\ 580 \\ 280 \end{bmatrix}$$

$$= \begin{bmatrix} 2232 \\ 580 \\ 280 \end{bmatrix}$$

Prob. 73. Population is divided into four age classes.

For finding the values at time = 1

$$N(1) = \begin{bmatrix} 0 & 0 & 4.6 & 3.7 \\ 0.7 & 0 & 0 & 0 \\ 0 & 0.5 & 0 & 0 \\ 0 & 0 & 0.1 & 0 \end{bmatrix} \begin{bmatrix} 1500 \\ 500 \\ 250 \\ 50 \end{bmatrix}$$

$$= \begin{bmatrix} (4.6)(250) + (3.7)(50) \\ (0.7)(1500) \\ (0.5)(500) \\ (0.1)(250) \end{bmatrix}$$

$$= \begin{bmatrix} 1150 + 185 \\ 1050 \\ 250 \\ 25 \end{bmatrix}$$

$$N(1) = \begin{bmatrix} 1335 \\ 1050 \\ 250 \\ 25 \end{bmatrix}$$

For finding the values at time $= 2$

$$N(2) = \begin{bmatrix} 0 & 0 & 4.6 & 3.7 \\ 0.7 & 0 & 0 & 0 \\ 0 & 0.5 & 0 & 0 \\ 0 & 0 & 0.1 & 0 \end{bmatrix} \begin{bmatrix} 1335 \\ 1050 \\ 250 \\ 25 \end{bmatrix}$$

$$= \begin{bmatrix} (4.6)(250) + (3.7)(25) \\ (0.7)(1325) \\ (0.5)(1050) \\ (0.1)(250) \end{bmatrix}$$

$$= \begin{bmatrix} 1150 + 92.5 \\ 934.5 \\ 525 \\ 25 \end{bmatrix}$$

$$N(2) = \begin{bmatrix} 1242.5 \\ 934.5 \\ 525 \\ 25 \end{bmatrix}$$

Prob. 75.

$$
L = \begin{bmatrix} 2 & 3 & 2 & 1 \\ 0.4 & 0 & 0 & 0 \\ 0 & 0.6 & 0 & 0 \\ 0 & 0 & 0.8 & 0 \end{bmatrix}
$$

(i) There are four age classes.

(ii) 60% of the one year olds are surviving until the end of the following breeding section.

(iii) The average number of female offspring of a two-year-old is 2.

Prob. 77.

$$
L = \begin{bmatrix} 1 & 2.5 & 3 & 1.5 \\ 0.9 & 0 & 0 & 0 \\ 0 & 0.3 & 0 & 0 \\ 0 & 0 & 0.2 & 0 \end{bmatrix}
$$

(i) 20% of the one year olds are surviving until the end of the following breeding section.

(ii) The average number of female offspring of one-year-old female is 3.

Prob. 79.

$$
L = \begin{bmatrix} 1.2 & 3.2 \\ 0.8 & 0 \end{bmatrix}
$$

$$
N_0(0) = 100
$$

$$
N_1(0) = 0
$$

From the above matrix

$$
N_0(t+1) = 1.2N_0(t) + 3.2N_1(t)
$$

$$
N_1(t+1) = 0.8N_0(t)
$$

$$
\begin{bmatrix} N_0(1) \\ N_1(1) \end{bmatrix} = \begin{bmatrix} 1.2 & 3.2 \\ 0.8 & 0 \end{bmatrix} \begin{bmatrix} 100 \\ 0 \end{bmatrix}
$$

$$N(1) = \begin{bmatrix} (1.2)(100) + (3.2)(0) \\ (0.8)(100) + 0 \end{bmatrix}$$

$$N(1) = \begin{bmatrix} 120 \\ 80 \end{bmatrix}$$

At t = 2

$$N(2) = \begin{bmatrix} (1.2)(120) + (3.2)(80) \\ (0.8)(120) + 0 \end{bmatrix}$$

$$N(2) = \begin{bmatrix} 400 \\ 96 \end{bmatrix}$$

At t = 3

$$N(3) = \begin{bmatrix} (1.2)(400) + (3.2)(96) \\ (0.8)(400) + 0 \end{bmatrix}$$

$$N(3) = \begin{bmatrix} 787.2 \\ 320 \end{bmatrix}$$

At t = 4

$$N(4) = \begin{bmatrix} (1.2)(787.2) + (3.2)(320) \\ (0.8)(787.2) + 0 \end{bmatrix}$$

$$N(4) = \begin{bmatrix} 1968 \\ 630 \end{bmatrix}$$

At t = 5

$$N(5) = \begin{bmatrix} (1.2)(1968) + (3.2)(630) \\ (0.8)(1968) + 0 \end{bmatrix}$$

$$N(5) = \begin{bmatrix} 4377 \\ 1574 \end{bmatrix}$$

At t = 6

$$N(6) = \begin{bmatrix} (1.2)(4377) + (3.2)(1574) \\ (0.8)(4377) + 0 \end{bmatrix}$$

$$N(6) = \begin{bmatrix} 10288 \\ 3501 \end{bmatrix}$$

At t = 7

$$N(7) = \begin{bmatrix} (1.2)(10288) + (3.2)(3501) \\ (0.8)(10288) + 0 \end{bmatrix}$$

$$N(7) = \begin{bmatrix} 23548 \\ 8230 \end{bmatrix}$$

At t = 8

$$N(8) = \begin{bmatrix} (1.2)(23548) + (3.2)(8230) \\ (0.8)(23548) + 0 \end{bmatrix}$$

$$N(8) = \begin{bmatrix} 54593 \\ 18838 \end{bmatrix}$$

At t = 9

$$N(9) = \begin{bmatrix} (1.2)(54593) + (3.2)(18838) \\ (0.8)(54593) + 0 \end{bmatrix}$$

$$N(9) = \begin{bmatrix} 125792 \\ 43674 \end{bmatrix}$$

At t = 10

$$N(10) = \begin{bmatrix} (1.2)(125792) + (3.2)(43674) \\ (0.8)(125792) + 0 \end{bmatrix}$$

$$N(10) = \begin{bmatrix} 290706 \\ 100633 \end{bmatrix}$$

t	1	2	3	4	5	6	7	8	9
$g_0(t)$	3.3	1.9	2.5	2.2	2.3	2.3	2.3	2.3	2.3

t	1	2	3	4	5	6	7	8	9
$g_1(t)$	1.2	3.3	2	2.5	2.2	2.3	2.3	2.3	2.3

The stable age distribution is $q_0(t) = q_1(t) = 2.3$

Prob. 81.

$$L_0 - \begin{bmatrix} 5 \\ 1 \end{bmatrix}, \qquad L_1 \begin{bmatrix} 2 \\ 3 \end{bmatrix}, \qquad L_2 = \begin{bmatrix} 6 \\ 1.2 \end{bmatrix},$$

$$L_3 = \begin{bmatrix} 2.4 \\ 3.6 \end{bmatrix}, \qquad L_4 = \begin{bmatrix} 7.2 \\ 1.44 \end{bmatrix}, \qquad L_5 = \begin{bmatrix} 2.88 \\ 4.32 \end{bmatrix},$$

$$L_6 = \begin{bmatrix} 8.64 \\ 1.73 \end{bmatrix}, \qquad L_7 = \begin{bmatrix} 3.46 \\ 5.18 \end{bmatrix}, \qquad L_8 = \begin{bmatrix} 10.34 \\ 2.07 \end{bmatrix},$$

$$L_9 = \begin{bmatrix} 4.15 \\ 6.22 \end{bmatrix}, \qquad L_{10} = \begin{bmatrix} 12.44 \\ 2.49 \end{bmatrix}.$$

Now,

t	1	2	3	4	5	6	7	8	9	10
$g_0(t)$	0.4	3	0.4	3	0.4	3	0.4	3	0.4	3

t	1	2	3	4	5	6	7	8	9	10
$g_1(t)$	3	0.4	3	0.4	3	0.4	3	0.4	3	0.4

Both $q_0(t)$ and $q_1(t)$ assilitate between 0.4 and 3.

9.3   Linear Maps, Eigenvectors and Eigenvalues

Prob. 1.

(a)

$$A = \begin{bmatrix} 2 & 1 \\ 3 & 4 \end{bmatrix}$$

$$x = \begin{bmatrix} x_1 \\ x_2 \end{bmatrix}$$

$$y = \begin{bmatrix} y_1 \\ y_2 \end{bmatrix}$$

To Prove $A(x + y) = Ax + Ay$

LHS

$$x + y = \begin{bmatrix} x_1 + y_1 \\ x_2 + y_2 \end{bmatrix}$$

$$A(x+y) = \begin{bmatrix} 2 & 1 \\ 3 & 4 \end{bmatrix} \begin{bmatrix} x_1 + y_1 \\ x_2 + y_2 \end{bmatrix}$$

$$= \begin{bmatrix} 2(x_1 + y_1) + 1(x_2 + y_2) \\ 3(x_1 + y_1) + 4(x_2 + y_2) \end{bmatrix}$$

$$= \begin{bmatrix} 2x_1 + 2y_1 + x_2 + y_2 \\ 3x_1 + 3y_1 + 4x_2 + 4y_2) \end{bmatrix}$$

RHS

$$Ax = \begin{bmatrix} 2 & 1 \\ 3 & 4 \end{bmatrix} \begin{bmatrix} x_1 \\ x_2 \end{bmatrix}$$

$$= \begin{bmatrix} 2x_1 + x_2 \\ 3x_1 + 4x_2 \end{bmatrix}$$

$$Ay = \begin{bmatrix} 2 & 1 \\ 3 & 4 \end{bmatrix} \begin{bmatrix} y_1 \\ y_2 \end{bmatrix}$$

$$= \begin{bmatrix} 2y_1 + y_2 \\ 3y_1 + 4y_2 \end{bmatrix}$$

Adding the above and rearranging

$$Ax + Ay = \begin{bmatrix} 2x_1 + 2y_1 + x_2 + y_2 \\ 3x_1 + 3y_1 + 4x_2 + 4y_2) \end{bmatrix}$$

so,

$$\text{LHS} = \text{RHS}$$

(b)

$$A = \begin{bmatrix} 2 & 1 \\ 3 & 4 \end{bmatrix}$$

$$x \;=\; \begin{bmatrix} x_1 \\ x_2 \end{bmatrix}$$

$$y \;=\; \begin{bmatrix} y_1 \\ y_2 \end{bmatrix}$$

To Prove $A(\lambda x) = \lambda(Ax)$ **LHS**

$$\lambda x \;=\; \begin{bmatrix} \lambda x_1 \\ \lambda x_2 \end{bmatrix}$$

$$A\lambda x \;=\; \begin{bmatrix} 2 & 1 \\ 3 & 4 \end{bmatrix} \begin{bmatrix} \lambda x_1 \\ \lambda x_2 \end{bmatrix}$$

$$=\; \begin{bmatrix} 2\lambda x_1 + \lambda x_2 \\ 3\lambda x_1 + 4\lambda x_2 \end{bmatrix}$$

LHS

$$Ax \;=\; \begin{bmatrix} 2 & 1 \\ 3 & 4 \end{bmatrix} \begin{bmatrix} x_1 \\ x_2 \end{bmatrix}$$

$$=\; \begin{bmatrix} 2x_1 + x_2 \\ 3x_1 + 4x_2 \end{bmatrix}$$

$$\lambda(Ax) \;=\; \begin{bmatrix} 2\lambda x_1 + \lambda x_2 \\ 3\lambda x_1 + 4\lambda x_2 \end{bmatrix}$$

so,   (9.3.1)

LHS = RHS   (9.3.2)

Prob. 3. Length: $|x| = \sqrt{2^2 + 2^2} = \sqrt{8} = 2\sqrt{2}$

Angle: $\tan \alpha = \frac{x_2}{x_1} = \frac{2}{2} = 1$ so $\alpha = \tan^{-1}$, that is $\alpha = 45^o$

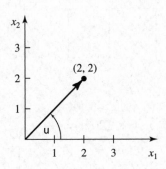

Prob. 5. Length: $|x| = \sqrt{(0)^2 + 3^2} = 3$.

Angle: $\tan \alpha = \frac{x_2}{x_1} = \frac{3}{(0)}$ which is not defined, so $\alpha = 90^o$.

Prob. 7. Length: $|x| = \sqrt{(\sqrt{3})^2 + (1)^2} = \sqrt{4} = 2$.

Angle: $\tan \alpha = \frac{x_2}{x_1} = \frac{-1}{\sqrt{3}}$, so $\alpha = \tan^{-1}(\frac{-1}{\sqrt{3}}) = 30^o$

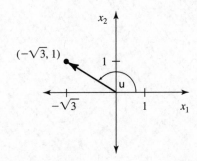

Prob. 9.

$$x_1 = 2\cos(30) = 2\frac{\sqrt{3}}{2} = \sqrt{3}$$

$$x_2 = 2\sin(30) = 2\frac{1}{2} = 1$$

$$\begin{bmatrix} x_1 \\ x_2 \end{bmatrix} = \begin{bmatrix} \sqrt{3} \\ 1 \end{bmatrix}$$

Prob. 11.

$$x_1 = 4\cos\alpha = 1 \cdot \cos(120^0) = -\frac{1}{2}$$

$$x_2 = 4\sin\alpha = 1 \cdot \sin(120^0) = \frac{\sqrt{3}}{2}, \quad \begin{bmatrix} x_1 \\ x_2 \end{bmatrix} = \begin{bmatrix} -\frac{1}{2} \\ \frac{\sqrt{3}}{2} \end{bmatrix}$$

Prob. 13. $\alpha = 360 - 15^o = 345^o$ so, $x_1 = 3\cos(345) = 3(0.966) = 2.898$ and $x_2 = 3\sin(345) = 3(-0.2588) = -0.7764$.

$$\begin{bmatrix} x_1 \\ x_2 \end{bmatrix} = \begin{bmatrix} 2.898 \\ -0.7764 \end{bmatrix}$$

Prob. 15.

$$x_1 = 5\cos(90 + 25) = -2.11$$

$$x_2 = 5\sin(90 + 25) = 4.53,$$

$$\begin{bmatrix} x_1 \\ x_2 \end{bmatrix} = \begin{bmatrix} -2.121 \\ 4.53 \end{bmatrix}$$

Prob. 17.

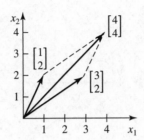

To add $\begin{bmatrix} 1 \\ 2 \end{bmatrix}$ to $\begin{bmatrix} 3 \\ 2 \end{bmatrix}$, move the vector $\begin{bmatrix} 1 \\ 2 \end{bmatrix}$ to the tip of $\begin{bmatrix} 3 \\ 2 \end{bmatrix}$ without changing the length of the former. The sum of two vectors start at (0,0). The sum is the diagonal in the parallelogram formed by these two vectors i.e., $\begin{bmatrix} 4 \\ 4 \end{bmatrix}$

Prob. 19.

To add $\begin{bmatrix} 0 \\ -2 \end{bmatrix}$ to $\begin{bmatrix} 1 \\ -1 \end{bmatrix}$, move the vector $\begin{bmatrix} 0 \\ -2 \end{bmatrix}$ to the tip of $\begin{bmatrix} 1 \\ -1 \end{bmatrix}$ without changing the length of the former. The sum of two vectors start at (0,0). The sum is the diagonal in the parallelogram formed by these two vectors i.e., $\begin{bmatrix} 1 \\ -3 \end{bmatrix}$.

Prob. 21.

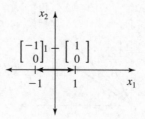

To add $\begin{bmatrix} 1 \\ 0 \end{bmatrix}$ to $\begin{bmatrix} -1 \\ 0 \end{bmatrix}$, move the vector $\begin{bmatrix} 1 \\ 0 \end{bmatrix}$ to the tip of $\begin{bmatrix} -1 \\ 0 \end{bmatrix}$ without changing the length of the former. The sum of two vectors start at $(0,0)$. The sum is the diagonal in the parallelogram formed by these two vectors i.e., $\begin{bmatrix} 0 \\ 0 \end{bmatrix}$.

Prob. 23.

$$a = 2$$

$$x = \begin{bmatrix} -2 \\ 1 \end{bmatrix}$$

$$ax = \begin{bmatrix} -4 \\ 2 \end{bmatrix}$$

If we multiply the vector with the scalar, there will not be any change in direction but in the length. The length will be increased by a factor 'a' (here it is 2). Since 'a' is positive here, the resulting vector points in the same direction as of $\begin{bmatrix} -2 \\ 1 \end{bmatrix}$.

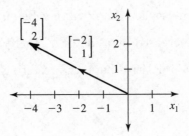

Prob. 25.

$$a = 0.5$$

$$x = \begin{bmatrix} 0 \\ -2 \end{bmatrix}$$

$$ax = \begin{bmatrix} 0 \\ -1 \end{bmatrix}$$

If we multiply the vector with the scalar, there will not be any change in direction but in the length. The length will be increased by a factor 'a' (here it is 0.5, so the length will be less than the original). Since 'a' is positive here, the resulting vector points in the same direction as of $\begin{bmatrix} 0 \\ -2 \end{bmatrix}$.

Prob. 27.

$$a = \frac{1}{4}$$

$$x = \begin{bmatrix} -4 \\ 1 \end{bmatrix}$$

$$ax = \begin{bmatrix} -1 \\ \frac{1}{4} \end{bmatrix}$$

If we multiply the vector with the scalar, there will not be any change in direction but in the length. The length will be increased by a factor 'a' (here it is $\frac{1}{4}$, so the length will be less than the original). Since 'a' is positive here, the resulting vector points in the same direction as of $\begin{bmatrix} -4 \\ 1 \end{bmatrix}$.

Prob. 29.

$$u = \begin{bmatrix} 3 \\ 4 \end{bmatrix}$$

$$v = \begin{bmatrix} -1 \\ -2 \end{bmatrix}$$

$$u + v = \begin{bmatrix} 2 \\ 2 \end{bmatrix}$$

To add $\begin{bmatrix} 3 \\ 4 \end{bmatrix}$ to $\begin{bmatrix} -1 \\ -2 \end{bmatrix}$, move the vector $\begin{bmatrix} 3 \\ 4 \end{bmatrix}$ to the tip of $\begin{bmatrix} -1 \\ -2 \end{bmatrix}$ without changing the length of the former. The sum of two vectors start at (0,0). The sum is the diagonal in the parallelogram formed by these two vectors i.e., $\begin{bmatrix} 2 \\ 2 \end{bmatrix}$

Prob. 31.

$$w = \begin{bmatrix} 1 \\ -2 \end{bmatrix}$$

$$u = \begin{bmatrix} 3 \\ 4 \end{bmatrix}$$

$$w - u = \begin{bmatrix} -2 \\ -6 \end{bmatrix}$$

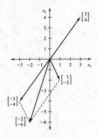

Prob. 33.

$$u = \begin{bmatrix} 3 \\ 4 \end{bmatrix}$$

$$v = \begin{bmatrix} -1 \\ -2 \end{bmatrix}$$

$$w = \begin{bmatrix} 1 \\ -2 \end{bmatrix}$$

$$u + v + w = \begin{bmatrix} 3 \\ 0 \end{bmatrix}$$

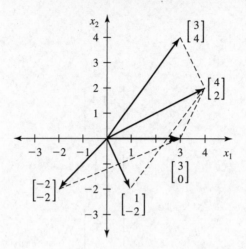

Prob. 35.

$$A = \begin{bmatrix} 1 & 0 \\ 0 & 1 \end{bmatrix}$$

$$x \to Ax$$

$$\begin{bmatrix} x_1 \\ x_2 \end{bmatrix} \to \begin{bmatrix} 1 & 0 \\ 0 & 1 \end{bmatrix} \begin{bmatrix} x_1 \\ x_2 \end{bmatrix}$$

$$= \begin{bmatrix} x_1 + 0 \\ 0 + x_2 \end{bmatrix}$$

$$= \begin{bmatrix} x_1 \\ x_2 \end{bmatrix}$$

So the vector is unchanged.

Prob. 37.

$$A = \begin{bmatrix} 0 & -1 \\ 1 & 0 \end{bmatrix}$$

$$x \to Ax$$

$$\begin{bmatrix} x_1 \\ x_2 \end{bmatrix} \to \begin{bmatrix} 0 & -1 \\ 1 & 0 \end{bmatrix} \begin{bmatrix} x_1 \\ x_2 \end{bmatrix}$$

$$= \begin{bmatrix} 0 - x_2 \\ x_1 \end{bmatrix}$$

$$= \begin{bmatrix} -x_2 \\ x_1 \end{bmatrix}$$

This map stretches or contracts each coordinate separately "counterclockwise rotation by $\theta = \frac{\pi}{2}$."

Prob. 39.

$$A = \frac{1}{2} \begin{bmatrix} \sqrt{3} & -1 \\ 1\sqrt{3} \end{bmatrix}$$

$$x \to Ax$$

$$\begin{bmatrix} x_1 \\ x_2 \end{bmatrix} \to \frac{1}{2} \begin{bmatrix} \sqrt{3} & -1 \\ 1 & \sqrt{3} \end{bmatrix} \begin{bmatrix} x_1 \\ x_2 \end{bmatrix}$$

$$= \begin{bmatrix} \frac{sqrt3}{2} x_1 - \frac{1}{2} x_2 \\ \frac{1}{2} x_1 + \frac{\sqrt{3}}{2} x_2 \end{bmatrix}$$

$$= \begin{bmatrix} \frac{\sqrt{3}}{2} x_1 - \frac{1}{2} x_2 \\ \frac{1}{2} x_1 + \frac{\sqrt{3}}{2} x_2 \end{bmatrix}$$

This map stretches or contracts each coordinate separately "counterclockwise rotation by $\theta = \frac{\pi}{6}$."

Prob. 41.

$$R_{\pi/6} = \begin{bmatrix} \cos(\pi/6) & -\sin(\pi/6) \\ \sin(\pi/6) & \cos(\pi/6) \end{bmatrix}$$

$$= \begin{bmatrix} \cos(30) & -\sin(30) \\ \sin(30) & \cos(30) \end{bmatrix} \begin{bmatrix} -1 \\ 2 \end{bmatrix}$$

$$= \begin{bmatrix} \sqrt{3}/2 & 1/2 \\ 1/2 & \sqrt{3}/2 \end{bmatrix} \begin{bmatrix} -1 \\ 2 \end{bmatrix}$$

$$= \begin{bmatrix} -\sqrt{3}/2 - 1 \\ -1/2 + \sqrt{3} \end{bmatrix}$$

$$= 1/2 \begin{bmatrix} -\sqrt{3} - 2 \\ -1 + 2\sqrt{3} \end{bmatrix}$$

Prob. 43.

$$R_{\pi/12} = \begin{bmatrix} \cos(\frac{\pi}{12}) & -\sin(\frac{\pi}{12}) \\ \sin(\frac{\pi}{12}) & \cos(\frac{\pi}{12} \end{bmatrix},$$

$$R_{\pi/12} \begin{bmatrix} 5 \\ 2 \end{bmatrix} = \begin{bmatrix} 5\cos(\frac{\pi}{12}) - 2\sin(\frac{\pi}{12}) \\ 5\sin(\frac{\pi}{12}) + 2\cos(\frac{\pi}{12}) \end{bmatrix}$$

Prob. 45.

$$R_{\pi/4} = \begin{bmatrix} \cos(\pi/4) & -\sin(\pi/4) \\ \sin(\pi/4) & \cos(\pi/4) \end{bmatrix}$$

$$= \begin{bmatrix} \cos(45) & -\sin(45) \\ \cos(45) & \sin(45) \end{bmatrix} \begin{bmatrix} -2 \\ -3 \end{bmatrix}$$

$$= \begin{bmatrix} 1/\sqrt{2} & -1/\sqrt{2} \\ 1/\sqrt{2} & 1/\sqrt{2} \end{bmatrix} \begin{bmatrix} -2 \\ -3 \end{bmatrix}$$

$$= \begin{bmatrix} \sqrt{2} - 1/\sqrt{2} \\ \sqrt{2} + 1/\sqrt{2} \end{bmatrix}$$

Prob. 47.

$$R_{\pi/7} = \begin{bmatrix} \cos(-\frac{\pi}{7}) & -\sin(-\frac{\pi}{7}) \\ \sin(-\frac{\pi}{7}) & \cos(-\frac{\pi}{7}) \end{bmatrix}$$

$$R_{\pi/7} \begin{bmatrix} 5 \\ 3 \end{bmatrix} = \begin{bmatrix} 5\cos(-\frac{\pi}{7}) - 3\sin(-\frac{\pi}{7}) \\ 5\sin(-\frac{\pi}{7}) + 3\cos(-\frac{\pi}{7}) \end{bmatrix}$$

Prob. 49.

$$A = \begin{bmatrix} 2 & 3 \\ 0 & -1 \end{bmatrix}$$

$$Ax = \lambda I x$$

$$(A - \lambda I)x = 0$$

$$[A - \lambda I] = 0$$

To get the above,

$$[A - \lambda I] = \begin{bmatrix} 2 & 3 \\ 0 & -1 \end{bmatrix} - \lambda \begin{bmatrix} 1 & 0 \\ 0 & 1 \end{bmatrix}$$

$$= \begin{bmatrix} 2-\lambda & 0 \\ 0 & -1-\lambda \end{bmatrix}$$

$$(2-\lambda)(-1-\lambda) = 0$$

So,

$$\lambda = 2 \;\&\; -1$$

The eigenvalues are 2 & -1.

To find the eigenvector associated with the eigenvalue $\lambda = 2$, we must determine x_1 and x_2 (not both equal to 0), so that

$$\begin{bmatrix} 2 & 3 \\ 0 & -1 \end{bmatrix} \begin{bmatrix} x_1 \\ x_2 \end{bmatrix} = 2 \begin{bmatrix} x_1 \\ x_2 \end{bmatrix}$$

Writing this as a system of linear equations,

$$2x_1 + 3x_2 = 2x_1 \qquad (9.3.3)$$

$$-2x_2 = 2x_2$$

So,

$$x_2 = 0$$

If we apply $x_2 = 0$, in the eqn (9.4.37), will give

$x_1 = x_1$. x_1 can take any value but not zero. So $x_1 = 0$

The eigenvector is (1,0)

To find the eigenvector associated with the eigenvalue $\lambda = -1$, we must determine x_1 and x_2 (not both equal to 0), so that

$$\begin{bmatrix} 2 & 3 \\ 0 & -1 \end{bmatrix} \begin{bmatrix} x_1 \\ x_2 \end{bmatrix} = -1 \begin{bmatrix} x_1 \\ x_2 \end{bmatrix}$$

Writing this as a system of linear equations,

$$2x_1 + 3x_2 = -x_1 \qquad (9.3.4)$$

$$-2x_2 = -x_2$$

Eqn (9.4.38), gives

$3x_1 + 3x_2 = 0$

i.e., in the eqn (9.4.37), will give

$3x_1 + 3x_2 = 0$

so,

$x_1 = -x_2.$

If $x_1 = 1$ then $x_2 = -1$

The eigenvector is (1,-1)

$$Av_1 = 2 \begin{bmatrix} 1 \\ 0 \end{bmatrix} = \begin{bmatrix} 2 \\ 0 \end{bmatrix}$$

$$Av_2 = -1 \begin{bmatrix} 1 \\ -1 \end{bmatrix} = \begin{bmatrix} -1 \\ 1 \end{bmatrix}$$

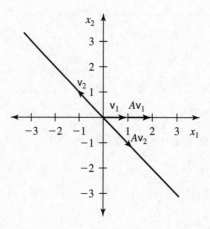

Prob. 51.

$$A = \begin{bmatrix} 1 & 0 \\ 1 & -1 \end{bmatrix}$$

$$Ax = \lambda I x$$

$$(A - \lambda I)x = 0$$

$$[A - \lambda I] = 0$$

To get the above,

$$[A - \lambda I] = \begin{bmatrix} 1 & 0 \\ 1 & -1 \end{bmatrix} - \lambda \begin{bmatrix} 1 & 0 \\ 0 & 1 \end{bmatrix}$$

$$= \begin{bmatrix} 1 - \lambda & 0 \\ 0 & -1 - \lambda \end{bmatrix}$$

$$(1 - \lambda)(-1 - \lambda) = 0$$

So,

$$\lambda = 1 \ \& \ -1$$

The eigenvalues are 1 & -1.

To find the eigenvector associated with the eigenvalue $\lambda = 1$, we must determine x_1 and x_2 (not both equal to 0), so that

$$\begin{bmatrix} 1 & 0 \\ 1 & -1 \end{bmatrix} \begin{bmatrix} x_1 \\ x_2 \end{bmatrix} = 0 \begin{bmatrix} x_1 \\ x_2 \end{bmatrix}$$

Writing this as a system of linear equations,

$$x_1 = x_1$$

$$-x_2 = x_2$$

So,

$$x_2 = 0$$

From $x_1 = x_1$ we can conclude that x_1 can take any value other than zero. The eigenvector is (1,0)

To find the eigenvector associated with the eigenvalue $\lambda = -1$, we must determine x_1 and x_2 (not both equal to 0), so that

$$\begin{bmatrix} 1 & 0 \\ 1 & -1 \end{bmatrix} \begin{bmatrix} x_1 \\ x_2 \end{bmatrix} = 0 \begin{bmatrix} x_1 \\ x_2 \end{bmatrix}$$

Writing this as a system of linear equations,

$$x_1 = -x_1$$

$$-x_2 = -x_2$$

$$x_2 = x_2$$

$$So,$$

$$x_1 = 0$$

From $x_2 = x_2$ we can conclude that x_2 can take any value other than zero.

The eigenvector is (0,1)

Av = eigenvalue x corresponding eigenvector

$$Av_1 = 1 \begin{bmatrix} 1 \\ 0 \end{bmatrix} = \begin{bmatrix} 1 \\ 0 \end{bmatrix}$$

$$Av_2 = -1 \begin{bmatrix} 0 \\ 1 \end{bmatrix} = \begin{bmatrix} 0 \\ -1 \end{bmatrix}$$

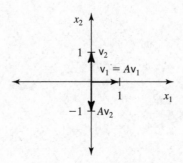

Prob. 53.

$$A = \begin{bmatrix} -4 & 2 \\ -3 & 1 \end{bmatrix}$$

$$Ax = \lambda I x$$

$$(A - \lambda I)x = 0$$

$$[A - \lambda I] = 0$$

To get the above,

$$[A - \lambda I] = \begin{bmatrix} -4 & 2 \\ -3 & 1 \end{bmatrix} - \lambda \begin{bmatrix} 1 & 0 \\ 0 & 1 \end{bmatrix}$$

$$= \begin{bmatrix} -4 - \lambda & 2 \\ -3 & 1 - \lambda \end{bmatrix}$$

$$[A - \lambda I] = (-4 - \lambda)(1 - \lambda) + 6$$

$$(-4 - \lambda)(1 - \lambda) + 6 = 0$$

$$\lambda^2 + 4\lambda + -\lambda + 6 - 4 = 0$$

$$\lambda^2 + 3\lambda + 2 = 0$$

$$(1 + \lambda)(2 + \lambda) = 0$$

So,

$$\lambda = -1 \ \& \ -2$$

The eigenvalues are -1 & -2.

To find the eigenvector associated with the eigenvalue $\lambda = -1$, we must determine x_1 and x_2 (not both equal to 0), so that

$$\begin{bmatrix} -4 & 2 \\ -3 & 1 \end{bmatrix} \begin{bmatrix} x_1 \\ x_2 \end{bmatrix} = 0 \begin{bmatrix} x_1 \\ x_2 \end{bmatrix}$$

Writing this as a system of linear equations,

$$-4x_1 + 2x_2 = -x_1$$

$$-3x_1 + x_2 = -x_2$$

$$3x_1 + 2x_2 = 0$$

$$3x_1 = 2x_2$$

So,

$$x_2 = \frac{3}{2}$$

If $x_1 = 1$, then $x_2 = \frac{3}{2}$.

The eigenvector is $(1, \frac{3}{2})$

To find the eigenvector associated with the eigenvalue $\lambda = -2$, we must determine x_1 and x_2 (not both equal to 0), so that

$$\begin{bmatrix} -4 & 2 \\ -3 & 1 \end{bmatrix} \begin{bmatrix} x_1 \\ x_2 \end{bmatrix} = 0 \begin{bmatrix} x_1 \\ x_2 \end{bmatrix}$$

Writing this as a system of linear equations,

$$-4x_1 + 2x_2 = -2x_1$$
$$-3x_1 + x_2 = -2x_2$$
$$x_1 + x_2 = 0$$

So,

$$x_1 = -x_2$$

If $x_1 = 1$, then $x_2 = -1$.

The eigenvector is (1,-1)

Av = eigenvalue x corresponding eigenvector

$$Av_1 = -2 \begin{bmatrix} 1 \\ 1 \end{bmatrix} = \begin{bmatrix} -2 \\ -2 \end{bmatrix}$$

$$Av_2 = -1 \begin{bmatrix} -1 \\ -\frac{3}{2} \end{bmatrix} = \begin{bmatrix} -1 \\ -\frac{3}{2} \end{bmatrix}$$

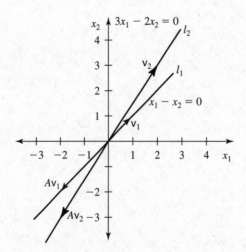

Prob. 55.

$$A = \begin{bmatrix} 2 & 1 \\ 2 & 3 \end{bmatrix}$$

$$Ax = \lambda I x$$

$$(A - \lambda I)x = 0$$

$$[A - \lambda I] = 0$$

To get the above,

$$[A - \lambda I] = \begin{bmatrix} 2 & 1 \\ 2 & 3 \end{bmatrix} - \lambda \begin{bmatrix} 1 & 0 \\ 0 & 1 \end{bmatrix}$$

$$= \begin{bmatrix} 2 - \lambda & 1 \\ 2 & 3 - \lambda \end{bmatrix}$$

$$[A - \lambda I] = (2 - \lambda)(3 - \lambda) - 2$$

$$(2 - \lambda)(3 - \lambda) - 2 = 0$$

$$\lambda^2 - 2\lambda + -3\lambda + 6 - 2 = 0$$

$$\lambda^2 - 5\lambda + 4 = 0$$

$$(\lambda - 1)(\lambda - 4) = 0$$

So,

$$\lambda = 1 \ \& \ 4$$

The eigenvalues are 1 & 4.

To find the eigenvector associated with the eigenvalue $\lambda = 1$, we must determine x_1 and x_2 (not both equal to 0), so that

$$\begin{bmatrix} 2 & 1 \\ 2 & 3 \end{bmatrix} \begin{bmatrix} x_1 \\ x_2 \end{bmatrix} = 0 \begin{bmatrix} x_1 \\ x_2 \end{bmatrix}$$

Writing this as a system of linear equations,

$$2x_1 + x_2 = x_1$$

$$2x_1 + 3x_2 = x_2$$

rearranging,

$$x_1 + x_2 = 0$$

$$x_1 = -x_2$$

If $x_1 = 1$, then $x_2 = -1$. The eigenvector is (1,-1)

To find the eigenvector associated with the eigenvalue $\lambda = 4$, we must determine x_1 and x_2 (not both equal to 0), so that

$$\begin{bmatrix} 2 & 1 \\ 2 & 3 \end{bmatrix} \begin{bmatrix} x_1 \\ x_2 \end{bmatrix} = 0 \begin{bmatrix} x_1 \\ x_2 \end{bmatrix}$$

Writing this as a system of linear equations,

$$2x_1 + x_2 = 4x_1$$

$$2x_1 + 3x_2 = 4x_2$$

rearranging,

$$-2x_1 + x_2 = 0$$

$$2x_1 - x_2 = 0$$

$$So,$$

$$2x_1 = x_2$$

If $x_1 = 1$, then $x_2 = 2$.

The eigenvector is (1,2)

Av = eigenvalue x corresponding eigenvector

$$Av_1 = 1 \begin{bmatrix} 1 \\ -1 \end{bmatrix} = \begin{bmatrix} 1 \\ -1 \end{bmatrix}$$

$$Av_2 = 4 \begin{bmatrix} 1 \\ 2 \end{bmatrix} = \begin{bmatrix} 4 \\ 8 \end{bmatrix}$$

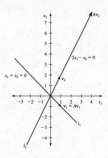

Prob. 57.

$$A = \begin{bmatrix} 4 & 0 \\ 0 & 3 \end{bmatrix}$$

$$Ax = \lambda I x$$

$$(A - \lambda I)x = 0$$

$$[A - \lambda I] = 0$$

To get the above,

$$[A - \lambda I] = \begin{bmatrix} 4 & 0 \\ 0 & 3 \end{bmatrix} - \lambda \begin{bmatrix} 1 & 0 \\ 0 & 1 \end{bmatrix}$$

$$= \begin{bmatrix} 4 - \lambda & 0 \\ 0 & 3 - \lambda \end{bmatrix}$$

$$[A - \lambda I] = (4 - \lambda)(3 - \lambda)$$

$$\lambda = 4 \ \& \ 3$$

The eigenvalues are 3 & 4.

To find the eigenvector associated with the eigenvalue $\lambda = 3$, we must determine x_1 and x_2 (not both equal to 0), so that

$$\begin{bmatrix} 4 & 0 \\ 0 & 3 \end{bmatrix} \begin{bmatrix} x_1 \\ x_2 \end{bmatrix} = 3 \begin{bmatrix} x_1 \\ x_2 \end{bmatrix}$$

Writing this as a system of linear equations,

$$4x_1 = 3x_1$$

$$3x_2 = 3x_2$$

rearranging,

$$x_1 = 0$$

$$x_2 = x_2$$

$x_1 = 0$, then x_2 can take any value other than zero. So, let us take $x_2 = 1$

The eigenvector is $(0, 1)$

To find the eigenvector associated with the eigenvalue $\lambda = 4$, we must determine x_1 and x_2 (not both equal to 0), so that

$$\begin{bmatrix} 4 & 0 \\ 0 & 3 \end{bmatrix} \begin{bmatrix} x_1 \\ x_2 \end{bmatrix} = 4 \begin{bmatrix} x_1 \\ x_2 \end{bmatrix}$$

Writing this as a system of linear equations,

$$4x_1 = 4x_1$$

$$3x_2 = 4x_2$$

rearranging,

$$x_2 = 0$$

$$x_1 = x_1$$

$$(9.3.5)$$

So, $x_2 = 0$, and then x_1 can have any value other than 0. So let us take $x_1 = 1$.

The eigenvector is $(1, 0)$

Prob. 59.

$$A = \begin{bmatrix} 1 & -3 \\ 0 & 2 \end{bmatrix}$$

$$Ax = \lambda Ix$$

$$(A - \lambda I)x = 0$$

$$[A - \lambda I] = 0$$

To get the above,

$$[A - \lambda I] = \begin{bmatrix} 1 & -3 \\ 0 & 2 \end{bmatrix} - \lambda \begin{bmatrix} 1 & 0 \\ 0 & 1 \end{bmatrix}$$

$$= \begin{bmatrix} 1-\lambda & -3 \\ 0 & 2-\lambda \end{bmatrix}$$

$$[A - \lambda I] = (1-\lambda)(2-\lambda)$$

$$\lambda = 1 \ \& \ 2$$

The eigenvalues are 1 & 2.

If one or both off-diagonal elements of 2x2 matrix are equal to zero, then the eigenvalues are that matrix is equal to the diagonal elements.

To find the eigenvector associated with the eigenvalue $\lambda = 1$, we must determine x_1 and x_2 (not both equal to 0), so that

$$\begin{bmatrix} 1 & -3 \\ 0 & 2 \end{bmatrix} \begin{bmatrix} x_1 \\ x_2 \end{bmatrix} = 1 \begin{bmatrix} x_1 \\ x_2 \end{bmatrix}$$

Writing this as a system of linear equations,

$$x_1 - 3x_2 = 2x_1$$

$$2x_2 = 2x_2$$

rearranging,

$$x_2 = 0$$

$$x_1 = x_1$$

$x_2 = 0$, then x_1 can take any value other than zero. So, let us take $x_1 = 1$

The eigenvector is (1, 0)

To find the eigenvector associated with the eigenvalue $\lambda = 2$, we must determine x_1 and x_2 (not both equal to 0), so that

$$\begin{bmatrix} 1 & -3 \\ 0 & 2 \end{bmatrix} \begin{bmatrix} x_1 \\ x_2 \end{bmatrix} = 2 \begin{bmatrix} x_1 \\ x_2 \end{bmatrix}$$

Writing this as a system of linear equations,

$$x_1 - 3x_2 = 2x_1$$

$$2x_2 = 2x_2$$

rearranging,

$$x_1 + 3x_2 = 0$$
$$x_2 = \frac{1}{3}x_1$$

If $x_1 = 1$, and then $x_2 = \frac{1}{3}$

The eigenvector is $(1, \frac{1}{3})$

Prob. 61. Eigenvalues satisfy:

$$\det(A - \lambda I) = 0$$

$$A - \lambda I = \begin{bmatrix} a & 0 \\ c & b \end{bmatrix} - \begin{bmatrix} \lambda & 0 \\ 0 & \lambda \end{bmatrix}$$

$$= \begin{bmatrix} a - \lambda & 0 \\ c & b - \lambda \end{bmatrix}.$$

So, we have: $(a - \lambda)(b - \lambda) = 0$, $\lambda_1 = a$. $\lambda_2 = b$.

Prob. 63. The real part of λ_1 and λ_2 will be negative, if and only if, trace $A < 0$ and $determinant A > 0$

$$tr\ A = 2 - 3 = -1 < 0$$
$$\det A = (2)(-3) - (4)(-2) = 2 > 0$$

So, the real parts of both eigenvalues are negative.

Prob. 65. The real part of λ_1 & λ_2 will be negative, if and only if, trace $A < 0$ and $determinant A > 0$

$$tr\ A = 4 - 3 = 1 > 0$$
$$\det A = (4)(-3) - (4)(-4) = 4 > 0$$

Though determinant A is positive, the trace of A is not negative.

As it is not satisfying the lemma, the real parts of both eigenvalues are not negative.

Prob. 67. The real part of λ_1 & λ_2 will be negative, if and only if, trace $A < 0$ and $determinant A > 0$

$$tr\ A = 2 - 3 = -1 < 0$$
$$\det A = (2)(-3) - (2)(-5) = 4 > 0$$

So, the real parts of both eigenvalues are negative.

Prob. 69.

(a) To be a linearly independent, if $\lambda_1 \neq \lambda_2$, then the eigenvector associated with λ_1 and the eigenvector associated with λ_2 are linearly independent.

$$A = \begin{bmatrix} -1 & 1 \\ 0 & 2 \end{bmatrix}$$

$$Ax = \lambda I x$$

$$(A - \lambda I)x = 0$$

$$[A - \lambda I] = 0$$

To get the above,

$$[A - \lambda I] = \begin{bmatrix} -1 & 1 \\ 0 & 2 \end{bmatrix} - \lambda \begin{bmatrix} 1 & 0 \\ 0 & 1 \end{bmatrix}$$

$$= \begin{bmatrix} -1 - \lambda & 1 \\ 0 & 2 - \lambda \end{bmatrix}$$

$$(-1 - \lambda)(2 - \lambda) = 0$$

So,

$$\lambda = 2 \ \& \ -1$$

The eigenvalues are -1 and 2. To find the eigenvector associated with the eigenvalue $\lambda = -1$, we must determine x_1 and x_2 (not both equal to 0), so that

$$\begin{bmatrix} -1 & 1 \\ 0 & 2 \end{bmatrix} \begin{bmatrix} x_1 \\ x_2 \end{bmatrix} = -1 \begin{bmatrix} x_1 \\ x_2 \end{bmatrix}$$

Writing this as a system of linear equations,

$$-x_1 + x_2 = -x_1$$

$$x_2 = 0$$

then

$$x_1 = x_1$$

So,

$$x_2 = 0$$

From $x_1 = x_1$ we can conclude that x_1 can take any value other than zero. The eigenvector is (1,0)

To find the eigenvector associated with the eigenvalue $\lambda = 2$, we must determine x_1 and x_2 (not both equal to 0), so that

$$\begin{bmatrix} -1 & 1 \\ 0 & 2 \end{bmatrix} \begin{bmatrix} x_1 \\ x_2 \end{bmatrix} = 2 \begin{bmatrix} x_1 \\ x_2 \end{bmatrix}$$

Writing this as a system of linear equations,

$$-x_1 + x_2 = 2x_1$$
$$-3x_1 + x_2 = 0$$
$$x_2 = 3x_1$$

So, if $x_1 = x_1$, x_2 will be equal to 3. The eigenvector is (1,3).

(b) Representing $\begin{bmatrix} 1 \\ -3 \end{bmatrix}$ as a linear combination of u_1 and u_2. We get:

$$\begin{bmatrix} 1 \\ -3 \end{bmatrix} = 2 \begin{bmatrix} 1 \\ 0 \end{bmatrix} - \begin{bmatrix} 1 \\ 3 \end{bmatrix}$$

(c) To compute A^{20}

$\lambda = -1,\ 2$

$$A^{20} = A^{20} \begin{bmatrix} 1 \\ -3 \end{bmatrix}$$

$$= A^{20} \left[2 \begin{pmatrix} 1 \\ 0 \end{pmatrix} - \begin{pmatrix} 1 \\ 3 \end{pmatrix} \right]$$

$$= 2A^{20} \begin{pmatrix} 1 \\ 0 \end{pmatrix} - A^{20} \begin{pmatrix} 1 \\ 3 \end{pmatrix}$$

$$= 2(-1)^{20} \begin{pmatrix} 1 \\ 0 \end{pmatrix} - (2)^{20} \begin{pmatrix} 1 \\ 3 \end{pmatrix}$$

$$= \begin{pmatrix} 2 \\ 0 \end{pmatrix} - 1048576 \begin{pmatrix} 1 \\ 3 \end{pmatrix}$$

$$= \begin{pmatrix} 2 \\ 0 \end{pmatrix} - \begin{pmatrix} 1048576 \\ 3145728 \end{pmatrix}$$

$$= \begin{pmatrix} -1048576 \\ -3145728 \end{pmatrix}$$

Prob. 71.

$$A = \begin{bmatrix} -1 & 0 \\ 3 & 1 \end{bmatrix}$$

To get the eigenvalues,

$$Ax = \lambda I x$$

$$(A - \lambda I)x = 0$$

$$[A - \lambda I] = 0$$

To get the above,

$$[A - \lambda I] = \begin{bmatrix} -1 & 0 \\ 3 & 1 \end{bmatrix} - \lambda \begin{bmatrix} 1 & 0 \\ 0 & 1 \end{bmatrix}$$

$$= \begin{bmatrix} -1 - \lambda & 0 \\ 3 & 1 - \lambda \end{bmatrix}$$

$$[A - \lambda I] = (-1 - \lambda)(1 - \lambda)$$

$$\lambda = 1 \ \& \ -1$$

The eigenvalues are 1 & -1.

To find the eigenvector associated with the eigenvalue $\lambda = 1$, we must determine x_1 and x_2 (not both equal to 0), so that

$$\begin{bmatrix} -1 & 0 \\ 3 & 1 \end{bmatrix} \begin{bmatrix} x_1 \\ x_2 \end{bmatrix} = 1 \begin{bmatrix} x_1 \\ x_2 \end{bmatrix}$$

Writing this as a system of linear equations,

$$-x_1 + 0 = -x_1$$

$$3x_1 + x_2 = -x_2$$

rearranging,

$$x_1 = x_1$$

$$3x_1 = -2x_2$$

So,

$$x_2 = -\frac{3}{2}x_1$$

If $x_1 = 1$, then $x_2 = -\frac{3}{2}$.

The eigenvector is $(1, -\frac{3}{2})$

To compute $A^{15}\begin{bmatrix} 2 \\ 0 \end{bmatrix}$,

By substituting u_1 & u_2 in the eqn

$x = a_1 u_1 + a_2 u_2$

Using the previous problem's method, we can find out the value for a_1 and a_2 and

applying the λ_1 and λ_2

we get,

$$A^{15}\begin{bmatrix} 2 \\ 0 \end{bmatrix} = \begin{bmatrix} -2 \\ 6 \end{bmatrix}$$

Prob. 73.

$$A = \begin{bmatrix} 5 & 7 \\ -2 & -4 \end{bmatrix}$$

To get the eigenvalues,

$$Ax = \lambda I x$$

$$(A - \lambda I)x = 0$$

$$[A - \lambda I] = 0$$

To get the above,

$$[A - \lambda I] = \begin{bmatrix} 5 & 7 \\ -2 & -4 \end{bmatrix} - \lambda \begin{bmatrix} 1 & 0 \\ 0 & 1 \end{bmatrix}$$

$$= \begin{bmatrix} 5 - \lambda & 7 \\ -2 & -4 - \lambda \end{bmatrix}$$

$$[A - \lambda I] = (5 - \lambda)(-4 - \lambda) + 14$$

$$-\lambda + \lambda^2 - 6 = 0$$

$$(\lambda - 3)(\lambda + 2) = 0$$

$$\lambda = 3 \ \& \ -2$$

The eigenvalues are 3 & -2.

To find the eigenvector associated with the eigenvalue $\lambda = 3$, we must determine x_1 and x_2 (not both equal to 0), so that

$$\begin{bmatrix} 5 & 7 \\ -2 & -4 \end{bmatrix} \begin{bmatrix} x_1 \\ x_2 \end{bmatrix} = 3 \begin{bmatrix} x_1 \\ x_2 \end{bmatrix}$$

Writing this as a system of linear equations,

$$5x_1 + 7x_2 = 3x_1$$

$$-2x_1 - 4x_2 = 3x_2$$

rearranging,

$$7x_2 = -2x_1$$

$$x_2 = -\frac{2}{7}x_1$$

If $x_1 = 1$, then $x_2 = -\frac{2}{7}$.

The eigenvector is $(1, -\frac{2}{7})$

To find the eigenvector associated with the eigenvalue $\lambda = -2$, we must determine x_1 and x_2 (not both equal to 0), so that

$$\begin{bmatrix} 5 & 7 \\ -2 & -4 \end{bmatrix} \begin{bmatrix} x_1 \\ x_2 \end{bmatrix} = -2 \begin{bmatrix} x_1 \\ x_2 \end{bmatrix}$$

Writing this as a system of linear equations,

$$5x_1 + 7x_2 = -2x_1$$

$$7x_1 = -7x_2$$

rearranging,

$$x_1 = -x_2$$

If $x_1 = 1$, then $x_2 = -1$.

The eigenvector is $(1, -1)$ To compute $A^{20} \begin{bmatrix} -3 \\ -2 \end{bmatrix}$,

use $x = a_1 u_1 + a_2 u_2$

Using the previous problem's method, we can find out the value for a_1 and a_2 and

applying the λ_1 and λ_2

we get,

$$A^{20} \begin{bmatrix} -3 \\ -2 \end{bmatrix} = -10^{10} \begin{bmatrix} -2.4 \\ 0.7 \end{bmatrix}$$

Prob. 75.

(a)

$$L = \begin{bmatrix} 2 & 4 \\ 0.3 & 0 \end{bmatrix}$$

$$Lx = \lambda I x$$

$$(L - \lambda I)x = 0$$

$$[L - \lambda I] = 0$$

To get the above,

$$[L - \lambda I] = \begin{bmatrix} 2 & 4 \\ 0.3 & 0 \end{bmatrix} - \lambda \begin{bmatrix} 1 & 0 \\ 0 & 1 \end{bmatrix}$$

$$= \begin{bmatrix} 2 - \lambda & 4 \\ 0.3 & -\lambda \end{bmatrix}$$

$$(2 - \lambda)(-\lambda) - 1.2 = 0$$

$$\lambda^2 - 2\lambda = 1.2$$

$$\lambda = \pm \frac{1}{5}(\sqrt{55} + 5)$$

i.e.,

$$\lambda = \frac{1}{5}(\sqrt{55} + 5)$$

$$\lambda = -\frac{1}{5}(\sqrt{55} - 5)$$

The eigenvalues are 2.48 & -0.48

(b) Biological Interpretation:

Among these two eigenvalues, the larger one determines the growth rate of the population. Here the larger one is 2.48.

(c) Stable age distribution:

The eigenvector corresponding to the larger eigenvalue is a stable age distribution.

To find the eigenvector associated with the larger eigenvalue $\lambda = 2.48$,

$$\begin{bmatrix} 2 & 4 \\ 0.3 & 0 \end{bmatrix} \begin{bmatrix} x_1 \\ x_2 \end{bmatrix} = 2 \begin{bmatrix} x_1 \\ x_2 \end{bmatrix}$$

Writing this as a system of linear equations,

$$2x_1 + 4x_2 = 2.48x_1$$

$$0.3x_1 = 2.48x_2$$

$$x_2 = \frac{0.3}{2.48}x_1$$

$$x_2 = 0.12x_1$$

If $x_1 = 1$ then $x_2 = .12$

The eigenvector is (1,0.12).

So the stable age distribution is $\begin{pmatrix} 1 \\ .12 \end{pmatrix}$

Prob. 77.

(a)

$$L = \begin{bmatrix} 7 & 3 \\ 0.1 & 0 \end{bmatrix}$$

$$Lx = \lambda I x$$

$$(L - \lambda I)x = 0$$

$$[L - \lambda I] = 0$$

To get the above,

$$[L - \lambda I] = \begin{bmatrix} 7 & 3 \\ 0.1 & 0 \end{bmatrix} - \lambda \begin{bmatrix} 1 & 0 \\ 0 & 1 \end{bmatrix}$$

$$= \begin{bmatrix} 7 - \lambda & 3 \\ 0.1 & -\lambda \end{bmatrix}$$

$$(7 - \lambda)(-\lambda) - 0.3 = 0$$

$$\lambda^2 - 7\lambda = 0.3$$

$$\lambda = \pm \frac{1}{10}(\sqrt{1255} + 35)$$

i.e.,

$$\lambda = \frac{1}{10}(\sqrt{1255} + 35)$$

$$\lambda = -\frac{1}{10}(\sqrt{1255} + 35)$$

The eigenvalues are 7.042 & -0.0426

(b) Biological Interpretation:

Among these two eigenvalues, the larger one determines the growth rate of the population. Here the larger one is 7.042.

(c) Stable age distribution:

The eigenvector corresponding to the larger eigenvalue is a stable age distribution.

To find the eigenvector associated with the larger eigenvalue $\lambda = 7.042$,

$$\begin{bmatrix} 7 & 3 \\ 0.1 & 0 \end{bmatrix} \begin{bmatrix} x_1 \\ x_2 \end{bmatrix} = 2.033 \begin{bmatrix} x_1 \\ x_2 \end{bmatrix}$$

Writing this as a system of linear equations,

$$7x_1 + 3x_2 = 7.042x_1$$

$$0.1x_1 = 7.042x_2$$

$$x_2 = \frac{0.1}{7.042}x_1$$

$$x_2 = 0.0142x_1$$

If $x_1 = 1$ then $x_2 = 0.0142$

The eigenvector is (1, 0.0142).

So the stable age distribution is $\begin{pmatrix} 1 \\ 0.0142 \end{pmatrix}$

Prob. 79.

(a)

$$L = \begin{bmatrix} 0 & 5 \\ 0.09 & 0 \end{bmatrix}$$

$$Lx = \lambda Ix$$

$$(L - \lambda I)x = 0$$

$$[L - \lambda I] = 0$$

To get the above,

$$[L - \lambda I] = \begin{bmatrix} 0 & 5 \\ 0.09 & 0 \end{bmatrix} - \lambda \begin{bmatrix} 1 & 0 \\ 0 & 1 \end{bmatrix}$$

$$= \begin{bmatrix} -\lambda & 5 \\ 0.09 & -\lambda \end{bmatrix}$$

$$(-\lambda)(-\lambda) - 0.45 = 0$$

$$\lambda^2 = 0.45$$

$$\lambda = \pm 0.67$$

i.e.,

$$\lambda = 0.67, -0.67$$

The eigenvalues are 0.67 & -0.67

(b) Biological Interpretation:

Among these two eigenvalues, the larger one determines the growth rate of the population. Here the larger one is 0.67.

(c) Stable age distribution:

The eigenvector corresponding to the larger eigenvalue is a stable age distribution. To find the eigenvector associated with the larger eigenvalue $\lambda = 0.67$,

$$\begin{bmatrix} 0 & 5 \\ 0.09 & 0 \end{bmatrix} \begin{bmatrix} x_1 \\ x_2 \end{bmatrix} = 2 \begin{bmatrix} x_1 \\ x_2 \end{bmatrix}$$

Writing this as a system of linear equations,

$$
\begin{aligned}
5x_2 &= 0.67x_1 \\
0.09x_1 &= 0.67x_2 \\
x_2 &= \frac{0.09}{0.67}x_1 \\
x_2 &= 1.34x_1
\end{aligned}
$$

If $x_1 = 1$ then $x_2 = 1.34$

The eigenvector is $(1,1.34)$.

So the stable age distribution is $\begin{pmatrix} 1 \\ 1.34 \end{pmatrix}$

9.4 Analytic Geometry

Prob. 1.

$$x = \begin{bmatrix} 1 \\ 4 \\ -1 \end{bmatrix}$$

$$y = \begin{bmatrix} -2 \\ 1 \\ 0 \end{bmatrix}$$

(a)

$$x + y = \begin{bmatrix} 1 - 2 \\ 4 + 1 \\ -1 + 0 \end{bmatrix}$$

$$= \begin{bmatrix} -1 \\ 5 \\ -1 \end{bmatrix}$$

(b)

$$2x = \begin{bmatrix} (2)(1) \\ (2)(4) \\ (2)(-1) \end{bmatrix}$$

$$= \begin{bmatrix} 2 \\ 8 \\ -2 \end{bmatrix}$$

(c)

$$-3y = \begin{bmatrix} (-3)(-2) \\ (-3)(1) \\ (-3)(0) \end{bmatrix}$$

$$= \begin{bmatrix} -6 \\ -3 \\ 0 \end{bmatrix}$$

Prob. 3.

$$A = (2, 3)$$

$$B = (4, 1)$$

$$\overrightarrow{AB} = \begin{bmatrix} b_1 - a_1 \\ b_2 - a_2 \end{bmatrix}$$

$$= \begin{bmatrix} 4 - 2 \\ 1 - 3 \end{bmatrix}$$

$$= \begin{bmatrix} 2 \\ -2 \end{bmatrix}$$

Prob. 5.

$$A = (0, -1, 3)$$

$$B = (-1, -1, 2)$$

$$\overrightarrow{AB} = \begin{bmatrix} b_1 - a_1 \\ b_2 - a_2 \\ b_3 - a_3 \end{bmatrix}$$

$$= \begin{bmatrix} -1 - 0 \\ -1 - 1 \\ 2 - (-3) \end{bmatrix}$$

$$= \begin{bmatrix} -1 - 0 \\ -1 - 1 \\ 2 - (-3) \end{bmatrix}$$

$$= \begin{bmatrix} -1 \\ -2 \\ 2 + 3 \end{bmatrix}$$

$$= \begin{bmatrix} -1 \\ -2 \\ 5 \end{bmatrix}$$

Prob. 7. Length of x

$$A = [1, 3]'$$

i.e.,

$$A = \begin{bmatrix} 1 \\ 3 \end{bmatrix}$$

$$|x| = \sqrt{(1)^2 + (3)^2}$$

$$= \sqrt{1+9}$$

$$= \sqrt{10}$$

Prob. 9. Length of x

$$A = [0,1,5]'$$

i.e.,

$$A = \begin{bmatrix} 0 \\ 1 \\ 5 \end{bmatrix}$$

$$|x| = \sqrt{(0)^2 + (1)^2 + (5)^2}$$

$$= \sqrt{0+1+25}$$

$$= \sqrt{26}$$

Prob. 11. Normalizing the vector

$$vector = [1,3,-1]'$$

i.e.,

$$A = \begin{bmatrix} 1 \\ 3 \\ -1 \end{bmatrix}$$

First calculate the length of the given vector

$$|x| = \sqrt{(1)^2 + (3)^2 + (-1)^2}$$

$$= \sqrt{1+9+1}$$

$$= \sqrt{11}$$

Normalizing the vector is finding out the unit vector.

$$\hat{x} = \frac{x}{|x|}$$

$$= \frac{1}{\sqrt{11}} \begin{bmatrix} 1 \\ 3 \\ -1 \end{bmatrix}$$

$$= \begin{bmatrix} \frac{1}{\sqrt{11}} \\ \frac{3}{\sqrt{11}} 3 \\ -\frac{1}{\sqrt{11}} \end{bmatrix}$$

Prob. 13. We find length: $\sqrt{36+0+0} = 6$. Hence, we get $\frac{1}{6}[6,0,0]' = [1,0,0]'$.

Prob. 15. Dot Product of $x = [1,2]'$ and $y = [3,-1]'$

$$x = \begin{bmatrix} 1 \\ 2 \end{bmatrix}$$

$$y = \begin{bmatrix} 3 \\ -1 \end{bmatrix}$$

$$x.y = \begin{bmatrix} 1 & 2 \end{bmatrix} \begin{bmatrix} 3 \\ -1 \end{bmatrix}$$

$$= \begin{bmatrix} 3 - 2 \end{bmatrix}$$

$$= 1$$

Prob. 17. Dot Product of $x = [0,-1,3]'$ and $y = [-3,1,1]'$

$$x = \begin{bmatrix} 0 \\ -1 \\ 3 \end{bmatrix}$$

$$y = \begin{bmatrix} -3 \\ 1 \\ 1 \end{bmatrix}$$

$$x.y = \begin{bmatrix} 0 & -1 & 3 \end{bmatrix} \begin{bmatrix} -3 \\ 1 \\ 1 \end{bmatrix}$$

$$= \begin{bmatrix} 0 - 1 + 3 \end{bmatrix}'$$

$$= 2$$

Prob. 19. Using the dot product to compute the length of $[0,-1,2]'$

$$|x| = \sqrt{x \cdot x} = \sqrt{(0)^2 + (-1)^2 + (2)^2} = \sqrt{5}$$

Prob. 21. Using the dot product to compute the length of $[1, 2, 3, 4]'$

$$|x| = \sqrt{x.x} = \sqrt{(1)^2 + (2)^2 + (3)^2 + (4)^2} = \sqrt{30}$$

Prob. 23.

$$x = \begin{bmatrix} 1 \\ 2 \end{bmatrix}$$

$$y = \begin{bmatrix} 3 \\ -1 \end{bmatrix}$$

To determine the angle θ between x & y

$$x.y = |x||y| \cos\theta \tag{9.4.1}$$

$$|x| = \sqrt{(1)^2 + (2)^2}$$

$$= \sqrt{1 + 4}$$

$$= \sqrt{5} \tag{9.4.2}$$

$$|y| = \sqrt{(3)^2 + (-1)^2}$$

$$= \sqrt{9 + 1}$$

$$= \sqrt{10} \tag{9.4.3}$$

$$x.y = \begin{bmatrix} 1 & 2 \end{bmatrix} \begin{bmatrix} 3 \\ -1 \end{bmatrix}$$

$$= [3 - 2]$$

$$= 1 \tag{9.4.4}$$

Substituting eqn (9.4.2), (9.4.3) and (9.4.4) in the eqn (9.4.1)

$$\cos\theta = \frac{1}{\sqrt{5}\sqrt{10}}$$

$$= \frac{1}{5\sqrt{2}}$$

$$\theta = \cos^{-1}\frac{1}{5\sqrt{2}}$$

Prob. 25.

$$x = \begin{bmatrix} 0 \\ -1 \\ 3 \end{bmatrix}$$

$$y = \begin{bmatrix} -3 \\ 1 \\ 1 \end{bmatrix}$$

To determine the angle θ between x & y

$$x.y = |x||y|\cos\theta \qquad (9.4.5)$$

$$|x| = \sqrt{(0)^2 + (-1)^2 + (3)^2}$$

$$= \sqrt{1+9}$$

$$= \sqrt{10} \qquad (9.4.6)$$

$$|y| = \sqrt{(-3)^2 + (1)^2 + (1)^2}$$

$$= \sqrt{9+1+1}$$

$$= \sqrt{11}$$

$$= \sqrt{11} \qquad (9.4.7)$$

$$x.y = \begin{bmatrix} 0 & -1 & 3 \end{bmatrix} \begin{bmatrix} -3 \\ 1 \\ 1 \end{bmatrix}$$

$$= [0 - 1 + 3]$$

$$= 2 \qquad (9.4.8)$$

Substituting eqn (9.4.6), (9.4.7) and (9.4.8) in the eqn (9.4.5)

$$cos\theta = \frac{2}{\sqrt{10}\sqrt{11}}$$

$$= -\frac{2}{\sqrt{110}}$$

$$\theta = \cos^{-1} -\frac{2}{\sqrt{110}}$$

Prob. 27. The vectors x & y are perpendicular if $x.y = 0$ $x = [1, -1]'$

Let us take

$$y = \begin{bmatrix} y_1 \\ y_2 \end{bmatrix}$$

$$x.y = \begin{bmatrix} 1 & -1 \end{bmatrix} \begin{bmatrix} y_1 \\ y_2 \end{bmatrix}$$

$$y_1 - y_2 = 0$$

So,

$$y_1 = y_2$$

From the above, we can assure that, any choice of numbers (y_1, y_2) that satisfies this equation would give a vector which will be perpendicular to x.

So,

if we give $y_1 = 1$ then $y_2 = 1$

$$y = \begin{bmatrix} 1 \\ 1 \end{bmatrix}$$

Prob. 29. Need to have $x \cdot y = 0$, meaning:

$$1 \cdot (y_1) + (-2) \cdot (y_2) + (4) \cdot (y_3) = 0.$$

We see that $y_1 = 2$, $y_2 = -1$, $y_3 = -1$ works. So, $y = [2, -1, -1]'$.

Prob. 31. The coordinates are $P = (0,0)$, $Q = (4,0)$ and $R = (4,3)$

$$\overrightarrow{PQ} = \begin{bmatrix} q_1 - p_1 \\ q_2 - p_2 \end{bmatrix}$$

$$= \begin{bmatrix} 4 - 0 \\ 0 - 0 \end{bmatrix}$$

$$= \begin{bmatrix} 4 \\ 0 \end{bmatrix}$$

$$\overrightarrow{QR} = \begin{bmatrix} r_1 - q_1 \\ r_2 - q_2 \end{bmatrix}$$

$$= \begin{bmatrix} 4 - 4 \\ 3 - 0 \end{bmatrix}$$

$$= \begin{bmatrix} 0 \\ 3 \end{bmatrix}$$

$$\overrightarrow{PR} = \begin{bmatrix} r_1 - p_1 \\ r_2 - p_2 \end{bmatrix}$$

$$= \begin{bmatrix} 4 - 0 \\ 3 - 0 \end{bmatrix}$$

$$= \begin{bmatrix} 4 \\ 3 \end{bmatrix}$$

To find out the length of the vector $\overrightarrow{PQ}$

$$\begin{aligned} |\overrightarrow{PQ}| &= \sqrt{(4)^2 + (0)^2} \\ &= \sqrt{16 + 0} \\ &= 4 \end{aligned} \tag{9.4.9}$$

To find out the length of the vector $\overrightarrow{QR}$

$$\begin{aligned} |\overrightarrow{QR}| &= \sqrt{(0)^2 + (3)^2} \\ &= \sqrt{0 + 9} \\ &= 3 \end{aligned} \tag{9.4.10}$$

To find out the length of the vector $\overrightarrow{PR}$

$$\begin{aligned} |\overrightarrow{PR}| &= \sqrt{(4)^2 + (3)^2} \\ &= \sqrt{16 + 9} \\ &= 5 \end{aligned} \tag{9.4.11}$$

Calculating the dot products

$$\begin{aligned} \overrightarrow{PQ}.\overrightarrow{QR} &= \begin{bmatrix} 4 & 0 \end{bmatrix} \begin{bmatrix} 0 \\ 3 \end{bmatrix} \\ &= 0 \end{aligned} \tag{9.4.12}$$

$$\begin{aligned} \overrightarrow{QR}.\overrightarrow{PR} &= \begin{bmatrix} 0 & 3 \end{bmatrix} \begin{bmatrix} 4 \\ 3 \end{bmatrix} \\ &= 9 \end{aligned} \tag{9.4.13}$$

$$\begin{aligned} \overrightarrow{PQ}.\overrightarrow{PR} &= \begin{bmatrix} 4 & 0 \end{bmatrix} \begin{bmatrix} 4 \\ 3 \end{bmatrix} \\ &= 16 + 0 \\ &= 0 \end{aligned} \tag{9.4.14}$$

To determine the angle θ between $\overrightarrow{PQ}$ & $\overrightarrow{QR}$

$$\overrightarrow{PQ}.\overrightarrow{QR} = |\overrightarrow{PQ}||\overrightarrow{PQ}|\cos\theta$$
$$cos\theta = \frac{\overrightarrow{PQ}.\overrightarrow{QR}}{|\overrightarrow{PQ}||\overrightarrow{PQ}|}$$

Substituting the values from the eqn (9.4.9), (9.4.10) & (9.4.12) in the above equation,

$$cos\theta = \frac{0}{(4)(3)}$$
$$= 0$$
$$\theta = \cos^{-1}(0) \qquad\qquad (9.4.15)$$

To determine the angle θ between $\overrightarrow{QR}$ & $\overrightarrow{PR}$

$$\overrightarrow{QR}.\overrightarrow{PR} = |\overrightarrow{QR}||\overrightarrow{PR}|\cos\theta$$
$$cos\theta = \frac{\overrightarrow{QR}.\overrightarrow{PR}}{|\overrightarrow{QR}||\overrightarrow{PR}|}$$

Substituting the values from the eqn (9.4.10), (9.4.11) & (9.4.13) in the above equation,

$$cos\theta = \frac{9}{(3)(5)}$$
$$\theta = \cos^{-1}\left(\frac{3}{5}\right) \qquad\qquad (9.4.16)$$

To determine the angle θ between $\overrightarrow{PQ}$ & $\overrightarrow{PR}$

$$\overrightarrow{PQ}.\overrightarrow{PR} = |\overrightarrow{PQ}||\overrightarrow{PR}|\cos\theta$$
$$cos\theta = \frac{\overrightarrow{PQ}.\overrightarrow{PR}}{|\overrightarrow{PQ}||\overrightarrow{PR}|}$$

Substituting the values from the eqn (9.4.9), (9.4.10) & (9.4.14) in the above equation,

$$cos\theta = \frac{16}{(4)(5)}$$
$$= \frac{4}{5}$$
$$\theta = \cos^{-1}\left(\frac{4}{5}\right) \qquad\qquad (9.4.17)$$

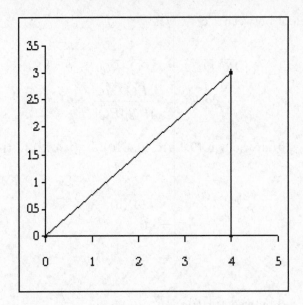

Prob. 33. The coordinates are $P = (1, 2, 3)$, $Q = (1, 5, 2)$ and $R = (2, 4, 2)$

$$\vec{PQ} = \begin{bmatrix} q_1 - p_1 \\ q_2 - p_2 \\ q_3 - p_3 \end{bmatrix} = \begin{bmatrix} 1 - 1 \\ 5 - 2 \\ 2 - 3) \end{bmatrix} = \begin{bmatrix} 0 \\ 3 \\ -1 \end{bmatrix}$$

$$\vec{QR} = \begin{bmatrix} r_1 - q_1 \\ r_2 - q_2 \\ r_3 - q_3 \end{bmatrix} = \begin{bmatrix} 2 - 1 \\ 4 - 5 \\ 2 - 2 \end{bmatrix} = \begin{bmatrix} 1 \\ -1 \\ 0 \end{bmatrix}$$

$$\vec{RP} = \begin{bmatrix} p_1 - r_1 \\ p_2 - r_2 \\ p_3 - r_3 \end{bmatrix} = \begin{bmatrix} 1 - 2 \\ 2 - 4 \\ 3 - 2 \end{bmatrix} = \begin{bmatrix} -1 \\ -2 \\ 1 \end{bmatrix}$$

To find out the length of the vector $\vec{PQ}$

$$|\vec{PQ}| = \sqrt{(0)^2 + (3)^2 + (-1)^2} = \sqrt{0 + 9 + 1} = \sqrt{10} \qquad (9.4.18)$$

To find out the length of the vector $\vec{QR}$

$$|\vec{QR}| = \sqrt{(1)^2 + (-1)^2 + (0)^2} = \sqrt{1 + 1} = \sqrt{2} \qquad (9.4.19)$$

To find out the length of the vector $\vec{RP}$

$$|\vec{RP}| = \sqrt{(-1)^2 + (-2)^2 + (1)^2} = \sqrt{1 + 4 + 1} = \sqrt{6} \qquad (9.4.20)$$

Calculating the dot products

$$\vec{PQ}.\vec{QR} = \begin{bmatrix} 0 & 3 & -1 \end{bmatrix} \begin{bmatrix} 1 \\ -1 \\ 0 \end{bmatrix} = 0 - 3 + 0 = -3 \tag{9.4.21}$$

$$\vec{QR}.\vec{RP} = \begin{bmatrix} 1 & -1 & 0 \end{bmatrix} \begin{bmatrix} -1 \\ -2 \\ 1 \end{bmatrix} = -1 + 2 + 0 = 1 \tag{9.4.22}$$

$$\vec{PQ}.\vec{RP} = \begin{bmatrix} -1 & -2 & 1 \end{bmatrix} \begin{bmatrix} 0 \\ 3 \\ -1 \end{bmatrix} = 0 - 6 - 1 = -7 \tag{9.4.23}$$

To determine the angle θ between $\vec{PQ}$ & $\vec{QR}$

$$\vec{PQ}.\vec{QR} = |\vec{PQ}||\vec{PQ}| \cos\theta$$
$$cos\theta = \frac{\vec{PQ}.\vec{QR}}{|\vec{PQ}||\vec{PQ}|}$$

Substituting the values from the eqn (9.4.18), (9.4.19) & (9.4.21) in the above equation,

$$cos\theta = \frac{-3}{(\sqrt{10})(\sqrt{2})} = cos\theta = \frac{-3}{2\sqrt{5}}$$
$$\theta = \cos^{-1}\left(\frac{-3}{2\sqrt{5}}\right) \tag{9.4.24}$$

To determine the angle θ between $\vec{QR}$ & $\vec{RP}$

$$\vec{QR}.\vec{RP} = |\vec{QR}||\vec{RP}| \cos\theta$$
$$cos\theta = \frac{\vec{QR}.\vec{RP}}{|\vec{QR}||\vec{RP}|}$$

Substituting the values from the eqn (9.4.19), (9.4.20) & (9.4.22) in the above equation,

$$cos\theta = \frac{1}{(\sqrt{2})(\sqrt{6})}$$
$$= \frac{1}{2\sqrt{3}}$$
$$\theta = \cos^{-1}\left(\frac{1}{2\sqrt{3}}\right) \tag{9.4.25}$$

To determine the angle θ between $\overrightarrow{PQ}$ & $\overrightarrow{RP}$

$$\overrightarrow{PQ}.\overrightarrow{RP} = |\overrightarrow{PQ}||\overrightarrow{RP}|\cos\theta$$
$$\cos\theta = \frac{\overrightarrow{PQ}.\overrightarrow{RP}}{|\overrightarrow{PQ}||\overrightarrow{RP}|}$$

Substituting the values from the eqn (9.4.18), (9.4.19) & (9.4.23) in the above equation,

$$\cos\theta = \frac{-7}{(\sqrt{6})(\sqrt{10})}$$
$$= \frac{-7}{2\sqrt{15}}$$
$$\theta = \cos^{-1}\left(\frac{-7}{2\sqrt{15}}\right) \qquad (9.4.26)$$

Prob. 35.

$$x_o = (2,1)$$
$$A = \begin{bmatrix} 1 \\ 2 \end{bmatrix}$$

To find the eqn. of the line

$$a(x - x_o) + b(y - y_o) = 0$$
$$1(x - 2) + 2(y - 1) = 0$$
$$x - 2 + 2y - 2 = 0$$
$$x + 2y - 4 = 0$$

Eqn. of the line is $x + 2y - 4 = 0$

Prob. 37.

$$x_o = (1,-2)$$
$$A = \begin{bmatrix} 4 \\ 1 \end{bmatrix}$$

To find the eqn. of the line

$$a(x - x_o) + b(y - y_o) = 0$$
$$(4)(x - 1) + 1(y + 2) = 0$$
$$4x - 4 + y + 2 = 0$$
$$4x + y - 2 = 0$$

Eqn. of the line is $4x + y - 2 = 0$

Prob. 39.

$$x_o = (-1, 2, 3)$$

$$A = \begin{bmatrix} 0 \\ -1 \\ 1 \end{bmatrix}$$

To find the eqn. of the line

$$
\begin{aligned}
a(x - x_o) + b(y - y_o) + c(z - z_o) &= 0 \\
0(x - 1) - 1(y - 2) + 1(z - 3) &= 0 \\
0 - 1(y - 2) + z - 3 &= 0 \\
-y + 2 + z - 3 &= 0 \\
-y + z - 1 &= 0 \\
y - z + 1 &= 0
\end{aligned}
$$

Eqn. of the line is:

$y - z + 1 = 0$

Prob. 41.

$$x_o = (0, 0, 0)$$

$$A = \begin{bmatrix} 1 \\ 0 \\ 0 \end{bmatrix}$$

To find the eqn. of the line

$$
\begin{aligned}
a(x - x_o) + b(y - y_o) + c(z - z_o) &= 0 \\
1(x + 0) - 0(y - 0) + 0(z - 0) &= 0 \\
x &= 0 \\
x &= 0
\end{aligned}
$$

Eqn. of the line is:

$x = 0$

Prob. 43. Parametric equation of the line in the x-y plane that goes through the point $(1,-1)$ in the direction of $\begin{bmatrix} 2 \\ 1 \end{bmatrix}$ To find the eqn. of the line

$$\begin{bmatrix} x \\ y \end{bmatrix} = \begin{bmatrix} 1 \\ -1 \end{bmatrix} + t \begin{bmatrix} 2 \\ -1 \end{bmatrix}$$

Writing the above matrix in the form of linear equations,

$$x = 1 + 2t \tag{9.4.27}$$

$$y = -1 - t \tag{9.4.28}$$

for eliminating t,

$$2t = x - 1$$

$$t = \frac{x}{2} - \frac{1}{2} \tag{9.4.29}$$

Substituting eqn (9.4.29) in eqn. (9.4.28)

$$y = -1 - \left(\frac{x}{2} - \frac{1}{2} \right)$$

$$= -1 - \frac{x}{2} + \frac{1}{2}$$

multiply by 2 on both sides,

$$2y = -x - 1$$

$$x + 2y + 1 = 0$$

$x + 2y + 1 = 0$ is the standard form of the equation of the line.

Prob. 45. Parametric equation of the line in the x-y plane that goes through the point $(3,-4)$ in the direction of $\begin{bmatrix} -1 \\ 2 \end{bmatrix}$ To find the eqn. of the line

$$\begin{bmatrix} x \\ y \end{bmatrix} = \begin{bmatrix} -1 \\ -2 \end{bmatrix} + t \begin{bmatrix} 1 \\ -3 \end{bmatrix}$$

$$\tag{9.4.30}$$

Writing the above matrix in the form of linear equations,

$$x = -1 + t \qquad (9.4.31)$$

$$y = -2 - 3t \qquad (9.4.32)$$

for eliminating t,

$$t = x + 1 \qquad (9.4.33)$$

Substituting eqn (9.4.33) in eqn. (9.4.32)

$$y = -2 - 3(x+1)$$

$$= -2 - 3x - 3$$

$$= -3x - 5$$

$$3x + y + 5 = 0$$

$3x + y + 5 = 0$ is the standard form of the equation of the line.

Prob. 47. Let us consider one of the two points say (-1,2) as the point P_o.

Then the direction of the indicated vector

$$\vec{U} = \begin{bmatrix} q_1 - p_1 \\ q_2 - p_2 \end{bmatrix}$$

$$= \begin{bmatrix} 3 - (-1) \\ 4 - 2 \end{bmatrix}$$

$$= \begin{bmatrix} 3 + 1 \\ 4 - 2 \end{bmatrix}$$

$$= \begin{bmatrix} 4 \\ 2 \end{bmatrix}$$

$$\begin{bmatrix} x \\ y \end{bmatrix} = \begin{bmatrix} -1 \\ 2 \end{bmatrix} + t \begin{bmatrix} 4 \\ 2 \end{bmatrix}$$

Writing the above matrix in the form of linear equations,

$$x = -1 + 4t \qquad (9.4.34)$$

$$y = 2 + 2t \qquad (9.4.35)$$

for eliminating t,

$$4t = x + 1$$

$$t = \frac{x+1}{4} \tag{9.4.36}$$

Substituting eqn (9.4.36) in eqn. (9.4.35)

$$y = 2 + 2\left(\frac{x+1}{4}\right)$$

$$= 2 + \frac{x}{2} + \frac{1}{2}$$

multiplying by 2 on both sides

$$2y = x + 5$$

$$x - 2y + 5 = 0$$

$x - 2y + 5 = 0$ is the standard form of the equation of the line.

Prob. 49. Let us consider one of the two points say (1, -3) as the point P_o.

Then the direction of the indicated vector

$$\vec{U} = \begin{bmatrix} q_1 - p_1 \\ q_2 - p_2 \end{bmatrix}$$

$$= \begin{bmatrix} 4 - 1 \\ 0 - (-3) \end{bmatrix}$$

$$= \begin{bmatrix} 3 \\ 3 \end{bmatrix}$$

$$\begin{bmatrix} x \\ y \end{bmatrix} = \begin{bmatrix} 1 \\ -3 \end{bmatrix} + t \begin{bmatrix} 3 \\ 3 \end{bmatrix}$$

Writing the above matrix in the form of linear equations,

$$x = 1 + 3t \tag{9.4.37}$$

$$y = 3 + 3t \tag{9.4.38}$$

for eliminating t,

$$3t = x - 1$$
$$t = \frac{x-1}{3} \tag{9.4.39}$$

Substituting eqn (9.4.39) in eqn. (9.4.38)

$$y = -3 + 3\left(\frac{x-1}{3}\right)$$
$$= -3 + x - 1$$
$$x - y - 4 = 0$$

$x - y - 4 = 0$ is the standard form of the equation of the line.

Prob. 51.

$$3x + 4y - 1 = 0$$

Eliminating the above equation

$$4y = -3x + 1$$
$$y = -\frac{3}{4}x + \frac{1}{4}$$

with x = t we can write the parametric equation as

$$x = t$$
$$y = -\frac{3}{4}t + \frac{1}{4}$$

Prob. 53.

$$2x + y - 3 = 0$$

Eliminating the above equation

$$y = -2x + 3$$

with x = t we can write the parametric equation as

$$x = t$$
$$y = -2t + 3$$

Prob. 55.

$$\begin{bmatrix} x \\ y \\ z \end{bmatrix} = \begin{bmatrix} 1 \\ -1 \\ 2 \end{bmatrix} + t \begin{bmatrix} 1 \\ -2 \\ 1 \end{bmatrix}$$

Writing the above matrix in the form of linear equations,

$$x = 1 + t \tag{9.4.40}$$

$$y = -1 - 2t \tag{9.4.41}$$

$$z = 2 + t \tag{9.4.42}$$

The above results itself tells about the parametric equation . But if we want to write the equation in the usual form we should solve the parametric equations as shown below.

To eliminate t, eqn (9.4.40) & eqn (9.4.42) can be rewritten as

$$t = x - 1 \tag{9.4.43}$$

$$t = z - 2 \tag{9.4.44}$$

We can write the eqn (9.4.41) by splitting it as

$$y = -1 - t - t$$

Substituting equation (9.4.43) & equation (9.4.44) in the above equation,

$$y = -1 - (x - 1) - (z - 2)$$
$$= -1 - x + 1 - z + 2)$$
$$= -x - z + 2)$$
$$x + y + z - 2 = 0$$

Prob. 57.

$$\begin{bmatrix} x \\ y \\ z \end{bmatrix} = \begin{bmatrix} -1 \\ 3 \\ -2 \end{bmatrix} + t \begin{bmatrix} -1 \\ -2 \\ 4 \end{bmatrix}$$

Writing the above matrix in the form of linear equations,

$$x = -1 - t \tag{9.4.45}$$

$$y = 3 - 2t \tag{9.4.46}$$

$$z = -2 + 4t \tag{9.4.47}$$

The eqn (9.4.45), eqn (9.4.46) & eqn (9.4.47) are the parametric equations of the line in the given x-y-z space

Prob. 59. Let us consider one of the two points say (5, 4, -1) as the point P_o.

Then the direction of the indicated vector

$$\vec{U} = \begin{bmatrix} q_1 - p_1 \\ q_2 - p_2 \\ q_3 - p_3 \end{bmatrix}$$

$$= \begin{bmatrix} 2 - 5 \\ 0 - 4 \\ 3 - (-1) \end{bmatrix}$$

$$= \begin{bmatrix} -3 \\ -4 \\ 4 \end{bmatrix}$$

$$\begin{bmatrix} x \\ y \\ z \end{bmatrix} = \begin{bmatrix} 5 \\ 4 \\ -1 \end{bmatrix} + t \begin{bmatrix} -3 \\ -4 \\ 4 \end{bmatrix}$$

Writing the above matrix in the form of linear equations,

$$x = 5 - 3t \tag{9.4.48}$$

$$y = 4 - 4t \tag{9.4.49}$$

$$y = -1 + 4t \tag{9.4.50}$$

The eqn (9.4.48), eqn (9.4.49) & eqn (9.4.50) are the parametric equations of the line in the given x-y-z space.

Prob. 61. The two points are (2, -3, 1) & (-5, 2, 1)

Let us consider one of the two points say (2, -3, 1) as the point P_o.

Then the direction of the indicated vector

$$\vec{U} = \begin{bmatrix} q_1 - p_1 \\ q_2 - p_2 \\ q_3 - p_3 \end{bmatrix}$$

$$= \begin{bmatrix} -5 - 2 \\ 2 - (-3) \\ 1 - 1 \end{bmatrix}$$

$$= \begin{bmatrix} -7 \\ 2 + 3 \\ 0) \end{bmatrix}$$

$$= \begin{bmatrix} -7 \\ 5 \\ 0 \end{bmatrix}$$

$$\begin{bmatrix} x \\ y \\ z \end{bmatrix} = \begin{bmatrix} 2 \\ -3 \\ 1 \end{bmatrix} + t \begin{bmatrix} -7 \\ 5 \\ 0 \end{bmatrix}$$

Writing the above matrix in the form of linear equations,

$$x = 2 - 7t \tag{9.4.51}$$

$$y = -3 + 5t \tag{9.4.52}$$

$$z = 1 \tag{9.4.53}$$

The eqn (9.4.51), eqn (9.4.52) & eqn (9.4.53) are the parametric equations of the line in the given x-y-z space.

Prob. 63. The given plane is (1, -1, 2)

The perpendicular line to the above mentioned plane is $\begin{bmatrix} 1 \\ 2 \\ 1 \end{bmatrix}$.

So, the equation of the plane in the three dimensional space is

$$x_o = (1, -1, 2)$$

$$A = \begin{bmatrix} 1 \\ 2 \\ 1 \end{bmatrix}$$

To find the eqn. of the line

$$a(x - x_o) + b(y - y_o) + c(z - z_o) = 0$$

$$1(x - 1) + 2(y + 1) + 1(z - 2) = 0$$

$$x - 1 + 2y + 2 + z - 2 = 0$$

$$x + 2y + z - 1 = 0$$

The line through the given points (0, -3, 2) and (-1, -2, 3)

$$\vec{U} = \begin{bmatrix} q_1 - p_1 \\ q_2 - p_2 \\ q_3 - p_3 \end{bmatrix}$$

$$= \begin{bmatrix} -1 - 0 \\ -2 + 3 \\ 3 - 2 \end{bmatrix}$$

$$= \begin{bmatrix} -1 \\ 1 \\ 1 \end{bmatrix}$$

To find out the angle of intersection:

We know that the given line is perpendicular to the given plane. So the angle the new line makes with the perpendicular line will be the angle of intersection.

$$u.a = |u||a|\cos\theta \tag{9.4.54}$$

$$|u| = \sqrt{(-1)^2 + (1)^2 + (1)^2}$$

$$= \sqrt{1 + 1 + 1}$$

$$= \sqrt{3} \tag{9.4.55}$$

$$|a| = \sqrt{(1)^2 + (2)^2 + (1)^2}$$

$$= \sqrt{1+4+1}$$

$$= \sqrt{6} \tag{9.4.56}$$

$$u.a = \begin{bmatrix} -1 & 1 & 1 \end{bmatrix} \begin{bmatrix} 1 \\ 2 \\ 1 \end{bmatrix}$$

$$= [-1 + 2 + 1]$$

$$= 2 \tag{9.4.57}$$

Substituting eqn (9.4.55), (9.4.56) and (9.4.57) in the eqn (9.4.54)

$$cos\theta = \frac{2}{\sqrt{3}\sqrt{6}}$$

$$= \frac{2}{3\sqrt{2}}$$

$$\theta = \cos^{-1} \frac{2}{3\sqrt{2}}$$

The angle of intersection is $\cos^{-1} \frac{2}{3\sqrt{2}}$

Prob. 65. The given plane is (0, -2, 1)

The perpendicular line to the above mentioned plane is $\begin{bmatrix} -1 \\ 1 \\ -1 \end{bmatrix}$.

So, the equation of the plane in the three dimensional space is

$$x_o = (0, -2, 1)$$

$$A = \begin{bmatrix} -1 \\ 1 \\ -1 \end{bmatrix}$$

To find the eqn. of the line

$$a(x - x_o) + b(y - y_o) + c(z - z_o) = 0$$

$$-1(x - 0) + 1(y + 2) - 1(z - 1) = 0$$

$$-x + 0 + y + 2 - z + 1 = 0$$

$$x - y + z - 3 = 0$$

The line through the point (5, -1, 0) and parallel to the given plane.

Parallel to the plane indicates, perpendicular to the given vector A.

Let us take the points of the unknown equation is (x, y, z)

$$\vec{U} = \begin{bmatrix} q_1 - p_1 \\ q_2 - p_2 \\ q_3 - p_3 \end{bmatrix}$$

$$= \begin{bmatrix} x - 5 \\ y + 1 \\ z + 0 \end{bmatrix}$$

If the line is parallel to the plane i.e., perpendicular ro the given line,

$$u.a = 0$$

i.e.,

$$\begin{bmatrix} x - 5 & y + 1 & z + 0 \end{bmatrix} \begin{bmatrix} -1 \\ 1 \\ -1 \end{bmatrix} = 0$$

$$(x - 5)(-1) + (y + 1) - z = 0$$

$$-x + 5 + y + 1 - z = 0$$

$$-x + y - z + 6 = 0$$

$$-x + y - z + 6 = 0$$

The equation of the line which is parallel to the given plane is $x - y + z - 6 = 0$

9.6 Review Problems

Prob. 1.

(a)

$$A = \begin{bmatrix} -1 & 1 \\ 0 & 2 \end{bmatrix}$$

$$x = \begin{bmatrix} 1 \\ 1 \end{bmatrix}$$

$$x \to Ax$$

$$\begin{bmatrix} x_1 \\ x_2 \end{bmatrix} \to \begin{bmatrix} -1 & 1 \\ 0 & 2 \end{bmatrix} \begin{bmatrix} 1 \\ 1 \end{bmatrix}$$

$$= \begin{bmatrix} -1+1 \\ 0+2 \end{bmatrix}$$

$$= \begin{bmatrix} 0 \\ 2 \end{bmatrix}$$

So this map stretches or contracts each coordinate separately. They are in the same quadrant.

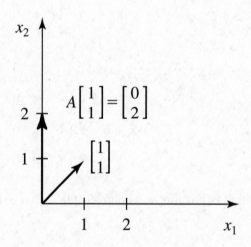

Length: $|x| = \sqrt{1^2 + 1^2} = \sqrt{1+1} = \sqrt{2}$

Angle: $\tan \alpha = \frac{x_2}{x_1} = \frac{1}{1} = 1$, so $\alpha = 45^o$.

(b) Eigenvalue & Eigenvector

$$A = \begin{bmatrix} -1 & 1 \\ 0 & 2 \end{bmatrix}$$

$$Ax = \lambda I x$$

$$(A - \lambda I)x = 0$$

$$[A - \lambda I] = 0$$

To get the above,

$$[A - \lambda I] = \begin{bmatrix} -1 & 1 \\ 0 & 2 \end{bmatrix} - \lambda \begin{bmatrix} 1 & 0 \\ 0 & 1 \end{bmatrix}$$

$$= \begin{bmatrix} -1 - \lambda & 1 \\ 0 & 2 - \lambda \end{bmatrix}$$

$$(-1 - \lambda)(2 - \lambda) = 0$$

So,

$$\lambda = -1 \ \& \ 2$$

The eigenvalues are 2 & -1.

To find the eigenvector associated with the eigenvalue $\lambda = -1$, we must determine x_1 and x_2 (not both equal to 0), so that

$$\begin{bmatrix} -1 & 1 \\ 0 & 2 \end{bmatrix} \begin{bmatrix} x_1 \\ x_2 \end{bmatrix} = -1 \begin{bmatrix} x_1 \\ x_2 \end{bmatrix}$$

Writing this as a system of linear equations,

$$-x_1 + x_2 = -x_1$$

$$-2x_2 = -x_2$$

$3x_2 = 0$

$x_2 = 0$

So substituting $x_2 = 0$ in the eqn $-x_1 + x_2 = -x_1$,

$-x_1 = -x_1$

so, x_1 can take any value other than zero. So, let us take $x_2 = 1$

The eigenvector is $(1, 0)$

To find the eigenvector associated with the eigenvalue $\lambda = 2$, we must determine x_1 and x_2 (not both equal to 0), so that

$$\begin{bmatrix} -1 & 1 \\ 0 & 2 \end{bmatrix} \begin{bmatrix} x_1 \\ x_2 \end{bmatrix} = 2 \begin{bmatrix} x_1 \\ x_2 \end{bmatrix}$$

Writing this as a system of linear equations,

$$\begin{aligned} -x_1 + x_2 &= 2x_1 \\ 2x_2 &= 2x_2 \end{aligned}$$

rearranging the above equations,

$$\begin{aligned} x_2 &= 3x_1 \\ x_1 &= x_2 \end{aligned}$$

If we apply $x_1 = 1$, in the equation

$x_2 = 3$. The eigenvector is $(1, 3)$

(c)

$$A = \begin{bmatrix} -1 & 1 \\ 0 & 2 \end{bmatrix}$$

$$u_i = \begin{bmatrix} 1 \\ 0 \end{bmatrix}$$

$$Au_i = \begin{bmatrix} -1 & 1 \\ 0 & 2 \end{bmatrix} \begin{bmatrix} -1+0 \\ 0+0 \end{bmatrix}$$

$$= \begin{bmatrix} -1+0 \\ 0+0 \end{bmatrix}$$

$$= \begin{bmatrix} -1 \\ 0 \end{bmatrix}$$

$$Au_{ii} = \begin{bmatrix} -1 & 1 \\ 0 & 2 \end{bmatrix} \begin{bmatrix} 1 \\ 3 \end{bmatrix}$$

$$= \begin{bmatrix} -1+3 \\ 0+6 \end{bmatrix}$$

$$= \begin{bmatrix} 2 \\ 6 \end{bmatrix}$$

That will lie on the same eigenvector.

$$x = a_1 u_1 + a_2 u_2$$

$$\begin{bmatrix} 1 \\ 1 \end{bmatrix} = a_1 \begin{bmatrix} 1 \\ 0 \end{bmatrix} + a_2 \begin{bmatrix} 1 \\ 1 \end{bmatrix}$$

$$a_1 + a_2 = 1 \qquad (9.6.1)$$

$$3a_2 = 1$$

$$a_2 = \frac{1}{3} \qquad (9.6.2)$$

Substituting eqn (9.6.2) in (9.6.1)

$$a_1 = 1 - \frac{1}{3}$$

$$= \frac{2}{3}$$

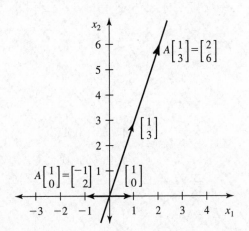

(d) u_1 and u_2 are both eigenvectors corresponding to A.

$$Au_1 = \lambda_1 u_1$$

$$Au_2 = \lambda_2 u_2$$

$$\begin{aligned}
Ax &= a_1\lambda_1 u_1 + a_2\lambda_2 u_2 \\
&= \frac{2}{3}(-1)\begin{pmatrix} 1 \\ 0 \end{pmatrix} + \frac{1}{3}(2)\begin{pmatrix} 1 \\ 3 \end{pmatrix} \\
&= \begin{pmatrix} -\frac{2}{3} \\ 0 \end{pmatrix} + \begin{pmatrix} \frac{2}{3} \\ 2 \end{pmatrix} \\
&= \begin{pmatrix} 0 \\ 2 \end{pmatrix}
\end{aligned}$$

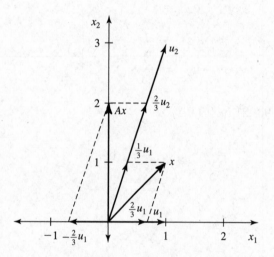

Prob. 3.

$$L = \begin{bmatrix} 1.5 & 0.875 \\ 0.5 & 0 \end{bmatrix}$$

$$Lx = \lambda I x$$

$$(L - \lambda I)x = 0$$

$$[L - \lambda I] = 0$$

To get the above,

$$[L - \lambda I] = \begin{bmatrix} 1.5 & 0.875 \\ 0.5 & 0 \end{bmatrix} - \lambda \begin{bmatrix} 1 & 0 \\ 0 & 1 \end{bmatrix}$$

$$= \begin{bmatrix} 1.5 - \lambda & 0.875 \\ 0.5 & -\lambda \end{bmatrix}$$

$$(1.5 - \lambda)(-\lambda) - (0.875)(0.5) = 0$$

$$\lambda^2 - 1.5\lambda - 0.4375 = 0$$

$$(\lambda - 1.75)(\lambda + 0.25) = 0$$

$$\lambda = 1.75, \ -0.25$$

The eigenvalues are 1.75 & -0.25

Among these two eigenvalues, the larger one determines the growth rate of the population. Here the larger one is 1.75.

Stable age distribution:

The eigenvector corresponding to the larger eigenvalue is a stable age distribution. To find the eigenvector associated with the larger eigenvalue $\lambda = 1.75$,

$$\begin{bmatrix} 1.5 & 0.875 \\ 0.5 & 0 \end{bmatrix} \begin{bmatrix} x_1 \\ x_2 \end{bmatrix} = 1.75 \begin{bmatrix} x_1 \\ x_2 \end{bmatrix}$$

Writing this as a system of linear equations,

$$1.5x_1 + 0.875x_2 = 1.75x_1$$

$$0.5x_1 = 1.75x_2$$

$$\frac{1}{2}x_1 = \frac{7}{4}x_2$$

$$x_2 = \frac{2}{7}x_1$$

If $x_1 = 1$ then $x_2 = \frac{2}{7}$

The eigenvector is $(1, \frac{2}{7})$.

So the stable age distribution is $\begin{pmatrix} 1 \\ \frac{2}{7} \end{pmatrix}$

Prob. 5.

$$AB = \begin{bmatrix} 0 & 1 \\ 2 & -1 \end{bmatrix}$$

$$A^{-1} = \begin{bmatrix} 4 & -1 \\ 8 & -1 \end{bmatrix}$$

$$B = A^{-1}AB$$

So

$$A^{-1}AB = \begin{bmatrix} 4 & -1 \\ 8 & -1 \end{bmatrix} \begin{bmatrix} 0 & 1 \\ 2 & -1 \end{bmatrix}$$

$$= \begin{bmatrix} 0-2 & 4+1 \\ 0-2 & 8+1 \end{bmatrix}$$

$$= \begin{bmatrix} -2 & 5 \\ -2 & 9 \end{bmatrix}$$

Prob. 7. The two different ways to solve the system of equations of the form

I Graphical Solution

Recall that the standard form of a linear equation in two variables is $Ax + By = C$ where A,B, and C are constants, A and B are not both equal to 0, and x and y are the two variables; its graph is a straight line. Any point (x, y) on this straight line satisfies (or solves) the equation Ax + By = C. We can extend this to more than one equation, to get the following system

$$a_{11}x_1 + a_{12}x_2 = b_1$$
$$a_{21}x_1 + a_{22}x2 = b_2$$

where a_{11}, a_{12}, a_{21}, a_{22} b_1, *and* b_2 are constants, x and y are the two variables. Finding an ordered pair (x, y) that satisfies each equation of the system given above. Because each equation in the system describes a straight line, we are therefore finding out the the point of intersection of these two lines.

The following three cases are possible:

1. The two lines have exactly one point of intersection. In this case, the system has exactly one solution.

2. The two lines are parallel and do not intersect. In this case, the system has no solution.

3. The two lines are parallel and intersect (that is, they are identical). In this case, the system has infinitely many solutions, namely each point on the line.

II Solution Method

$$a_{11}x_1 + a_{12}x_2 = b_1$$
$$a_{21}x_1 + a_{22}x2 = b_2$$

We will transform this system into an equivalent system in upper triangular form. To do so we will use the following three basic operations:

i. Multiplying an equation by a nonzero constant

ii. Adding one equation to another

iii. Rearranging the order of the equations

This method is named as Gaussian elimination.

The above mentioned three cases are identified such that:

1. If $a_{11}a_{22} - a_{12}a_{22} \neq 0$ Then the two lines have exactly one point of intersection. In this case, the system has exactly one solution.

2. If $a_{11}a_{22} - a_{12}a_{22} = 0$ and $b_1 \neq b_2$ then the two lines are parallel and do not intersect. In this case, the system has no solution.

3. If $a_{11}a_{22} - a_{12}a_{22} = 0$ and $b_1 = b_2$ then the two lines are parallel and intersect (that is, they are identical). In this case, the system has infinitely many solutions, namely each point on the line.

Prob. 9.

$$a_{11}x_1 + a_{12}x_2 = b_1$$
$$a_{21}x_1 + a_{22}x2 = b_2$$

The system will have infinitely many solutions if and only if $a_{11}a_{12} - a_{12}a_{22} = 0$ and $b_1 = b_2$

The given system is

$$
\begin{aligned}
ax + 3y &= 0 \\
x - y &= 0
\end{aligned}
$$

If we write it in the form of matrix (equivalent system)

$$
\begin{bmatrix} a & 3 \\ 1 & -1 \end{bmatrix}
\begin{bmatrix} x \\ y \end{bmatrix}
=
\begin{bmatrix} 0 \\ 0 \end{bmatrix}
$$

Here $b_1 = b_2$ i.e., 0, $-a - 3 = 0$ So, if $a = -3$ then the system will have infinitely many solutions. i.e., the two lines are parallel and intersect (that is, they are identical).

Prob. 11. We need to have the eigenvalues:

$$
\det(A - \lambda I) = 0
$$

$$
\det \begin{pmatrix} 0.5 - \lambda & 2.3 \\ a & -\lambda \end{pmatrix} = 0
$$

$$
-\lambda(0.5 - \lambda) - 2.3a = 0
$$

$$
\lambda_{1,2} = \frac{0.5 \pm \sqrt{0.25 + 9.2a}}{2}
$$

The largest eigenvalue must be greater than 1

$$
0.5 + \sqrt{0.25 + 9.2a} > 2
$$

we get $\frac{5}{23} < a \le 1$.

Prob. 13. Eigenvalues satisfy: $\det(A - I\lambda) = 0$

$$
\det \begin{pmatrix} a - \lambda & c \\ 0 & b - \lambda \end{pmatrix} = 0
$$

$$
(a - \lambda)(b - \lambda) = 0, \quad \lambda_1 = a, \quad \lambda_2 = b.
$$

Chapter 10

Multivariable Calculus

10.1   Functions of Two or More Independent Variables

Prob. 1. We have $CO = (HR) \cdot (SU)$. If NR is in $\frac{\text{beats}}{\text{minute}}$, SV is $\frac{\text{liters}}{\text{beat}}$, then

$$CO = \frac{\text{beats}}{\text{minute}} \cdot \frac{\text{liters}}{\text{beat}} = \frac{\text{liters}}{\text{minute}},$$

Also,

$$\texttt{domain} : \{(HR, SV) : HR \geq -. \quad SV \geq 0\}$$

$$\texttt{range} : \{CO : \quad CO \geq 0\}/$$

Prob. 3.

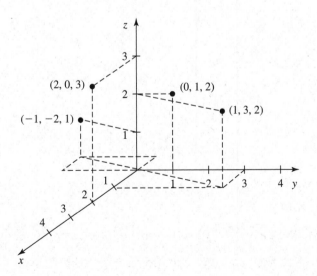

Prob. 5. $\frac{4}{13}$

Prob. 7.

a) $f_1(-1, 2) = 2(-1) - 3(2)^2 = -2 - 12 = -14$

b) $f_2(2, -1) = 2(2) - 3(-1)^2 = 4 - 3 = 1$

Prob. 9. 0.904837

Prob. 11. $h(2, -1) = (-1) \cdot e^2 = -e^2$.

Prob. 13. Domain is $\mathbb{R}^2$. Range is $\mathbb{R}_+$. The level curves are circles $x^2 + y^2 = c$ for $c \geq 0$ (for $c = 0$ the level curve is just a point).

Prob. 15. Domain is the set $\{(x, y) \in \mathbb{R}^2 : y > x^2\}$. Range is $\mathbb{R}$. Level curves are parabolas $y - x^2 = 2^c$, for every real c.

Prob. 17. For $f(x, y) = \frac{x-y}{x+y} = \frac{x+y-2y}{x+y} = 1 - \frac{2y}{x+y}$. We see that domain is: $\{(x, y) : x \neq -y\}$. Range is: all reals.

level curves: $\frac{x-y}{x+y} = c$, meaning $y = \frac{1-c}{1+c}x$. These are straight lines through the origin, $c \neq -1$.

Prob. 19. 10.23

Prob. 21. 10.24

Prob. 23.

(a) The level curves are circles.

Let $z = f_1(x, y)$. We have $f_1(x, 0) = x^2$ on the intersection of the surface with the plane xOz, therefore the curve has equation $z = x^2$. Similarly, $z = f_1(0, y) = y^2$ is the intersection with the plane yOz.

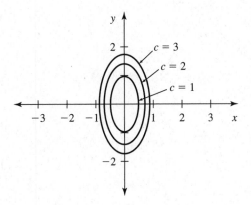

(b) For $z = f_4(x, y)$, we have $f_4(x, 0) = 4x^2$ on the intersection of the surface with the plane xOz, therefore the curve has equation $z = 4x^2$. Similarly, the surface intersects the plane yOz along the curve $z = y^2$.

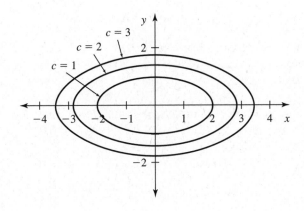

(c) For $z = f_{1/4}(x, y)$, we have $f_{1/4}(x, 0) = 0.25x^2$ on the intersection of the surface with the plane xOz, therefore the curve has equation $z = 0.25x^2$. Similarly, the surface intersects the plane yOz along the curve $z = y^2$.

Prob. 25. $23, 18, 15.$

10.2   Limits and Continuity

Prob. 1. 1.

Prob. 3. $2^2(-1)^3 - 3(2)(-1) = -4 + 6 = 2$

Prob. 5. $(-1)^2((3)^2 - 3(-1)(3)) = 9 + 9 = 18$

Prob. 7. $-\frac{1}{2}$.

Prob. 9. $\frac{(1)^2 + (0)^2}{(1)^2 - (0)^2} = 1$

Prob. 11. $-\frac{3}{2}$.

Prob. 13. $\frac{2}{3}$.

Prob. 15. We have $\lim_{x \to 0^+} \frac{x^2 - 2 \times 0^2}{x^2 + 0^2} = 1 \neq \lim_{y \to 0^+} \frac{0^2 - 2y^2}{0^2 + y^2} = -2$.

Prob. 17. We have $\lim_{x \to 0} \frac{4x \times 0}{x^2 + 0^2} = 0 = \lim_{y \to 0} \frac{4 \times 0 \times y}{0^2 + y^2}$, but $\lim_{x \to 0} \frac{4x^2}{2x^2} = 2$, so the limit does not exist.

Prob. 19. For $y = mx$ we get $\lim_{x \to 0} \frac{2mx^2}{x^3 + mx^2} = 2$. For $y = x^2$ we get $\lim_{x \to 0} \frac{2x^3}{2x^3} = 1$, so the given limit does not exist.

Prob. 21. Taking $\delta = \varepsilon$ it is easy to see that $\lim_{(x,y) \to (0,0)} = 0$. We have also $f(0,0) = 0$, therefore f is continuous at $(0,0)$.

Prob. 23. By problem 17 there is no limit at $(0,0)$.

Prob. 25. By problem 19 there is no limit at $(0,0)$.

Prob. 27.

(a) $h = f \circ g$ where $g(x,y) = x^2 + y^2$ and $h(z) = \sin z$.

(b) All $\mathbb{R}^2$.

Prob. 29.

(a) $h = f \circ g$ where $g(x,y) = xy$ and $h(z) = e^z$.

(b) All $\mathbb{R}^2$.

Prob. 31. $(x-1)^2 + (y+1)^2 < 4$.

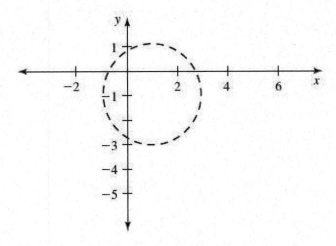

Prob. 33. Open disk centered at $(0,2)$ with radius 3.

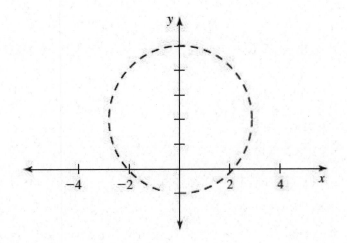

Prob. 35. As $2x^2 + y^2 \le 2(x^2 + y^2)$, given $\varepsilon > 0$ we can take $\delta = \varepsilon/2$. We have

$$0 < x^2 + y^2 < \delta \Rightarrow x^2 + y^2 < \varepsilon/2 \Rightarrow 2(x^2 + y^2) < \varepsilon \Rightarrow |f(x,y)| < \varepsilon.$$

10.3   Partial Derivatives

Prob. 1. $\frac{\partial f}{\partial x} = 2xy + y^2$; $\frac{\partial f}{\partial y} = x^2 + 2xy$.

Prob. 3. $\frac{\partial f}{\partial x} = \frac{3\sqrt{xy^3}}{2} - \frac{2\sqrt[3]{y^2}}{3\sqrt[3]{x}}$;

$\frac{\partial f}{\partial y} = \frac{3\sqrt{yx^3}}{2} - \frac{2\sqrt[3]{x^2}}{3\sqrt[3]{y}}$.

Prob. 5. $\frac{\partial f}{\partial x} = \frac{\partial f}{\partial y} = \cos(x + y)$.

Prob. 7.

$$\frac{\partial f}{\partial x} = 2\cos(x^2 - 2y) \cdot (-\sin(x^2 - 2y)) \cdot 2x$$

$$\frac{\partial f}{\partial y} = 2\cos(x^2 - 2y) \cdot (-\sin(x^2 - 2y)) \cdot (-2).$$

Prob. 9. $\frac{\partial f}{\partial x} = \frac{\partial f}{\partial y} = \frac{e^{\sqrt{x+y}}}{2\sqrt{x+y}}$.

Prob. 11. $\frac{\partial f}{\partial x} = e^x \sin(xy) + e^x y \cos(xy)$; $\frac{\partial f}{\partial y} = e^x x \cos(xy)$.

Prob. 13. $\frac{\partial t}{\partial x} = \frac{1}{2x+y} \cdot 2$, $\frac{\partial f}{\partial y} = \frac{1}{2x+y}$

Prob. 15. $\frac{\partial f}{\partial x} = \frac{-2x}{(y^2-x^2)\ln 3}$; $\frac{\partial f}{\partial y} = \frac{2y}{(y^2-x^2)\ln 3}$.

Prob. 17. 6.

Prob. 19. $3e^5$.

Prob. 21. $\frac{\partial f}{\partial z} = \frac{1}{xz}x$, so $f_z(e, 1) = 1$.

Prob. 23. $\frac{2}{9}$.

Prob. 25. $\frac{\partial f}{\partial x}(1,1) = \frac{\partial f}{\partial y}(1,1) = -2$. The fact that $\frac{\partial f}{\partial x}(1,1) = -2$ means that the slope of the tangent to the graph of the function $y \mapsto f(1, y)$ at the point $y = 1$ is -2. The fact that $\frac{\partial f}{\partial y}(1,1) = -2$ means that the slope of the tangent to the graph of the function $x \mapsto f(x, 1)$ at the point $x = 1$ is -2.

Prob. 27. $\frac{\partial f}{\partial x}(-2,1) = -4$; $\frac{\partial f}{\partial y}(-2,1) = 4$. The fact that $\frac{\partial f}{\partial x}(-2,1) = -4$ means that the slope of the tangent to the graph of the function $y \mapsto f(-2, y)$ at the point $y = 1$ is -4. The fact that $\frac{\partial f}{\partial y}(-2,1) = 4$ means that the slope of the tangent to the graph of the function $x \mapsto f(x, 1)$ at the point $x = -2$ is 4.

Prob. 29.

(a) $\frac{\partial P_e}{\partial a} = \frac{NT}{(1+aT_hN)^2} > 0$, therefore P_e increases when a increases.

(b) $\frac{\partial P_e}{\partial T} = \frac{aN}{1+T_hN} > 0$, therefore P_e increases when N increases.

Prob. 31. $\frac{\partial f}{\partial x} = 2xz - y$, $\frac{\partial f}{\partial y} = z^2 - x$, $\frac{\partial f}{\partial z} = x^2 + 2yz$

Prob. 33.

$$\frac{\partial f}{\partial x} = 3x^2 y^2 z + \frac{1}{yz}, \quad \frac{\partial f}{\partial y} = 2x^3 yz - \frac{x}{zy^2}$$

$$\frac{\partial f}{\partial z} = x^3 y^2 - \frac{x}{yz^2}$$

Prob. 35. $\frac{\partial f}{\partial x} = e^{x+y+z}$; $\frac{\partial f}{\partial y} = e^{x+y+z}$; $\frac{\partial f}{\partial z} = e^{x+y+z}$.

Prob. 37. $\frac{\partial f}{\partial x} = \frac{1}{x+y+z}$; $\frac{\partial f}{\partial y} = \frac{1}{x+y+z}$; $\frac{\partial f}{\partial z} = \frac{1}{x+y+z}$.

Prob. 39. $\frac{\partial^2 f}{\partial x^2} = 2y$.

Prob. 41. $\frac{\partial^2 f}{\partial x \partial y} = e^y$.

Prob. 43. $\frac{\partial^2 f}{\partial u^2} = 2\sec^2(u+w)\tan(u+w)$.

Prob. 45. $\frac{\partial^3 f}{\partial x^2 \partial y} = -6x\sin y$.

Prob. 47. We have $f(x,y) = xy$ with constraint $g(x,y) = x + y - c = 0$, applying Lagrange multipliers gives: $x = y$. So, we get a square.

Prob. 49. We have $f(x,y) = 3x + 2y$ with constraint $g(x,y) = x \cdot y - 384 = 0$, applying Lagrange multipliers gives: $x = 16$, $y = 24$. So, the perimeter is 96.

Prob. 51. We have $f(x,y) = \frac{1}{2}x \cdot y$ with constraint $g(x,y) = x^2 + y^2 - 16 = 0$, applying Lagrange multipliers gives: $c = 2\sqrt{2}$, $y = 4$. So, the area is 4.

Prob. 53. We have $f(x,y) = 2x + 2y$ with constraint $g(x,y) = y - \frac{1}{x} = 0$, applying Lagrange multipliers gives: $x = y = 1$. So, perimeter is 4.

10.4   Tangent Planes, Differentiability, and Linearization

Prob. 1. $z = 6x + 4y - 8$.

Prob. 3. $z = -2x - y + 2$

Prob. 5. $z = y$.

Prob. 7. $z = 2ex - e$.

Prob. 9. $z = x + y - 1$

Prob. 11. $\frac{\partial f}{\partial x} = y^2 + 2xy$, $\frac{\partial f}{\partial y} = 2xy + x^2$, as these are polynomials they are continuous everywhere, therefore f is differentiable at any point.

Prob. 13. $\frac{\partial f}{\partial x} = -\sin(x+y) = \frac{\partial f}{\partial y}$, as this function is continuous everywhere, f is differentiable at any point.

Prob. 15. $\frac{\partial f}{\partial x} = 1 - 2y$, $\frac{\partial f}{\partial y} = 2y - 2x$, as these are polynomials they are continuous everywhere, therefore f is differentiable at any point.

Prob. 17. $L(x, y) = x - 3y$.

Prob. 19. $L(x, y) = \frac{1}{2}x + 2y + \frac{1}{2}$.

Prob. 21. $L(x, y) = x + y$.

Prob. 23. $L(x, y) = x + \frac{1}{2}y + \ln 2 - \frac{3}{2}$.

Prob. 25. $L(x, y) = 1 + x + y$, $L(0.1, 0.05) = 1.15$, $f(0.1, 0.05) = 1.1618$.

Prob. 27. $L(x, y) = 2x - 3y - 2$, $L(1.1, 0.1) = -0.1$, $f(1.1, 0.1) = -0.0943$.

Prob. 29. $\begin{bmatrix} 1 & 1 \\ 2x & -2y \end{bmatrix}$.

Prob. 31. $\begin{bmatrix} e^{x-y} & -e^{x-y} \\ e^{x+y} & e^{x+y} \end{bmatrix}$

Prob. 33. $\begin{bmatrix} -\sin(x-y) & \sin(x-y) \\ -\sin(x+y) & -\sin(x+y) \end{bmatrix}$

Prob. 35. $\begin{bmatrix} 4xy + 12x^2 - 3 \\ e^x \sin y\, e^x \cos y \end{bmatrix}$.

Prob. 37. $L(x, y) = \begin{bmatrix} 4x + 2y - 4 \\ -x - y + 3 \end{bmatrix}$.

Prob. 39. $L(x, y) = \begin{bmatrix} y \\ 1 - y \end{bmatrix}$.

Prob. 41. $L(x, y) = \begin{bmatrix} 0 \\ x - y + 1 \end{bmatrix}$.

Prob. 43. $L(x, y) = \begin{bmatrix} x + y - 1 \\ 2x - 2y + 2 \end{bmatrix}$, $L(-0.9, 1.05) = \begin{bmatrix} -0.85 \\ -1.9 \end{bmatrix}$,

$f(-0.9, 1.05) = \begin{bmatrix} -0.8571 \\ -1.89 \end{bmatrix}$.

Prob. 45. $L(x, y) = \begin{bmatrix} \frac{x}{2} + \frac{y}{4} + 1 \\ x - 4y + 4 \end{bmatrix}$, $L(1.05, 2.05) = \begin{bmatrix} 2.0375 \\ -3.15 \end{bmatrix}$,

$f(1.05, 2.05) = \begin{bmatrix} 2.03715 \\ -3.1525 \end{bmatrix}$.

10.5 More About Derivatives

Prob. 1. $18 \ln 2 + 8$.

Prob. 3. $\frac{4\pi + 3\sqrt{3}}{2\sqrt{4\pi^2 + 9}}$.

Prob. 5. 0.

Prob. 7. $\frac{\partial f}{\partial x} \frac{du}{dt} + \frac{\partial f}{\partial y} \frac{dv}{dt}$.

Prob. 9. $\frac{-2x}{x^2 + 2y + y^2}$.

Prob. 11. $\frac{3x^2 y + 3y^3 - 2x}{2y - 3x^3 - 3xy^2}$.

Prob. 13. $\frac{-1}{\sqrt{1 - x^2}}$, for $-1 \leq x \leq 1$.

Prob. 15. We have $\frac{\partial r}{\partial t} = \frac{\partial r}{\partial F} \frac{\partial F}{\partial t} + \frac{\partial r}{\partial N} \frac{\partial N}{\partial t} < 0$, therefore the growth rate is negative.

Prob. 17. $\begin{bmatrix} 3x^2 y^2 \\ 2x^3 y \end{bmatrix}$.

Prob. 19. $\begin{bmatrix} \frac{3x^2 - 3y}{2\sqrt{x^3 - 3xy}} \end{bmatrix}$.

Prob. 21. $\begin{bmatrix} \frac{xe^{\sqrt{x^2 + y^2}}}{\sqrt{x^2 + y^2}} \end{bmatrix}$.

Prob. 23. $\begin{bmatrix} \frac{x^2 - y^2}{x(x^2 + y^2)} \\ \frac{y^2 - x^2}{y(x^2 + y^2)} \end{bmatrix}$.

Prob. 25. $\frac{2}{\sqrt{3}}$.

Prob. 27. $-\sqrt{2}$.

Prob. 29. $\frac{15}{\sqrt{10}}$.

Prob. 31. $\frac{13}{\sqrt{2}}$.

Prob. 33. $D_u f(1,6) = -\frac{1}{4\sqrt{29}}$.

Prob. 35. $\begin{bmatrix} 5 \\ -3 \end{bmatrix}$.

Prob. 37. $\begin{bmatrix} \frac{5}{4} \\ \frac{3}{4} \end{bmatrix}$.

Prob. 39. $\begin{bmatrix} 3 \\ 4 \end{bmatrix}$.

Prob. 41. We get $\begin{bmatrix} 2 \\ -27 \end{bmatrix}$, normalized is $\frac{1}{\sqrt{733}} \begin{bmatrix} 2 \\ -27 \end{bmatrix}$.

Prob. 43. $\begin{bmatrix} -\frac{4}{5} \\ -\frac{4}{5} \end{bmatrix}$.

10.6   Applications

Prob. 1. Minimum at $(1, 0)$.

Prob. 3. Saddle at $(\pm 2, 4)$.

Prob. 5. Saddle at $(0, 3)$.

Prob. 7. Maximum at $(0, 0)$.

Prob. 9. Saddle at $(0, (2k + 1)\frac{\pi}{2})$, $k \in \mathbb{Z}$.

Prob. 11.

(a) We have:

$$\nabla f_1(x, y) = \begin{bmatrix} 2x \\ 0 \end{bmatrix}, \quad \nabla f_1(0, 0) = \begin{bmatrix} 0 \\ 0 \end{bmatrix}$$

$$\nabla f_2(x, y) = \begin{bmatrix} 2x \\ 3y2 \end{bmatrix}, \quad \nabla f_2(0, 0) = \begin{bmatrix} 0 \\ 0 \end{bmatrix}$$

$$\nabla f_3(x, y) = \begin{bmatrix} 2x \\ 4y3 \end{bmatrix}, \quad \nabla f_3(0, 0) = \begin{bmatrix} 0 \\ 0 \end{bmatrix}.$$

(b) $0, 2$.

(c) f_1 has a minimum at $(0, 0)$ but there is a direction along which f_1 is constant; f_2 has a saddle at $(0, 0)$; f_3 has a minimum at $(0, 0)$.

Prob. 13. Absolute maximum: $(1, -1)$, absolute minimum: $(-1, 1)$.

Prob. 15. Maximum is 1, minimum is -1.

Prob. 17. Absolute maximum: $(0, 0)$, $(1, 0)$, $(1, -2)$, $(0, -2)$, Absolute minimum $(\frac{1}{2}, -1)$.

Prob. 19. Absolute maximum at $(\frac{2}{3}, \frac{2}{3})$, absolute minimum occur at all points along the boundary of the domain.

Prob. 21. Absolute maximum: $(3, 0)$, absolute minimum: $(-2, 0)$

Prob. 23. Absolute minimum: $(-frac12, \frac{1}{2})$, absolute maximum: $(\frac{1}{\sqrt{2}}, -\frac{1}{\sqrt{2}})$.

Prob. 25. Yes.

Prob. 27. $(1,1)$.

Prob. 29. $(2\sqrt{2})^3$.

Prob. 31. 172.

Prob. 33. $\frac{1}{3}$.

Prob. 35.

(a) We have

$H = p_1 \ln p_1 + p_2 \ln p_2 + p_3 \ln p_3 = p_1 \ln p_1 + p_2 \ln p_2 + (1 - p_1 - p_2) \ln(1 - p_1 - p_2)$. As the function $\ln x$ is only defined for $x > 0$, we must have $p_1 > 0$, $p_2 > 0$, $1 > p_1 + p_2$.

(b) H is differentiable on its domain. The limit on the boundary of the triangular domain is 0. H is positive, so it attains a maximum in the interior of the triangle. The only critical point is $(p_1, p_2) = (\frac{1}{3}, \frac{1}{3})$.

Prob. 37. Absolute maximum: $\left(\frac{-\sqrt{35}}{6}, \frac{1}{6}\right)$, $\left(\frac{\sqrt{35}}{6}, \frac{1}{6}\right)$ Absolute minimum: $(0, -1)$.

Prob. 39. Absolute minimum: $(\frac{1}{4}, -\frac{1}{8})$, no maximum

Prob. 41. Absolute minimum: $9\left(\frac{12}{13}, -\frac{8}{13}\right)$, no maximum.

Prob. 43. Local minimum: $(0, \frac{1}{3})$, no absolute maxima Absolute maxima: $(\frac{1}{\sqrt{2}}, \frac{1}{6})$, $(\frac{-1}{\sqrt{2}}, \frac{1}{6})$.

Prob. 45. The minimum of f is 0.

Prob. 47. We have $f(x, y) = xy$ with constraint $g(x, y) = x + y - c = 0$, applying Lagrange multipliers gives: $x = y$.

Prob. 49. We have $f(x, y) = 3x + 2y$ with constraint $g(x, y) = x \cdot y - 384 = 0$, applying Lagrange multipliers gives: $x = 16$, $y = 24$. So, the perimeter is 96.

Prob. 51. We have $f(x, y) = \frac{1}{2}x \cdot y$ with constraint $g(x, y) = x^2 + y^2 - 16 = 0$, applying Lagrange multipliers gives: $x = 2\sqrt{2}$, $y = 4$. So, the area is 4.

Prob. 53. We have $f(x, y) = 2x + 2y$ with constraint $g(x, y) = y - \frac{1}{x} = 0$, applying Lagrange multipliers gives: $x = y = 1$. So, perimeter is 4.

Prob. 55. We have $f(r, h) = \pi r^2 + \pi rh$ with constraint $y(r, h) = \pi r^2 h - 1 = 0$, applying Lagrange multipliers gives $r = \sqrt{\frac{0.5}{\pi}}$, $h = 2$.

Prob. 57.

(a) f has local minimum at $(2, 2)$ $(f(2, 2) = 4)$.

(b) No. $f(x, 1/(x - 1))$ can take arbitrarily large positive and negative values.

Prob. 59. P is the only point on $g(x, y) = 0$ where f attains the value c_2, so this point is a local maximum.

Prob. 61. (b) Absolute maximum at $\left(\frac{65 - 40\sqrt{2}}{3}, \frac{40\sqrt{2} - 55}{3} \right)$.

Prob. 63.

(a) ii) We have $\frac{\partial c}{\partial x} = \frac{-x}{2Dt} c(x, t)$, therefore c increases when x is negative and decreases when x is positive.

iv) We have $\frac{\partial^2 c}{\partial x^2} = \frac{\exp\left[-\frac{x^2}{4Dt}\right]}{2Dt\sqrt{4\pi Dt}} \left[\frac{x^2}{2Dt} - 1 \right]$, therefore $\frac{\partial^2 c}{\partial x^2} = 0$ when $|x| = \sqrt{2Dt}$, $\frac{\partial^2 c}{\partial x^2} > 0$ when $|x| > \sqrt{2Dt}$ and $\frac{\partial^2 c}{\partial x^2} < 0$ when $|x| < \sqrt{2Dt}$.

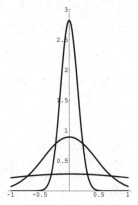

(b)

Prob. 65. The successive differentiation and algebra leads to the answer.

10.7 Systems of Difference Equations

Prob. 1. Refer to the table below:

t	N_t	P_t
0	5	0
1	7.5	0
2	11.25	0
3	16.87	0
4	25.31	0
5	37.96	0
6	56.95	0
7	85.42	0
8	128.14	0
9	192.21	0
10	288.32	0

Prob. 3. For any t we have $P_{t+1} = cN_t \times 0 = 0$, since $1 - \exp 0 = 0$. We have also $N_{t+1} = b^{t+1}N_0$.

Prob. 5. Refer to the table below:

t	N_t	P_t
0	5	5
1	6.78	1.42
2	9.89	0.57
3	14.67	0.33
4	21.85	0.29
5	32.59	0.38
6	48.50	0.75
7	71.67	2.18
8	102.91	9.18
9	128.47	51.81
10	68.36	248.67
11	0.70	203.69
12	0.01	2.09
13	0.02	0.002
14	0.03	3.48×10^{-6}
15	0.05	8.15×10^{-9}

Prob. 7. Refer to the table below:

t	N_t	P_t
0	5	0
1	7.5	0
2	11.25	0
3	16.87	0
4	25.31	0
5	37.96	0
6	56.95	0
7	85.42	0
8	128.14	0
9	192.21	0
10	288.32	0

Prob. 9. For any t we have $P_{t+1} = cN_t \times 0 = 0$. We have also $N_{t+1} = b^{t+1}N_0$.

Prob. 11. Refer to the table below:

t	N_t	P_t
0	100	50
1	79.45	141.09
2	36.96	164.42
3	15.68	79.52
4	10.02	27.01
5	10.00	10.05
6	12.56	4.89
7	17.18	3.31
8	24.19	3.17
9	34.14	4.28
10	47.21	7.98
11	61.27	19.10
12	67.49	48.83
13	54.16	94.14
14	31.68	99.14
15	18.01	59.00

Prob. 13.

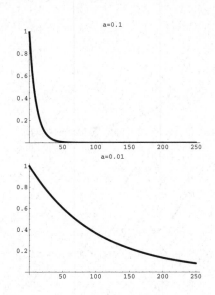

(a)

(b) The chances of escaping decrease when a increases.

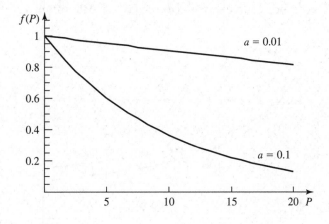

Prob. 15.

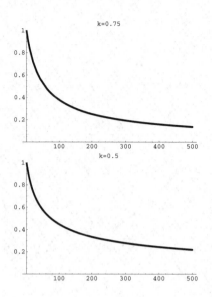

(a)

(b) The chances of escaping decrease when k increases. Maximum is 2, minimum is 0.

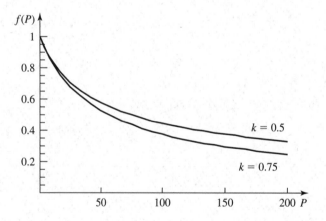

Prob. 17. We have $\begin{bmatrix} -0.7 & 0 \\ -0.3 & 0.2 \end{bmatrix} \begin{bmatrix} 0 \\ 0 \end{bmatrix} = \begin{bmatrix} 0 \\ 0 \end{bmatrix}$, so $\begin{bmatrix} 0 \\ 0 \end{bmatrix}$ is an equilibrium point. As the eigenvalues are -0.7 and 0.2, both with modulus less than 1, this equilibrium is stable.

Prob. 19. We have $\begin{bmatrix} -1.4 & 0 \\ -0.5 & 0.1 \end{bmatrix} = \begin{bmatrix} 0 \\ 0 \end{bmatrix}$, so $\begin{bmatrix} 0 \\ 0 \end{bmatrix}$ is an equilibrium point. As there is an eigenvalue, -1.4, with modulus larger than 1, this equilibrium is unstable.

Prob. 21. We have $\begin{bmatrix} 1 & 2 \\ 3 & 2 \end{bmatrix} \begin{bmatrix} 0 \\ 0 \end{bmatrix} = \begin{bmatrix} 0 \\ 0 \end{bmatrix}$, so $\begin{bmatrix} 0 \\ 0 \end{bmatrix}$ is an equilibrium point. As there is an eigenvalue, 4, with modulus larger than 1, this equilibrium is unstable.

Prob. 23. We have $\begin{bmatrix} -0.2 & -0.4 \\ 0.6 & 0.1 \end{bmatrix} \begin{bmatrix} 0 \\ 0 \end{bmatrix} = \begin{bmatrix} 0 \\ 0 \end{bmatrix}$, so $\begin{bmatrix} 0 \\ 0 \end{bmatrix}$ is an equilibrium point. As the eigenvalues are complex and the square of the modulus of each equals the determinant, which is 0.22, this equilibrium is stable.

Prob. 25. We have $\begin{bmatrix} 4.2 & -3.4 \\ 2.4 & -1.1 \end{bmatrix} \begin{bmatrix} 0 \\ 0 \end{bmatrix} = \begin{bmatrix} 0 \\ 0 \end{bmatrix}$, so $\begin{bmatrix} 0 \\ 0 \end{bmatrix} = \begin{bmatrix} 0 \\ 0 \end{bmatrix}$ is an equilibrium point. As the eigenvalues are complex and the square of the modulus of each equals the determinant, which is 3.54, this equilibrium is unstable.

Prob. 27. The corresponding Jacobi matrix is $\begin{bmatrix} 0 & 0.25 \\ 2 & 0 \end{bmatrix}$, which has eigenvalues $\pm 1/\sqrt{2}$, both with modulus less than 1, so the equilibrium is stable.

Prob. 29. The corresponding Jacobi matrix is $\begin{bmatrix} 0 & 1 \\ -0.5 & 1 \end{bmatrix}$, which has complex eigenvalues. As its determinant is $0.5 < 1$ the equilibrium is stable.

Prob. 31. We have $\begin{bmatrix} 0 \\ 0 \end{bmatrix} = \begin{bmatrix} a \times 0 \\ 2 \times 0 - \cos 0 + 1 \end{bmatrix}$, so $\begin{bmatrix} 0 \\ 0 \end{bmatrix}$ is an equilibrium point.

The Jacobi matrix for the corresponding linearization is $\begin{bmatrix} 0 & a \\ 2 & 0 \end{bmatrix}$ which has eigenvalues $\pm\sqrt{2a}$, so the equilibrium is stable when $a < \frac{1}{2}$.

Prob. 33. The nonnegative equilibrium is $\begin{bmatrix} 0 \\ 0 \end{bmatrix}$. The Jacobi matrix of the corresponding linearization is $\begin{bmatrix} 0 & 1 \\ \frac{1}{2} & \frac{2}{3} - 2y \end{bmatrix}$. One of the eigenvalues of the Jacobi matrix at $\begin{bmatrix} 0 \\ 0 \end{bmatrix}$ is 1.115, which has modulus larger than 1, so this equilibrium is unstable.

Prob. 35. The corresponding Jacobi matrix is $\begin{bmatrix} 0 & a \\ 1 & 0 \end{bmatrix}$. The corresponding eigenvalues are the solutions of the equation $\lambda^2 = a$, therefore the equilibrium is stable if $-1 < a < 1$.

Prob. 37.

(a) $\begin{bmatrix} r - \frac{1}{2} \\ r - \frac{1}{2} \end{bmatrix}$ is an equilibrium point, so if $r > 1/2$, then $(r - 1/2, r - 1/2)$ is an equilibrium.

(b) The Jacobi matrix of the corresponding linearization is $\begin{bmatrix} 0 & 1 \\ \frac{1}{2} & r - 2y \end{bmatrix}$. At the equilibrium the eigenvalues are given by $\frac{1 - r \pm \sqrt{r^2 - 2r + 3}}{2}$

The quantity inside the square root $(r^2 - 2r + 3)$ is always positive. To see that observer that $r^2 - 2r + 3 = (r - 1)^2 + 1$, so both eigenvalues are real.

Forcing the modulus of the eigenvalues to be less than 1, we see that the numerator should be bound by:

$$-2 < 1 - r \pm \sqrt{r^2 - 2r + 3} < 2$$

Solving the inequality on the right we obtain $r < 1/2$, which was already known and the one on the left produces $r > 3/2$, so for values of $1/2 < r < 3/2$, the modulus of the eigenvalues are all less then 1 and the equilibrium is stable.

Prob. 39. There is an unstable equilibrium at $(140 ln4)/3, 10 ln4)$.

Prob. 41. There is a stable equilibrium at $\begin{bmatrix} 1000 \\ 750 \end{bmatrix}$.

10.9   Review Problems

Prob. 1. See graphs below:

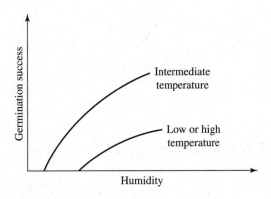

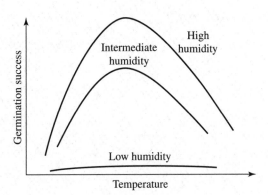

Prob. 3.

(a) $\frac{\partial A_i}{\partial F} > 0, \frac{\partial A_i}{\partial D} < 0.$

(b) The area covered by the introduced species increases with the amount of fertilizer used and also with the disturbance intensity.

(c) Both have areas that increase with the amount of fertilizer used, but the indigenous species area decreases when the disturbance intensity increases, while the introduced species area increases when the disturbance intensity increases.

Prob. 5. $\begin{bmatrix} 2x & -1 \\ 3x^2 & -2y \end{bmatrix}$.

Prob. 7.

(a)

(b) Since $r_{avg} = \sqrt{\pi Dt}$, $(r_{avg})^2 = \pi DT$. So, $D = \frac{(r_{avg})^2}{\pi t}$.

(c) Since r_{avg} = arithmetic average, we have: $r_{avg} = \frac{1}{N}\sum_{i=1}^{N} d_i$. But then we see that:

$$D = \frac{1}{\pi t} \cdot \left(\frac{1}{N} \cdot \sum_{i=1}^{N} di \right)^2.$$

Chapter 11

Systems of Differential Equations

11.1 Linear Systems: Theory

Prob. 1. $\frac{dx}{dt} = \begin{bmatrix} 2 & 3 \\ -4 & 1 \end{bmatrix} x(t)$

Prob. 3. $\frac{dx}{dt} = \begin{bmatrix} -2 & 0 & 1 \\ -1 & 0 & 0 \\ 1 & 1 & 1 \end{bmatrix} x(t)$

Prob. 5. $(1,0): \begin{bmatrix} -1 \\ 1 \end{bmatrix}$; $(0,1): \begin{bmatrix} 2 \\ 0 \end{bmatrix}$; $(-1,0): \begin{bmatrix} 1 \\ -1 \end{bmatrix}$; $(0,-1): \begin{bmatrix} -2 \\ 0 \end{bmatrix}$;

$(1,1): \begin{bmatrix} 1 \\ 1 \end{bmatrix}$; $(0,0): \begin{bmatrix} 0 \\ 0 \end{bmatrix}$; $(-1,1): \begin{bmatrix} 4 \\ -2 \end{bmatrix}$;

Prob. 7. $(1,0) \to \frac{dx}{dt} = \begin{pmatrix} 1 \\ -1 \end{pmatrix}$

$(0,1) \to \frac{dx}{dt} = \begin{pmatrix} 3 \\ 2 \end{pmatrix}$

$(-1, 1) \to \frac{dx}{dt} = \begin{pmatrix} 2 \\ 3 \end{pmatrix}$

$(0, -1) \to \frac{dx}{dt} = \begin{pmatrix} -3 \\ -2 \end{pmatrix}$

$(-3, 1) \to \frac{dx}{dt} = \begin{pmatrix} 0 \\ 5 \end{pmatrix}$

$(0, 0) \to \frac{dx}{dt} = \begin{pmatrix} 0 \\ 0 \end{pmatrix}$

$(-2, 1) \to \frac{dx}{dt} = \begin{pmatrix} 1 \\ 4 \end{pmatrix}$

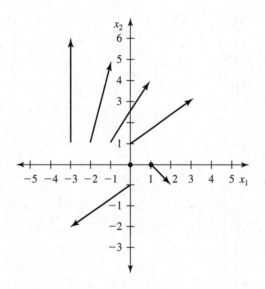

Prob. 9.

(a) Fig 11.21

(b) Fig 11.20

(c) Fig 11.19

(d) Fig 11.18

Prob. 11.

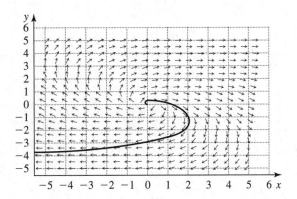

Prob. 13. $\frac{dy}{dt} = \begin{pmatrix} 1 & 3 \\ 5 & 3 \end{pmatrix} x(t)$

$A = \begin{pmatrix} 1 & 3 \\ 5 & 3 \end{pmatrix}$ and the characteristic equation is $\det(A - \lambda I) = 0$, or

$$\begin{vmatrix} 1 - \lambda & 3 \\ 5 & 3 - \lambda \end{vmatrix} = 0$$

$$(1 - \lambda)(3 - \lambda) - 15 = 0$$

$$\lambda^2 - 4\lambda - 12 = 0$$

and the eigenvalues are $\lambda = 6$ and $\lambda = -2$ Computating the eigenvector we have: $\lambda = G$; the system of equations reduce to

$$\begin{pmatrix} -5 & 3 \\ 5 & -3 \end{pmatrix} \begin{pmatrix} u_1 \\ u_2 \end{pmatrix} = \begin{pmatrix} 0 \\ 0 \end{pmatrix}$$

or $5u_1 = 3u_2$, and one eigenvalue is the vector $\begin{pmatrix} 3 \\ 5 \end{pmatrix}$.

$\lambda = -2$, the equation reduces to

$$\begin{pmatrix} 3 & 3 \\ 5 & 5 \end{pmatrix} \begin{pmatrix} u_1 \\ u_2 \end{pmatrix} = \begin{pmatrix} 0 \\ 0 \end{pmatrix}$$

and the eigenvalue will be $\begin{pmatrix} 1 \\ -1 \end{pmatrix}$, so the general solution is:

$$x(t) = c_1 e^{6t} \begin{pmatrix} 3 \\ 5 \end{pmatrix} + c_2 e^{-2t} \begin{pmatrix} 1 \\ -1 \end{pmatrix}$$

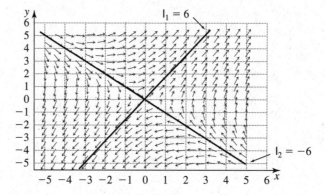

Prob. 15. $\frac{dx}{dt} = \begin{pmatrix} -3 & 3 \\ 6 & 4 \end{pmatrix} x(t)$

$A = \begin{pmatrix} -3 & 3 \\ 6 & 4 \end{pmatrix}$ and the characteristic equation is $\det(A - \lambda I) = 0$, or

$$\begin{vmatrix} -3 - \lambda & 3 \\ 6 & 4 - \lambda \end{vmatrix} = 0$$

$$(-3 - \lambda)(4 - \lambda) - 18 = 0$$

and the eigenvalues are $\lambda = 6$ and $\lambda = -5$. Computing the eigenvector we have

$\lambda = 6$;

$$\begin{pmatrix} -9 & 3 \\ 6 & -2 \end{pmatrix} \begin{pmatrix} u_1 \\ u_2 \end{pmatrix} = \begin{pmatrix} 0 \\ 0 \end{pmatrix}$$

or $6u_1 = 2u_2$ or $3u_1 = u_2$ and the eigenvector is $\begin{pmatrix} 1 \\ 3 \end{pmatrix}$.

Similarly for $\lambda = -5$

$$\begin{pmatrix} 2 & 3 \\ 6 & 9 \end{pmatrix} \begin{pmatrix} u_1 \\ u_2 \end{pmatrix} = \begin{pmatrix} 0 \\ 0 \end{pmatrix}$$

or $2u_1 + 3u_2 = 0$ and the eigenvector is $\begin{pmatrix} 3 \\ -2 \end{pmatrix}$ and the general solution is:

$$x(t) = c_1 e^{6t} \begin{pmatrix} 1 \\ 3 \end{pmatrix} + c_2 e^{-5t} \begin{pmatrix} 3 \\ -2 \end{pmatrix}$$

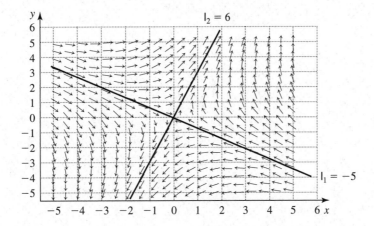

Prob. 17. $\frac{dx}{dt} = \begin{pmatrix} -2 & 0 \\ -3 & 1 \end{pmatrix} x(t)$

In this case $A = \begin{pmatrix} -2 & 0 \\ -3 & 1 \end{pmatrix}$, and the characteristic equation is

$$\begin{vmatrix} -2 - \lambda & 0 \\ -3 & 1 - \lambda \end{vmatrix} = 0$$

$$(-2 - \lambda)(1 - \lambda) = 0$$

and the characteristic values can be read straight out of the equation and are $\lambda = -2$ and $\lambda = 1$. Computing the eigenvector

$\lambda = -2$;

$$\begin{pmatrix} 0 & 0 \\ -3 & 3 \end{pmatrix} \begin{pmatrix} u_1 \\ u_2 \end{pmatrix} = \begin{pmatrix} 0 \\ 0 \end{pmatrix}$$

and $u_1 = u_2$, that is, the eigenvector is $\begin{pmatrix} 1 \\ 1 \end{pmatrix}$

$\lambda = 1$;

$$\begin{pmatrix} -3 & 0 \\ -3 & 0 \end{pmatrix} \begin{pmatrix} u_1 \\ u_2 \end{pmatrix} = \begin{pmatrix} 0 \\ 0 \end{pmatrix}$$

and $3u_1 = 0$, or $u_1 = 0$ and an eigenvector is $\begin{pmatrix} 0 \\ 1 \end{pmatrix}$, so the general solution is

$$x(t) = c_1 e^{-2t} \begin{pmatrix} 1 \\ 1 \end{pmatrix} + c_2 e^{t} \begin{pmatrix} 0 \\ 1 \end{pmatrix}$$

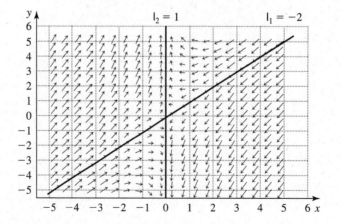

Prob. 19. $\frac{dx}{dt} = \begin{pmatrix} -3 & 0 \\ 4 & 2 \end{pmatrix} x(t)$

$A = \begin{pmatrix} -3 & 0 \\ 4 & 2 \end{pmatrix}$, and the characteristic equation is

$$\begin{vmatrix} -3 - \lambda & 0 \\ 4 & 2 - \lambda \end{vmatrix} = 0$$

or $(-3 - \lambda)(2 - \lambda) = 0$ which was roots or characteristic values $\lambda = -3$ and $\lambda = 2$.

Computing the eigenvectors, we have:

$\lambda = -3$;

$$\begin{pmatrix} 0 & 0 \\ 4 & 5 \end{pmatrix} \begin{pmatrix} u_1 \\ u_2 \end{pmatrix} = \begin{pmatrix} 0 \\ 0 \end{pmatrix}$$

$4u_1 = -5u_2$, and the eigenvector is $\begin{pmatrix} 5 \\ -4 \end{pmatrix}$

$\lambda = 2$;

$$\begin{pmatrix} -5 & 0 \\ 4 & 0 \end{pmatrix} \begin{pmatrix} u_1 \\ u_2 \end{pmatrix} = \begin{pmatrix} 0 \\ 0 \end{pmatrix}$$

and $u_1 = 0$, in this case, and the eigenvector is $\begin{pmatrix} 0 \\ 1 \end{pmatrix}$. So the general solution is

$$x(t) = c_1 e^{-3t} \begin{pmatrix} 5 \\ -4 \end{pmatrix} + c_2 e^{2t} \begin{pmatrix} 0 \\ 1 \end{pmatrix}$$

Since $x_1(0) = -5$ and $x_2(0) = 5$. We end up with the equation, substituting for $t = 0$.

$$\begin{pmatrix} -5 \\ 5 \end{pmatrix} = c_1 \begin{pmatrix} 5 \\ -4 \end{pmatrix} + c_2 \begin{pmatrix} 0 \\ 1 \end{pmatrix}$$

or

$$\begin{cases} 5c_1 = -5 \\ -4c_1 + c_2 = 5 \end{cases}$$

whose solution is $c_1 = -1$ and $c_2 = 1$, so the solution is:

$$x(t) = -ce^{-3t} \begin{pmatrix} 5 \\ -4 \end{pmatrix} + e^{2t} \begin{pmatrix} 0 \\ 1 \end{pmatrix}.$$

Prob. 21. $A = \begin{pmatrix} 3 & -2 \\ 0 & 1 \end{pmatrix}$, so the characteristic equation is:

$$\begin{vmatrix} 3 - \lambda & -2 \\ 0 & 1 - \lambda \end{vmatrix} = 0$$

or $(5 - \lambda)(6 - \lambda) = 0$ with eigenvectors $\lambda = 1$ and $\lambda = 3$. Computing the eigenvalues, we have:

$\lambda = 1$;

$$\begin{pmatrix} 2 & -2 \\ 0 & 0 \end{pmatrix} \begin{pmatrix} u_1 \\ u_2 \end{pmatrix} = \begin{pmatrix} 0 \\ 0 \end{pmatrix}$$

$2u_1 = u_2$, and the eigenvector is $\begin{pmatrix} 1 \\ 1 \end{pmatrix}$

$\lambda = 3$;

$$\begin{pmatrix} 0 & -2 \\ 0 & -2 \end{pmatrix} \begin{pmatrix} u_1 \\ u_2 \end{pmatrix} = \begin{pmatrix} 0 \\ 0 \end{pmatrix}$$

and $u_2 = 0$, giving us an eigenvector $\begin{pmatrix} 1 \\ 0 \end{pmatrix}$, so the general solution is

$$x(t) = c_1 e^t \begin{pmatrix} 1 \\ 1 \end{pmatrix} + c_2 e^{3t} \begin{pmatrix} 1 \\ 0 \end{pmatrix}.$$

with the initial conditions, we obtain the equations

$$\begin{pmatrix} 1 \\ 1 \end{pmatrix} = c_1 \begin{pmatrix} 1 \\ 1 \end{pmatrix} + c_2 \begin{pmatrix} 1 \\ 0 \end{pmatrix}$$

whose solution is $c_1 = 1$ and $c_2 = 0$, so the solution is

$$x(t) = e^t \begin{pmatrix} 1 \\ 1 \end{pmatrix}$$

or better

$$x_1(t) = e^t = x_2(t)$$

Prob. 23. $A = \begin{pmatrix} 4 & -7 \\ 2 & -5 \end{pmatrix}$, we have: $\begin{vmatrix} 4 - \lambda & -7 \\ 2 & -5 - \lambda \end{vmatrix} = 0$

$$(4 - \lambda)(-5 - \lambda) + 14 = 0 \text{ gives } \lambda = -3, \quad \lambda = 2.$$

For $\lambda = -3$:

$$\begin{bmatrix} 7 & -7 \\ 2 & -2 \end{bmatrix} \begin{pmatrix} u_1 \\ u_2 \end{pmatrix} = \begin{pmatrix} 0 \\ 0 \end{pmatrix} \text{ gives } \begin{bmatrix} 1 \\ 1 \end{bmatrix}.$$

For $\lambda = 2$:

$$\begin{bmatrix} 2 & -7 \\ 2 & -7 \end{bmatrix} \begin{pmatrix} u_1 \\ u_2 \end{pmatrix} = \begin{pmatrix} 0 \\ 0 \end{pmatrix} \text{ gives } \begin{bmatrix} 7 \\ 2 \end{bmatrix}.$$

So, general solution is: $x(t) = c_1 e^{-3t} \begin{bmatrix} 1 \\ 1 \end{bmatrix} + c_2 e^{2t} \begin{bmatrix} 7 \\ 2 \end{bmatrix}.$

Now, using initial conditions:

$$\begin{pmatrix} 13 \\ 3 \end{pmatrix} = c_1 \begin{bmatrix} 1 \\ 1 \end{bmatrix} + c_2 \begin{bmatrix} 7 \\ 2 \end{bmatrix} \text{ gives } c_1 = -1, \quad c_2 = 2.$$

So, $x(t) = -e^{-3t} \begin{bmatrix} 1 \\ 1 \end{bmatrix} + 2e^{2t} \begin{bmatrix} 7 \\ 2 \end{bmatrix}.$

Prob. 25. $A = \begin{pmatrix} 4 & 7 \\ 1 & -2 \end{pmatrix}$, and the characteristic equation is:

$$\begin{vmatrix} 4 - \lambda & 7 \\ 1 & -2 - \lambda \end{vmatrix} = 0$$

$$(4 - \lambda)(-2 - \lambda) - 7 = 0$$

$$\lambda^2 - 2\lambda - 15 = 0$$

and the eigenvalues are $\lambda = -3$ and $\lambda = 5$, so computing the eigenvectors we have:

$\lambda = -3$;

$$\begin{pmatrix} 7 & 7 \\ 1 & 1 \end{pmatrix} \begin{pmatrix} u_1 \\ u_2 \end{pmatrix} = \begin{pmatrix} 0 \\ 0 \end{pmatrix}$$

or $u_2 = -u_1$, and the eigenvector is $\begin{pmatrix} 1 \\ -1 \end{pmatrix}$

$\lambda = 5$;

$$\begin{pmatrix} -1 & 7 \\ 1 & -7 \end{pmatrix} \begin{pmatrix} u_1 \\ u_2 \end{pmatrix} = \begin{pmatrix} 0 \\ 0 \end{pmatrix}$$

or $u_1 = 7u_2$ and the eigenvector is $\begin{pmatrix} 7 \\ 1 \end{pmatrix}$ and the general solution is

$$x(t) = c_1 e^{-3t} \begin{pmatrix} 1 \\ -1 \end{pmatrix} + c_2 e^{5t} \begin{pmatrix} 7 \\ 1 \end{pmatrix}.$$

Using the initial conditions we end up with

$$\begin{pmatrix} -1 \\ -2 \end{pmatrix} = c_1 \begin{pmatrix} 1 \\ 1 \end{pmatrix} + c_2 \begin{pmatrix} 7 \\ 1 \end{pmatrix}$$

or

$$\begin{cases} c_1 + 7c_2 = -1 \\ c_1 + c_2 = -2 \end{cases}$$

whose solution is $c_1 = -\frac{13}{6}$ and $c_2 = \frac{1}{6}$ so the solution is

$$x(t) = -\frac{13}{6} e^{-3t} \begin{pmatrix} 1 \\ 1 \end{pmatrix} + \frac{1}{6} e^{5t} \begin{pmatrix} 7 \\ 1 \end{pmatrix}.$$

or

$$x_1(t) = -\frac{13}{e}^{-3t} + \frac{7}{6} e^{5t}$$

$$x_2(t) = -\frac{13}{6} e^{-3t} + \frac{1}{6} e^{5t}$$

Prob. 27.

(a) $A = \begin{pmatrix} 1 & 0 \\ 0 & 1 \end{pmatrix}$, and a straight computation shows that

$$\begin{vmatrix} 1-\lambda & 0 \\ 0 & 1-\lambda \end{vmatrix} = 0$$

that is, $(1-\lambda)^2 = 0$ whose only root is $\lambda = 1$, so A has repested eigenvalues.

(b) Let's call the two vectors e_1 and e_2, respectively, then it is a trivial confirmation that

$$Ae_1 = e_1 \quad \text{and} \quad Ae_2 = e_2$$

so e_1 and e_2 are eigenvectors of $A = Id$.

Now any vector $\begin{pmatrix} c_1 \\ c_2 \end{pmatrix}$ can be written as $c_1 e_1 + c_2 e_2$, just use the two coordenates of the vector as coeficients of the decomposition

(c) Now let $x(t) = c_1 e^t \begin{pmatrix} 1 \\ 0 \end{pmatrix} + c_2 e^t \begin{pmatrix} 0 \\ 1 \end{pmatrix}$, by substituting $t = 0$ we can easily see that it satisfies the initial conditions $x_1(0) = c_1$ and $x_2(0) = c_2$, let's now differentiate it individually

$$x_1 = c_1 e^t$$

$$x_2 = c_2 e^t$$

we get

$$\frac{dx_1}{dt} = c_1 e^t = x_1$$

$$\frac{dx_2}{dt} = c_2 e^t = x_2,$$

so

$$\begin{pmatrix} \frac{dx_1}{dt} \\ \frac{dx_2}{dt} \end{pmatrix} = \begin{pmatrix} 1 & 0 \\ 0 & 1 \end{pmatrix} \begin{pmatrix} x_1(t) \\ x_2(t) \end{pmatrix}$$

satisfying the equation.

Prob. 29. $\det(A - \lambda I) = \begin{vmatrix} 2 - \lambda & -1 \\ 0 & 3 - \lambda \end{vmatrix} = (2 - \lambda)(3 - \lambda) = 0$

Both eigenvalues are positive, hence source.

Prob. 31. $\det(A - \lambda I) = \begin{vmatrix} -2 - \lambda & 2 \\ 2 & 1 - \lambda \end{vmatrix} = (-2 - \lambda)(1 - \lambda) - 4 = 0$

We get $\lambda = -3$, $\lambda = 2$, hence saddle point.

Prob. 33. $\det(A - \lambda I) = \begin{vmatrix} -4 - \lambda & 2 \\ -5 & 3 - \lambda \end{vmatrix} = (-4 - \lambda)(3 - \lambda) + 10 = 0$

Eigenvalues are of opposite signs, hence saddle point.

Prob. 35.

$$\det(A - \lambda I) = \begin{vmatrix} 6 - \lambda & -4 \\ -3 & 5 - \lambda \end{vmatrix} = (6 - \lambda)(5 - \lambda) - 12$$

$$= \lambda^2 - 11\lambda + 18 = 0$$

and the roots are 9 and 2, both positive, so a source.

Prob. 37.

$$\det(A - \lambda I) = \begin{vmatrix} -3 - \lambda & -1 \\ 1 & -6 - \lambda \end{vmatrix} = (3 + \lambda)(6 + \lambda) + 1$$

$$= \lambda^2 + 9\lambda + 19 = 0$$

and the roots are $\frac{-9 \pm \sqrt{5}}{2}$, both negative, so a sink.

Prob. 39.

$$\det(A - \lambda I) = \begin{vmatrix} -\lambda & -2 \\ -1 & 3 - \lambda \end{vmatrix} = (-\lambda)(3 - \lambda) - 2$$

$$= \lambda^2 - 3\lambda - 2$$

and the roots are $\frac{3 \pm \sqrt{17}}{2}$, one positive, one negative, a saddle.

Prob. 41.

$$\det(A - \lambda I) = \begin{vmatrix} -2 - \lambda & -3 \\ 1 & 3 - \lambda \end{vmatrix} = (-2 - \lambda)(3 - \lambda) + 3$$

$$= \lambda^2 - \lambda - 3 = 0$$

and the roots are $\frac{1 \pm \sqrt{13}}{2}$, one positive, one negative, so a saddle.

Prob. 43. $\det(A - \lambda I) = \begin{vmatrix} 2 - \lambda & -1 \\ 3 & -\lambda \end{vmatrix} = -\lambda(2 - \lambda) + 3 = 0$

Eigenvalues have positive real part, so unstable spiral.

Prob. 45. $\det(A - \lambda I) = \begin{vmatrix} -2 - \lambda & 4 \\ -3 & -2 - \lambda \end{vmatrix} = (-2 - \lambda)^2 + 12 = 0.$

Eigenvalues have negative real part, so stable spiral.

Prob. 47.

$$\det(A - \lambda I) = \begin{vmatrix} 1 - \lambda & 3 \\ -2 & -2 - \lambda \end{vmatrix} = (1 - \lambda)(-2 - \lambda) + 6$$

$$= \lambda^2 + \lambda + 4$$

and the roots are $\frac{-1\pm\sqrt{-15}}{2}$, or $-\frac{1}{2} \pm \frac{\sqrt{15}}{2}i$ which have negative real part, so it is a stable spiral. Observe that all our equations are of the type

$$\lambda^2 + b\lambda + c = 0$$

with complex roots, so the real part will always be $-b/2$, and so the sign will always be the apposite sign of b, and *ih* the case $b = 0$ we have a center.

In fact, we can go further and observe that the term b in the second degree equation of these matrices is equal to minus the sum of the two diagonal elements. So, for these matrices that have imaginary roots, if the sum of two diagonal elements is positive it will be a unstable spiral, if the sum is zero a center, and if the sum is negative a stable spiral.

Prob. 49. Positive diagonal sum, by the Solution to Problem 47, an unstable spiral.

Prob. 51. Diagonal sum zero, by the Solution of Problem 47, a center.

Prob. 53. Diagonal sum is zero, so by the Solution to Problem 47, a center.

Prob. 55. Negative diagonal sum, so by the Solution to Problem 47, a stable spiral.

Prob. 57. $\det(A - \lambda I) = \begin{vmatrix} -1-\lambda & -2 \\ 1 & 3-\lambda \end{vmatrix} = (-1-\lambda)(3-\lambda) + 2 = 0.$

Eigenvalues are real and opposite signs, hence saddle.

Prob. 59.

$$\det(A - \lambda I) = \begin{vmatrix} -1-\lambda & -1 \\ 5 & -3-\lambda \end{vmatrix} = (\lambda+1)(\lambda+3) + 5$$

$$= \lambda^2 + 4\lambda + 8 = 0$$

and the roots are $\lambda = \frac{-41\pm\sqrt{-16}}{2} = -2 \pm 2i$, so a stable spiral.

Prob. 61. $\det(A - \lambda I) = \begin{vmatrix} 1-\lambda & 3 \\ 2 & 3-\lambda \end{vmatrix} = (1-\lambda)(3-\lambda) - 6 = 0$

Eigenvalues are real and opposite signs, hence saddle.

Prob. 63.

$$\det(A - \lambda I) = \begin{vmatrix} -2-\lambda & 3 \\ 1 & -4-\lambda \end{vmatrix} = (\lambda+4)(\lambda+2) - 3$$

$$= \lambda^2 + 6\lambda + 5 = 0$$

with roots -1 and -5, so a sink.

Prob. 65. $\det(A - \lambda I) = \begin{vmatrix} 3 - \lambda & -5 \\ 2 & -1 - \lambda \end{vmatrix} = (1 + \lambda)(\lambda - 3) + 10 = 0$

Eigenvalues are imaginary with positive real part, hence unstable spiral.

Prob. 67.

(a) $\det(A - \lambda I) = 0$ produces the equation

$$\begin{vmatrix} 4 - \lambda & 8 \\ 1 & 2 - \lambda \end{vmatrix} = 0$$

$$(\lambda - 2)(\lambda - 4) - 8 = 0$$

$$\lambda^2 - 6\lambda = 0$$

so the eigenvalues are $\lambda = 0$ and $\lambda = 6$

$\lambda = 0$;

$$\begin{pmatrix} 4 & 8 \\ 1 & 2 \end{pmatrix} \begin{pmatrix} u_1 \\ u_2 \end{pmatrix} = \begin{pmatrix} 0 \\ 0 \end{pmatrix}$$

which reduces to $u_1 + 2u_2 = 0$ and the associated eigenvector $\begin{pmatrix} -2 \\ 1 \end{pmatrix}$.

$\lambda = 6$;

$$\begin{pmatrix} -2 & 8 \\ 1 & -4 \end{pmatrix} \begin{pmatrix} u_1 \\ u_2 \end{pmatrix} = \begin{pmatrix} 0 \\ 0 \end{pmatrix}$$

which reduces to $u_1 = 4u_2$ and the associated eigenvector $\begin{pmatrix} 4 \\ 1 \end{pmatrix}$. The general

solution is given by

$$x(t) = c_1 e^{0t} \begin{pmatrix} -2 \\ 1 \end{pmatrix} + c_2 e^{6t} \begin{pmatrix} 4 \\ 1 \end{pmatrix}.$$

or

$$x(t) = \begin{pmatrix} -2c_1 \\ c_1 \end{pmatrix} + c_2 e^{6t} \begin{pmatrix} 4 \\ 1 \end{pmatrix}.$$

(b) and the individual solutions are:

$$x_1(t) = 4c_2e^{6t} - 2c_1$$

$$x_2(t) = c_2e^{6t} + c_1$$

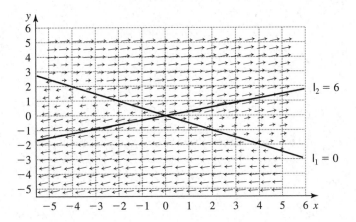

(c)

$$\frac{dx_1}{dt} = 4c_2 6e^{6t}$$

$$\frac{dx_2}{dt} = 6c_2 e^{6t}$$

so $\frac{dx_2}{dx_1} = \frac{1}{4}$ which is the direction parallel to eigenvector associated with the non-zero eigenvalue.

Solutions diverge always from the direction of the eigenvector associate with the zero eigenvalue.

11.2 Linear Systems: Applications

Prob. 1. The system is

$$\frac{dx_1}{dt} = -(0.5 + 0.05)x_1 + 0.1x_2$$

$$\frac{dx_2}{dt} = 0.5x_1 - (0.1 + 0.02)x_2$$

The determinant in this case is

$$\Delta = (0.12)(0.55) - (0.5)(0.1) > 0$$

which is closely positive, then here eigenvalues are strictly negative and the equilibrium is a stable node.

Prob. 3. The system is

$$\frac{dx_1}{dt} = -(2.5)x_1 + 0.7x_2$$
$$\frac{dx_2}{dt} = 2.5x_2 - (0.7 + 0.1)x_2$$

The determinant is

$$\Delta = (2.5)(0.8) - (2.5)(0.7) > 0$$

which is strictly positive, then the eigenvalues are strictly negative and the equilibrium is a stable node.

Prob. 5. The system is

$$\frac{dx_1}{dt} = -0.1x_1 + 0x_2$$
$$\frac{dx_2}{dt} = 0x_1 - 0.3x_2$$

The determinant is $\Delta = (0.1)(0.3) = 0.3 > 0$ which is strictly positive, so the eigenvalues are strictly negative and the equilibrium is a stable node.

Prob. 7. The systems is:

$$\frac{dx_1}{dt} = -0.6x_1 + 1.2x_2$$
$$\frac{dx_2}{dt} = 0.1x_1 - 1.25x_2$$

The determinant is: $(-0.6)(-1.25) - (0.1)(1.2) > 0$, stable node.

Prob. 9. Setting the system to find the values of a, b, c and d we have

$$a + c = 0.4$$
$$b = 0.3$$
$$a = 0.1$$
$$b + d = 0.5$$

so $a = 0.1$, $b = 0.3$, $c = 0.3$ and $d = 0.2$.

Prob. 11.

$$a + c = 0.2$$

$$b = 0.1, \text{ we get } a = 0, \quad b = 0.1, \quad c = 0.2, \quad d = 0$$

$$b + d = 0.1$$

$$a = 0$$

Prob. 13. Set up the system to find a, b, c, d same way as Problem 12

Prob. 15. Set up the system to find a, b, c, d same way as Problem 12

Prob. 17. Set up the system to find a, b, c, d same way as Problem 12

Prob. 19. Since $\frac{dx_1}{dt} = -0.3x_1(t)$ we can integrate this equation to obtain

$$x_1(t) = x(0)e^{-0.3t}$$

for the expression for $x_1(t)$, or better

$$x_1(t) = 4e^{-0.3t}$$

since the amount in the drive will be the complement of what is in the tissue

$$x_2(t) = 4 - x_1(t)$$

$$= -4e^{-03t}$$

Another Solution: We can also look at the problem as a system of two linear differential equations, the first one being given by

$$\frac{dx_1}{dt} = -0.3x_1(t)$$

and the second being exactly the same equation with the opposite sign; because the total amount of drug in the body is constant, so

$$\frac{dx_2}{dt} = 0.3x_1(t)$$

which gives us the system $\frac{dx}{dt} = Ax(t)$, where

$$A = \begin{pmatrix} -0.3 & 0 \\ 0.3 & 0 \end{pmatrix}.$$

Computing the eigenvalues we have

$$\det(A - \lambda I) = \begin{vmatrix} -0.3 - \lambda & 0 \\ 0.3 & -\lambda \end{vmatrix} = \lambda(\lambda + 0.3) = 0$$

which are $\lambda = 0$ and $\lambda = -0.3$, then proceeding to compute the eigenvectors, we have
$\lambda = 0$;

$$\begin{pmatrix} -0.3 & 0 \\ 0.3 & 0 \end{pmatrix} \begin{pmatrix} u_1 \\ u_2 \end{pmatrix} = \begin{pmatrix} 0 \\ 0 \end{pmatrix}$$

and it is easy to see that $\begin{pmatrix} 0 \\ 1 \end{pmatrix}$ is an eigenvector

$\lambda = 0$;

$$\begin{pmatrix} 0 & 0 \\ 0.3 & +0.3 \end{pmatrix} \begin{pmatrix} u_1 \\ u_2 \end{pmatrix} = \begin{pmatrix} 0 \\ 0 \end{pmatrix}$$

which gives $\begin{pmatrix} 1 \\ -1 \end{pmatrix}$ is an eigenvector, so the general solution is

$$x(t) = c_1 \begin{pmatrix} 0 \\ 1 \end{pmatrix} + c_2 e^{-0.3t} \begin{pmatrix} 1 \\ -1 \end{pmatrix}$$

and if we use the initial conditions

$$\begin{cases} 4 = c_2 \\ 0 = c_1 - c_2 \end{cases}$$

so the solution to the system is $c_1 = 4$, $c_2 = 4$, and the final equations for $x_1(t)$ and $x_2(t)$ are:

$$x_1(t) = 4e^{-0.3t}$$

$$x_2(t) = 4 - 4e^{-0.3t}$$

Prob. 21. (a) The system that gives each component of the diagram is

$$a + c = 0.2$$

$$b = 0.1$$

$$a = 0.2$$

$$b + d = 0.1$$

so separating each component we have $a = 0.2$, $b = 0.1$, $c = 0$ and $d = 0$.

(b)

$$\frac{d}{dt}(x_1 + x_2) = \frac{dx_1}{dt} + \frac{dx_2}{dt}$$

$$= -0.2x_1 + 0.1x_2 + 0.2x_1 - 0.1x_2$$

$$= 0$$

so the total quantity $x_1 + x_2$ is constant, $x_1 + x_2 = A$. A here denotes the total Area, the one occupied by adlt trees plus the ones occupied by the gaps.

(c) The function $A(t) = x_1(t) + x_2(t)$, is constant, because the derivative is zero, so if $A(0) = 20$, so it if for all values of $t > 0$.

(d) Since $x_1(t) + x_2(t) = 20$, we can write

$$x_2(t) = 20 - x_1(t)$$

and substituting on the second equation of the system

$$\frac{dx_2(t)}{dt} = 0 - \frac{dx_1(t)}{dt} = 0.2x_1 - 0.1x_2$$

$$\frac{dx_1(t)}{dt} = 0.2x_1 - 0.1(20 - x_1(t))$$

$$\frac{dx_1}{dt} = 2 + 0.3x_1$$

which is what we needed.

(e) To solve the system we first compute the eigenvalues of

$$A = \begin{pmatrix} -0.2 & 0.1 \\ 0.2 & -0.1 \end{pmatrix}$$

$$\det(A - \lambda I) = \begin{vmatrix} -0.2 - \lambda & 0.1 \\ 0.2 & -0.1 - \lambda \end{vmatrix} = (\lambda + 0.1)(\lambda + 0.2) - 0.02$$

$$= \lambda^2 + 0.3\lambda = \lambda(\lambda + 0.3) = 0$$

so the eigenvalues are $\lambda = 0$ and $\lambda = -0.3$. Computing the eigenvectors we have $\lambda = 0$;

$$\begin{pmatrix} -0.2 & 0.1 \\ 0.2 & -0.1 \end{pmatrix} \begin{pmatrix} u_1 \\ u_2 \end{pmatrix} = \begin{pmatrix} 0 \\ 0 \end{pmatrix}$$

which reduces to $u_2 = 2u_1$ and the eigenvector is $\begin{pmatrix} 1 \\ 2 \end{pmatrix}$ is an eigenvector

$\lambda = 0.3$;

$$\begin{pmatrix} 0.1 & 0.1 \\ 0.2 & 0.2 \end{pmatrix} \begin{pmatrix} u_1 \\ u_2 \end{pmatrix} = \begin{pmatrix} 0 \\ 0 \end{pmatrix}$$

which reduces to $u_2 = -u_1$ and the eigenvector is $\begin{pmatrix} 1 \\ -1 \end{pmatrix}$ so the general solution is

$$x(t) = c_1 \begin{pmatrix} 1 \\ 2 \end{pmatrix} + c_2 e^{-0.3t} \begin{pmatrix} 1 \\ -1 \end{pmatrix}$$

Using the initial conditions, we have

$$\begin{cases} 2 = c_1 + c_2 \\ 18 = 2c_1 - c_2 \end{cases}$$

and the solution is $c_1 = \frac{20}{3}$ and $c_2 = \frac{-14}{3}$, that is

$$x_1(t) = \frac{20}{3} - \frac{14}{3} e^{-0.3t}$$

$$x_2(t) = \frac{40}{3} + \frac{14}{3} e^{-0.3t}.$$

So the fraction of the forest occupied by adult trees is $x_2(t)/A$, that is

$$\left(\frac{2}{3} + \frac{7}{30} e^{-0.3t} \right) \times 100$$

or

$$\frac{200}{3} + \frac{70}{3} e^{-0.3t}$$

when $t \to \infty$, the second term of the sum will approach 0 and the amount covered by adult trees will approach 200/3%, or approximately 66.6%

Prob. 23. The general solution is

$$x(t) = c_1 \sin 2t + c_2 \cos 2t$$

using the initial condition $x(0) = 0$, we get

$$0 = x(0) = c_2 \cos 0 = c_2$$

which implies that $c_2 = 0$, reducing the solution to

$$x(t) = c_1 \sin 2t$$

derivating once and using the initial condition on the derivative, we get:

$$G = \frac{dx}{dt}(0) = c_1(\cos 2t)2|_{t=0}$$

implying that $c_1 = 3$ and the solution is

$$x(t) = 3 \sin 2t$$

Prob. 25. Let $y = \frac{dx}{dt}$, then $\frac{dy}{dt} = \frac{d^2x}{dt^2} = 3x$, so the system becomes

$$\begin{cases} \frac{dx}{dt} = y \\ \frac{dy}{dt} = 3x \end{cases}$$

Prob. 27. Let $y = \frac{dx}{dt}$, then $\frac{dy}{dt} = \frac{d^2x}{dt^2} = x - \frac{dx}{dt} = x - y$, and the system is:

$$\begin{cases} \frac{dx}{dt} = y \\ \frac{dy}{dt} = x - y \end{cases}$$

11.3 Nonlinear Autonomous Systems: Theory

Prob. 1. The Jacobian matrix will be given by

$$Df = \begin{pmatrix} 1 + x_2 & -2 + x_1 \\ -1 & 1 \end{pmatrix}$$

and at the origin it is

$$Df(0) = \begin{pmatrix} 1 & -2 \\ -1 & 1 \end{pmatrix}$$

computing the trace and determinant we have, $\tau = 2$ and $\Delta = -1$, so it is a saddle.

Prob. 3. The Jacobian matrix will be given by

$$Df = \begin{pmatrix} 1 + 2x_1 - 2x_2 & -2x_1 + 1 \\ 1 & 0 \end{pmatrix}$$

at the origin it is,

$$Df(0) = \begin{pmatrix} 1 & 1 \\ 1 & 0 \end{pmatrix}$$

computing the trace and determinant we have, $\tau = 1$ and $\Delta = -1$, so it is a saddle.

Prob. 5. The Jacobian is:

$$\begin{bmatrix} e^{-x_2} & -x_1 e^{-x_2} \\ 2x_2 e^{x_1} & 2e^{x_1} \end{bmatrix}$$

At the origin: $\begin{bmatrix} 1 & 0 \\ 0 & 2 \end{bmatrix}$, $\Delta = 2$, $\tau = 3$. $\tau^2 - 4\Delta = 9 - 8 = 1 > 0$.

Unstable node

Prob. 7. First let's find the equilibrium points, which are given by the solutions to the system:

$$\begin{cases} -x_1 + 2x_1(1 - x_1) = 0 \\ -x_2 + 5x_2(1 - x_1 - x_2) = 0 \end{cases}$$

or rewriting it:

$$\begin{cases} x_1 = 2x_1(1 - x_1) \\ x_2 = 5x_2(1 - x_1 - x_2) \end{cases}$$

so one obvious solution is $x_1 = x_2 = 0$. To find the other ones let's assume $x_1 \neq 0$, then solving the first equation we get $1 = 2(1 - x_1)$, that is, $x_1 = 1/2$. Substituting that on the second equation we have $x_2 = 5x_2(\frac{1}{2} - x_2)$, whose solutions are $x_2 = 0$, or $x_2 = \frac{3}{10}$, giving two more solutions.

Now let's assume that $x_1 = 0$, then the first equation is easily verified and the second one reduces to the second degree $x_2 = 5x_2(1 - x_2)$ whose solutions $x_2 = 0$ and $x_2 = 4/5$. So the equilibrium points are $(0,0)$, $(1/2,0)$, $(1/2,3/10)$ and $0,4/5)$. Now the Jacobian is given by the expression

$$Df = \begin{pmatrix} 1 - 4x_1 & 0 \\ -5x_2 & 4 - 5x_1 - 10x_2 \end{pmatrix}$$

and the determinant is $\Delta = (1 - 4x_1)(4 - 5x_1 - 10x_2)$ and the trace is $\tau = 5 - 9x_1 - 10x_2$, so analyzing it on each of the equilibrium points we have:

$(0, 0) \rightarrow \Delta = 4$, $\tau = 5 \rightarrow$ unstable node ($\tau^2 > 4\Delta$)

$(1/2, 0) \rightarrow \Delta = -3/2$, $\tau = 1/2 \rightarrow$ saddle ($\Delta < 0$)

$(1/2, 3/10) \rightarrow \Delta = 3/2$, $\tau = -5/2 \rightarrow$ stable node ($\tau^2 > 4\Delta$)

$(0, 4/5) \rightarrow \Delta = -4$, $\tau = -3 \rightarrow$ saddle ($\Delta < 0$)

Prob. 9. The equilibrium points are given by the equations:

$$\begin{cases} 4x_1(1 - x_1) - 2x_1x_2 = 0 \\ x_2(2 - x_2) - x_2 = 0 \end{cases}$$

or other

$$\begin{cases} 2x_1(1 - x_1) = x_1x_2 \\ x_2(2 - x_2) = x_2 \end{cases}$$

The second equation has degree 2 and the roots are $x_2 = 0$ or $x_2 = 1$. In the frist case if we substitute on the first equation we get

$$x_1(1 - x_1) = 0$$

which has solutions $x_1 = 0$ and $x_1 = 1$, generating two equilibrium points $(0, 0)$ and $(0, 1)$. Now if we take $x_2 = 1$ on the first equation we end up with

$$2x_1(1 - x_1) = x_1$$

which has the obvious solution $x_1 = 0$ which is already noted as one of the equilibrium points, and the other solution (simplyfing x_1) is

$$2(1 - x_1) = 1$$

or $x_1 = 1/2$, which gives the last equilibrium point as $(1/2, 1)$.

The Jacobian is given by the expression

$$Df = \begin{pmatrix} 4 - 8x_1 - 2x_2 & -2x_1 \\ 0 & 1 - 2x_2 \end{pmatrix}$$

and the determinant and trace are given by:

$$\Delta = (4 - 8x_1 - 2x_1)(1 - 2x_2)$$

$$\tau = 5 - 8x_1 - 4x_2$$

So at the equilibrium points we have:

$(0,0) \to \Delta = 4,\ \tau = 5 \to \tau^2 > 4\Delta \to$ unstable node

$(0,1) \to \Delta = -4,\ \tau = 1 \to \Delta < 0 \to$ saddle

$(1/2,1) \to \Delta = 2,\ \tau = -3 \to \tau^2 > 4\Delta \to$ stable node.

Prob. 11. The equilibrium points are given by the equations:

$$\begin{cases} x_1 - x_2 = 0 \\ x_1 x_2 - x_2 = 0. \end{cases}$$

So from the first equation we know that $x_1 = x_2$, and substituting on the second, we get

$$x_2^2 - x_2 = 0$$

or

$$x_2(x_2 - 1) = 0$$

with solutions $x_2 = 0$ and $x_2 = 1$, so the two equilibrium points are $(0,0)$ and $(1,1)$.

Looking at the Jacobian at these points we have

$$Df = \begin{pmatrix} 1 & -1 \\ x_2 & x_1 - 1 \end{pmatrix}$$

so determinant is $\Delta = x_1 + x_2 - 1$ and the trace $\tau = x_1$, so at the equilibrium points

$(0,0) \to \Delta = -1 \to$ saddle

$(1,1) \to \Delta = 1,\ \tau = 1 \to \tau^2 < 4\Delta \to$ unstable spiral

Prob. 13. We have:

$$x_2(x_1 + a) = 0,$$
$$x_2 = 0,$$
$$x_1 = -a,$$
$$x_2^2 + x_2 - x_1 = 0,$$
$$x_2 = \frac{-1 \pm \sqrt{1 + 4x_1}}{2}$$

$$1 + 4x_1 \le 0$$

To be unique, we must have: $1 - 4a \le 0$.

$$a \ge \tfrac{1}{4}$$

Now,

$$D = \begin{bmatrix} x_2 & a \\ -1 & 2x_2 + 1 \end{bmatrix}, \quad \Delta = (2x_2 + 1)x_2 + a, \quad \tau = 3x_2 + 1.$$

We have unstable spiral for $a > \frac{1}{4}$.

Prob. 15.

(a) The isoclines are the curves

$$x_1(10 - 2x_1 - x_2) = 0$$

and

$$x_2(10 - x_1 - 2x_2) = 0$$

so solving each one of the equations we get

$$10x_1 - 2x_1^2 - x_1x_2 = 0$$

$$x_1x_2 = 10x_1 - 2x_1^2$$

so for $x_1 \neq 0$, we have $x_2 = 10 - 2x_1$, for the first curve and solving the second we have

$$10x_2 - x_1x_2 - 2x_2^2 = 0$$

$$x_1x_2 = 10x_2 - 2x_2^2$$

and in a similar way, for $x_2 \neq 0$

$$x_1 = 10 - 2x_2$$

so the two graphs are

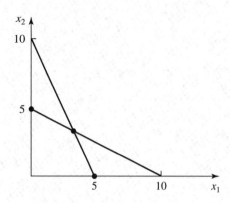

(b) Looking for the intersection point of two isoclines is the same as solving the system

$$\begin{cases} x_1 = 10 - 2x_2 \\ x_2 = 10 - 2x_1 \end{cases}$$

and substituting one equation into another we get

$$x_1 = 10 - 2(10 - x_1)$$

whose solution is $x_1 = 10/3$, and substituting back on the second equation we find $x_2 = 10/3$, so $\left(\frac{10}{3}, \frac{10}{3}\right)$ is an equilibrium.

The Jacobian of the system is given by

$$Df = \begin{pmatrix} 10 - 4x_1 - x_2 & -x_1 \\ -x_2 & 10 - x_1 - 4x_2 \end{pmatrix}$$

so at the equilibrium point

$$Df = \begin{pmatrix} 220/3 & -10/3 \\ -10/3 & -20/3 \end{pmatrix}$$

so $\Delta = 100/3$ and $\tau = -40/3$, since $\tau^2 > 4\Delta$, the equilibrium is a stable node.

Prob. 17. The matrix of the signs of elements of the Jacobian is

$$\begin{bmatrix} - & - \\ - & - \end{bmatrix}$$

so the determinant is unknown and the trace has negative sign, so the equilibrium can be one of three, a stable spiral, a stable node or a saddle.

Prob. 19. The matrix of signs of the Jacobian is

$$\begin{bmatrix} + & - \\ - & - \end{bmatrix}$$

so the trace is unknown and determinant has negative sign, so the equilibrium is a saddle.

Prob. 21. The matrix of signs of the Jacobian

$$\begin{bmatrix} - & 0 \\ - & - \end{bmatrix}$$

so determinant is positive and trace negative and the equilibrium is a stable node or a stable spiral.

Prob. 23.

(a) The first isocline will be given by the equation

$$x_1(2 - x_1) - x_1 x_2 = 0$$

so for $x_1 \neq 0$, it solves to $x_2 = 2 - x_1$.

The second isocline is given by

$$x_1 x_2 - x_2 = 0$$

and for $x_2 \neq 0$ it is the vertical line $x_1 = 1$, so the graph looks like:

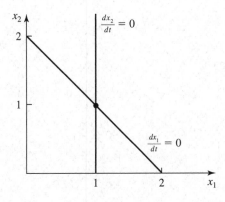

(b) The two lines intersect at $(1, 1)$ in the first quadrant, by trivial substitutive and this is the equilibrium point.

Looking at the signs of both functions on the side of each line we have

and the matrix of signs of the Jacobian is

$$\begin{bmatrix} - & - \\ + & 0 \end{bmatrix}$$

so the determinant is positive and the trace negative, making it a stable node or stable spiral.

11.4 Nonlinear Systems: Applications

Prob. 1. $N_2 = 20$ and $\alpha_{12}N_2 = 4$ so $\alpha_{12} = 1/5$ and by the same token $N_1 = 30$ and $\alpha_{21}N_1 = 6$ so $\alpha_{21} = 1/5$, so the equation are

$$\frac{dN_1}{dt} = 2N_1 \left(1 - \frac{N_1}{20} - \frac{N_2}{100} \right)$$

$$\frac{dN_2}{dt} = 3N_2 \left(1 - \frac{N_2}{15} - \frac{N_1}{75} \right)$$

Prob. 3. In this case $K_1 = 10$, $K_2 = 15$, $\alpha_{12} = 0.7$ and $\alpha_{21} = 0.3$, and we clearly see that

$$K_1 = 10 < 10.5 = \alpha_{12}K_2$$

and

$$K_2 = 15 > 3 = \alpha_{21}K_1$$

and species 2 excludes species 1.

Prob. 5. In this case $K_1 = 20$, $K_2 = 15$, $\alpha_{12} = 4$ and $\alpha_{21} = 5$, so we can clearly see that

$$K_1 = 20 < 60 = \alpha_{12}K_2$$

$$K_2 = 15 < 100 = \alpha_{21}K_1$$

and this is a case of founder control.

Prob. 7. The four equilibrium points will be $(0,0)$, $(K_1, 0)$, $(0, K_2)$ and the positive solutions to the system of equations

$$\begin{cases} N_1 + 1.3N_2 = 18 \\ 0.6N_1 + N_2 = 20 \end{cases}$$

a simple substitution argument shows that the systems solution is negative, so we only have to analyze the other three equilibrium points.

The equilibrium at $(0,0)$ has eigenvalues $\lambda_1 = r_1 = 3$ and $\lambda_2 = r_2 = 2$, making it unstable.

The equilibrium at $(k_1, 0) = (18, 0)$ has eigenvalues $\lambda_1 = -r_1 = -3$ and

$$\lambda_2 = r_2 \left(1 - \alpha_{21}\frac{k_1}{k_2}\right) = 2 \left(1 - 0.6\frac{18}{20}\right) > 0$$

making it unstable.

The equilibrium at $(0, k_2) = (0, 20)$ has eigenvalues $\lambda_2 = -r_1 = -2$ and

$$\lambda_1 = r_1 \left(1 - \alpha_{12}\frac{k_2}{k_1}\right) = 3 \left(1 - 1.3\frac{20}{18}\right) < 0,$$

so it is locally stable.

Prob. 9. The equilibrium points are $(0,0)$, $(k_1, 0)$, $(0, k_2)$ and the positive solutions to the system of equations

$$\begin{cases} N_1 + 3N_2 = 35 \\ 4N_2 + N_1 = 40 \end{cases}$$

multiplying the second by 3 and subtracting out of the first equation we get

$$-11N_1 = -85$$

guaranteeing a positive solution for N_1; and with substitution back into the first equation we see the positive solution for N_2 as well.

The equilibrium at $(0,0)$ is always unstable with eigenvalues $\lambda_1 = r_1 = 1$ and $\lambda_2 = r_2 = 3$.

The equilibrium at $(k_1, 0) = (35, 0)$ has eigenvalues $\lambda_1 = -r_1 = -1$ and

$$\lambda_2 = r_2 \left(1 - \alpha_{21}\frac{k_1}{k_2}\right) = 3 \left(1 - 4.\frac{35}{40}\right) < 0$$

so the equilibrium is stable.

The equilibrium at $(0, K_2) = (0, 40)$ has eigenvalues $\lambda_2 = -r_2 = -3$ and

$$\lambda_1 = r_1 \left(1 - \alpha_{12}\frac{k_2}{k_1}\right) = 1\left(1 - 3.\frac{40}{35}\right) < 0$$

so the equilibrium is stable.

Now, for the non-trivial point, $\alpha_{12}\alpha_{21} > 1$ and the equilibrium is unstable.

Prob. 11. The difference among the two situations for the N_1 population is 20, so

$$\alpha_{12} = \frac{20}{N_2} = \frac{20}{80} = \frac{1}{4}$$

The difference among the two situations for the N_2 population is 70, so

$$\alpha_{21} = \frac{70}{N_1} = \frac{70}{180} = \frac{7}{18}$$

Prob. 13.

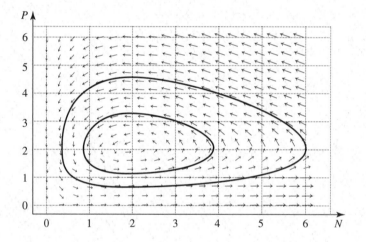

Prob. 15.

(a) The two isoclines are $N - 4PN = 0$ and $2PN - 3P = 0$

so with the first equation we have

$$N(1 - 4P) = 0$$

with solutions $N = 0$ and $P = 1/4$. As for the second isocline we have

$$2PN - 3P = 0$$

or factoring $P(2N - 3) = 0$ with solutions $P = 0$, and $N = 3/2$, so there are two

points of equilibrium $(0, 0)$ and $(3/2, 1/4)$.

(b) The Jacobi matrix at $(0, 0)$ is given by

$$Df(0, 0) = \begin{bmatrix} 1 & 0 \\ 0 & -3 \end{bmatrix}$$

a saddle, unstable equilibrium.

(c) At the non-trivial equilibrium point

$$Df(3/2, 1/4) = \begin{bmatrix} 0 & -1 \\ 5 & 0 \end{bmatrix}$$

and the eigenvalues can be computed and are $\lambda_1 = i\sqrt{5}$ and $\lambda_2 = -i\sqrt{5}$, which does not allows us to infer anything about the stability of the equilibrium.

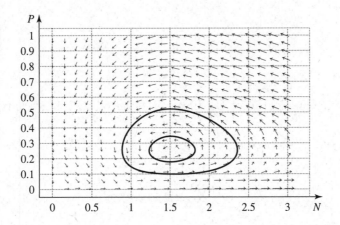

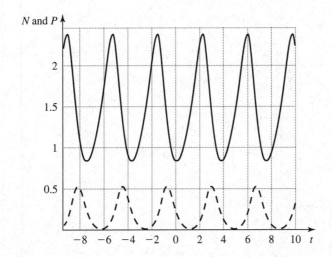

(d)

Prob. 17.

(a) If the predator is absent $P = 0$ and the equation simply becomes

$$\frac{dN}{dt} = 5N$$

Separating the variables we have

$$\frac{dN}{N} = 5dt$$

and integrating both sides

$$\int \frac{dN}{N} = \int 5dt$$

or

$$\ln N = 5t + c$$

or

$$N(t) = ce^{5t}$$

so in the absence of a predator the insect population will increase exponentially.

(b) When the predator is introduced the growth rate for $N(t)$ decreases, according to the equation

$$\frac{dN}{dt} = 5N - 3PN$$

which will help control the density of the insect.

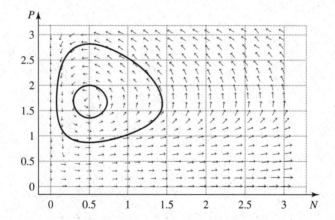

(c)

Prob. 19.

(a) In the absence of a predator $P = 0$ and the first equation becomes

$$\frac{dN}{dt} = 3N\left(1 - \frac{N}{10}\right)$$

(b) The isoclines are

$$3N\left(1 - \frac{N}{10}\right) - 2PN = 0$$

and

$$PN - 4P = 0$$

The first one solves to $N = 0$ and $P = \frac{3}{2}\left(1 - \frac{N}{10}\right)$ and the second to $P = 0$ and $N = 4$, so the equilibrium points are $(0,0)$ and $(4, 9/10)$.

The Jacobian matrix is given by

$$Df(N, P) = \begin{pmatrix} 3 - \frac{3N}{5} - 2P & -2N \\ P & N - 4 \end{pmatrix}.$$

So at $(0,0)$ it is

$$Df(0,0) = \begin{pmatrix} 3 & 0 \\ 0 & -4 \end{pmatrix}$$

which is a diagonal matrix with eigenvalues 3 and -4 so the equilibrium is unstable.

For the other point

$$Df(4, 9/10) = \begin{pmatrix} -6/5 & -8 \\ 9/10 & 0 \end{pmatrix}$$

so with $\Delta = 36/5$ and $\tau = -6/5$ it is a stable spiral.

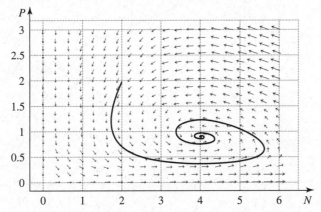

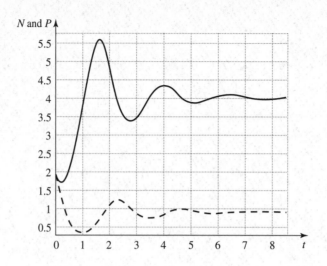

(c)

Prob. 21.

(a) The isoclines will be given by the equations

$$N\left(1 - \frac{N}{20}\right) = 5PN$$

$$PN = 4P$$

so they are $N = 0$ and $P = \frac{1}{5} - \frac{N}{100}$ for the first equation and $P = 0$ and $N = 4$ for the second.

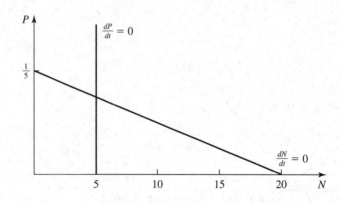

(b) The signs of the Jacobian matrix are

$$Df = \begin{bmatrix} - & - \\ + & 0 \end{bmatrix}$$

with positive determinant and negative trace, so it is a stable equilibrium.

Prob. 23. The intersection of th zero-isoclines will happen at the point

$$N = \frac{d}{c} \quad \text{and} \quad P = \frac{a}{b}\left(1 - \frac{d}{ck}\right)$$

so the density of the prey does not depend on the value of "a" and will increase the density of the predator.

Prob. 25. The equilibrium point of Problem 22 is given by

$$N = \frac{d}{c} \quad \text{and} \quad P = \frac{a}{b}\left(1 - \frac{d}{ck}\right)$$

so an increase in the value of "c" will increase the denominator of $N = \frac{d}{c}$ and then reduce the abundance of the prey. At the same time increasing the value of "c" will decrease the amount that gets subtracted in P, increasing the abundance of the predator at the equilibrium point.

Prob. 27. It is a $(+-)$ case of *Predation*, the determinant is positive, trace negative, so stable.

Prob. 29. It is a $(++)$ case of *Mutualism*, or *Symbiosis*, the determinant $\Delta > 0$ and trace negative, so stable.

Prob. 31. It is a $(++)$ case of *Mutualism* or *Symbiosis*, the determinant $\Delta = 1.5 - 2.6 = -1.1 < 0$, so unstable.

Prob. 33. It is a $(--)$ case of *Competition*, the determinant negative, so unstable.

Prob. 35. The signs of the Jacobian matrix are

$$\begin{bmatrix} - & - \\ - & - \end{bmatrix}$$

so the determinant is unknown and the equilibrium cannot be determined from the data.

Prob. 37. The signs of the Jacobian matrix are

$$\begin{bmatrix} - & - \\ - & - \end{bmatrix}$$

so the determinant is unknown and the equilibrium cannot be determined.

Prob. 39. The signs of the Jacobian matrix are

$$\begin{bmatrix} - & - \\ - & - \end{bmatrix}$$

so the determinant is unknown and the type of equilibrium cannot de determined.

Prob. 41. The diagonal element a_{ii} measures the effect species i has on itself, and the value of the derivative

$$a_{ii} = \frac{af_i(N_1, N_2)}{aN_i}$$

so if they are negative the growth of the number of individuals will imply in a lower growth rate.

Prob. 43.

(a) The zero-isoclines are given by the equations

$$aN = bNP$$

$$cNP = dP$$

with solutions $N = 0$ and $P = a/b$ for the first and $P = 0$ and $N = d/c$ for the second, so the non-trivial equilibrium happens at the point:

$$\left(\frac{d}{c}, \frac{a}{b} \right)$$

(b) The community matrix is given by

$$A = \begin{bmatrix} \frac{af_1}{aN} & \frac{af_1}{aP} \\ \frac{af_2}{aN} & \frac{af_2}{aP} \end{bmatrix} = \begin{bmatrix} a - bP & -bN \\ cP & cN - d \end{bmatrix}$$

(c) $a_{12} = -bN < 0$, so the growth rate of the prey is decreased if the abundance of the predator increase.

$a_{21} = cP > 0$, so the grouwth rate of the predator increases if the abundance of the prey increase.

Prob. 45.

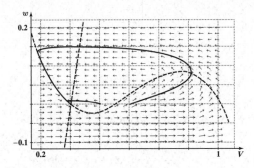

Prob. 47. We have first to verify both eigenvalues are real and negative, for one to observe an action potential. In this particular case the Jacobian

$$Df = \begin{bmatrix} -3V^2 + 2.6V - 0.3 & -1 \\ 0.01 & -0.004 \end{bmatrix}$$

so at $(0,0)$ it reduces to

$$Df = \begin{bmatrix} -0.3 & -1 \\ 0.01 & -0.004 \end{bmatrix}$$

so computing the eigenvalues

$$\begin{aligned} \det(A - \lambda I) &= \begin{vmatrix} -0.3 - \lambda & -1 \\ 0.01 & -0.004 - \lambda \end{vmatrix} \\ &= (\lambda + 0.004)(\lambda + 0.3) + 0.01 \\ &= \lambda^2 + 0.304\lambda + 0.0012 + 0.01 \\ &= \lambda^2 + 0.304\lambda + 0.0112 \end{aligned}$$

which can easily verified to have roots which are real and negative, so we will observe an action potential for any value of $V \in (0.3, 1)$.

Prob. 49.

$$\begin{aligned} \frac{db}{dt} &= -kab \\ \frac{dc}{dt} &= kab \end{aligned}$$

Prob. 51.

$$\frac{ds}{dt} = -k_1 se$$

$$\frac{de}{dt} = k_2 c - k_1 se$$

$$\frac{dc}{dt} = k_1 se - k_2 c$$

$$\frac{dp}{dt} = k_2 c$$

Prob. 53. The conserved quantity is $x + y$, because

$$\frac{d}{dt}(x+y) = \frac{dx}{dt} + \frac{dy}{dt} = 2x - 3y + 3y - 2x = 0$$

Prob. 55. The conserved quantity is $x + y + z$, because

$$\frac{d}{dt}(x+y+z) \qquad = \frac{dx}{dt} + \frac{dy}{dt} + \frac{dz}{dt}$$

$$= (-x + 2y + z) + (-2xy) + (x - z) = 0$$

Prob. 57.

(a)

$$\lim_{s \to \infty} f(s) = \lim_{s \to \infty} \frac{\vartheta_m s}{k_m + s} = \lim_{s \to \infty} \frac{\frac{\vartheta_m s}{s}}{\frac{k_m + s}{s}}$$

$$= \lim_{s \to \infty} \frac{\vartheta_m}{\frac{k_m}{s} + 1}$$

$$= \vartheta_m$$

because the denominator tends to 1.

(b)

$$f(k_m) = \frac{\vartheta_m \cdot k_m}{k_m + k_m} = \frac{\vartheta_m \cdot k_m}{2k_m} = \frac{\vartheta_m}{2}$$

(c) (i) Both numerator and denominator of $f(s)$ are positive, so $f(s) > 0$

(ii) Assume that s_1 and s_2 are on the line and that $s_1 < s_2$, then taking the inverses and multiplying by k_m (positive) we get

$$\frac{k_m}{s_1} > \frac{k_m}{s_2}$$

adding 1 and taking inverses again changes the direction of the inequality

$$\frac{1}{\frac{k_m}{s_1} + 1} < \frac{1}{\frac{k_m}{s_2} + 1}$$

so multiplying both sides ϑ_m we get

$$\frac{\vartheta_m}{\frac{k_m}{s_1}+1} < \frac{\vartheta_m}{\frac{k_m}{s_2}+1}$$

multiplying both numerator and denominator on the left by s_1 and on the right by s_2, should not change the value of the fractions and

$$\frac{\vartheta_m}{k_m+s_1} < \frac{\vartheta_m}{k_m+s_2}$$

showing that $f(s_1) < f(s_2)$, that is, $f(s)$ is increasing.

Another way to see this result is to compute the derivative of $f(s)$ and show that it is positive everywhere

$$f'(s) = \frac{(k_m+s)(\vartheta_m) - \vartheta_m s}{(k_m+s)^2} = \frac{k_m \vartheta_m}{(k_m+s)^2} > 0$$

so $f(s)$ is increasing.

(iii) To show that $f(s)$ is concave down we compute the second derivative

$$f''(s) = -\frac{k_m \vartheta_m}{(k_m+s)^3} < 0$$

so $f(s)$ is concave down

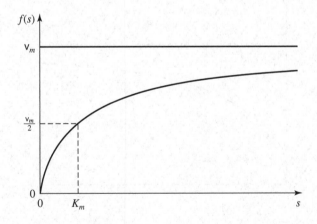

(d) The reaction rate $\frac{dp}{dt} = \frac{\vartheta_m s}{k_m+s}$ is limited above by ϑ_m and according to the graph depends on s, the total amount of substrate.

Prob. 59.

(a) Setting the equations equal to zero and adding algebraically we see that the equilibrium point is

$$\hat{s} = \frac{Dk_m}{y\vartheta_m - D}$$

which is an increasing function of D, to see this just compute the derivative of

$$f(x) = \frac{xk_m}{y\vartheta_m - x}$$

with respect to the variable x, or observe the fact that $1/f(x)$ can be easily seen as a decreasing function of x.

(b) To find $\hat{s}$, just mark $\frac{D}{y}$ on the q-axis of the graph and find the inverse point (mark the horizontal until the graph and then the vertical until the s-axis). The following calculation show why this is true

$$q\left(\frac{Dk_m}{y\vartheta_m - D}\right) = \frac{\vartheta_m \frac{Dk_m}{y\vartheta_m - D}}{k_m + \frac{Dk_m}{y\vartheta_m - D}} = \frac{D}{y}$$

As it can be seen from the graph the higher value of D, the higher will be the value of $\hat{s}$

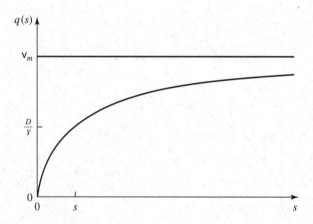

Prob. 61. In this case $D = 2$, $s_0 = 4$, $\vartheta_m = 3$, $k_m = 1$ and $Y = 1$. We can easily see that the inequality

$$0 < D/Y < \vartheta_m$$

is verified and we will have two equilibrium points

$$(s_0, 0) = (4, 0)$$

and

$$\left(\frac{Dk_m}{y\vartheta_m - D}, y\left(s_0 - \frac{Dk_m}{y\vartheta_m - D}\right) \right) = (2, 2)$$

At the first point the Jacobi matrix is

$$Df(s_0, 0) = \begin{bmatrix} -D & -q(s_0) \\ 0 & yq(s_0) - D \end{bmatrix} = \begin{bmatrix} -2 & -12/5 \\ 0 & 2/5 \end{bmatrix}$$

with determinant $\Delta = -4/5$ negative making it an unstable point.

From the text we see that when the nontrivial equilibrium exists (our case here) it is always stable.

11.6 Review Problems

Prob. 1. Since $Z(t) = N_1(t)/N_2(t)$, we can apply logarithms

$$\ln Z(t) = \ln N_1(t) - \ln N_2(t)$$

and computing the derivative

$$\begin{aligned} \frac{d}{dt} \ln Z(t) &= \frac{d}{dt} \ln N_1(t) - \frac{d}{dt} \ln N_2(t) \\ &= \frac{N_1'(t)}{N_1(t)} - \frac{N_2'(t)}{N_2(t)} \\ &= r_1 - r_2 \end{aligned}$$

Now solving the differential equation

$$\frac{d}{dt} \ln Z(t) = r_1 - r_2$$

we get

$$\ln Z(t) = (r_1 - r_2)t + c$$

and taking exponentiation on both sides

$$Z(t) = Z_0 e^{(r_1 - r_2)t}$$

so if $r_1 > r_2$, the coefficient for t is positive and $\lim\limits_{t \to \infty} Z(t) = \infty$, so if $r_1 > r_2$, population 1 becomes numerically dominant.

Prob. 3.

(a) The zero isoclines are given by the equations

$$2N(1 - \frac{N}{10}) = 3PN$$

and

$$PM = 3P$$

with solutions $N = 0$ and $P = \frac{2}{3}(1 - \frac{N}{10})$ for the first and $P = 0$ and $N = 3$ for the second

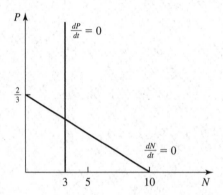

(b) The matrix of signs of the Jacobian are

$$\begin{bmatrix} + & 0 \\ - & - \end{bmatrix}$$

so the determinant is negative and the equilibrium unstable.

Prob. 5.

(a) In the absence of species 2, $p_2 = 0$, so the two equations reduceto

$$\frac{dp_1}{dt} = c_1 p_1(1 - p_1) - m_1 p_1$$

and the equilibrium points are given by

$$c_1 p_1 (1 - p_1) = m_1 p_1$$

with solutions $p_1 = 0$ and $p_1 = \frac{m_1}{c_1} - 1$, so a nontrivial equilibrium will happen with $o < p_1 \leq 1$, if and only if $1 < \frac{m_1}{c_1} \leq 2$, that is

$$c_1 < m_1 \leq 2c_1$$

(b) We can re-write the equations in the Lotka-Volterra format to obtain

$$
\begin{aligned}
\frac{dp_1}{dt} &= c_1 p_1 (1 - p_1 - p_2) - c_1 p_1 \frac{m_1}{c_1} \\
&= c_1 p_1 (1 - \frac{m_1}{c_1} - p_1 - p_2) \\
&= c_1 p_1 \left(\frac{c_1 - m_1}{c_1} - p_1 - p_2 \right) \\
&= p_1 (c_1 - m_1 - c_1 p_1 - c_1 p_2) \\
&= (c_1 - m_1) p_1 \left(1 - \frac{c_1}{c_1 - m_1} p_1 - \frac{c_1}{c_1 - m_1} p_2 \right)
\end{aligned}
$$

which is our standard equation with

$$r_1 = c_1 - m_1, \quad K_1 = 1 - \frac{m_1}{c_1} \quad \text{and} \quad \alpha_{12} = 1$$

and in the same fashion we rearrange the second equation to obtain

$$r_2 = c_2 - m_2, \quad K_2 = 1 - \frac{m_2}{c_2} \quad \text{and} \quad \alpha_{21} = 1.$$

Assuming that $\frac{c_1}{m_1} > \frac{c_2}{m_2}$ we can invert to obtain

$$\frac{m_1}{c_1} < \frac{m_2}{c_2}$$

and

$$1 - \frac{m_1}{c_2} > 1 - \frac{m_2}{c_2}$$

that is $K_1 > K_2$. Since $\alpha_{12} = \alpha_{21} = 1$, we have both $K_1 > \alpha_{12} K_2$ and $K_2 < \alpha_{21} K_1$, and species 1 will exclude species 2, if they start on a positive parte of the plane.

Prob. 7.

(a) The zero-isoclines of the system is given by the equations

$$D(s_0-) = q(s)x$$

and

$$Yq(s)x = Dx$$

so the nontrivial solutions are $q(\hat{x}) = \frac{D}{Y}$, and using this on the first equation we get

$$D(s_0 - \hat{s}) = \frac{D}{Y}\hat{x}$$

so at the equilibrium point: $\hat{x} = Y(s_0 - \hat{s})$, which is positive when $\hat{s} < s_0$.

(b) Since $q(\hat{s}) = \frac{D}{Y}$, we can write out the equation

$$\frac{v_m\hat{s}}{K_m + \hat{s}} = \frac{D}{Y}$$

and solving for $\hat{s}$ we get: $\hat{s} = \frac{DK_m}{Yv_m-D}$ so the equilibrium abundance of the microbe is

$$\hat{x} = Y(x_0 - \hat{s}) = \frac{YDK_m}{Yv_m - D}$$

Consider now the ivnerse of $\hat{x}$, that is

$$\frac{1}{\hat{x}} = \frac{Yv_m - D}{YDK_m} = \frac{Yv_m}{YDK_m} - \frac{D}{YDK_m} = \frac{v_m}{DK_m} - \frac{1}{YK_m}$$

so we can clearly see that $\frac{1}{\hat{x}}$ is a decreasing function of D and an increasing function of Y, so $\hat{x}$ will be increasing with D and decreasing with Y.

Chapter 12

Probability and Statistics

12.1 Counting

Prob. 1. One experiment (trial) consists of using one fertilizer with one light level. There are 5 different fertilizers and each is to be tested with every one of the four light levels. Thus one has to conduct $5 \cdot 4 = 20$ trials. But each trial is required to have 4 replicates. Hence the total number of replicates is given by $5 \cdot 2 \cdot 4$, which is equal to 40.

Prob. 3. The beetle is to be tested with 2 satiation levels. For each satiation level, it is to be given two (out of three) items of food. Since there are 3 ways in which two food items can be selected out of 3 and since 2 satiation levels are to be tested, six experiments have to be conducted on the beetle to complete all possibilities. However, we are told that to ensure accuracy of results, each experiment is to be repeated 20 times. Thus there are a total number of 120 $(20 \cdot 6)$ replicates.

Prob. 5. For a breakfast a person takes, as per the problem, one beverage, one cereal and one fruit. Since there is a choice of 3 beverages, 7 cereals and 4 fruits, a person has a choice of $(3 \cdot 7 \cdot 4 = 84)$ different breakfasts.

Prob. 7. We have 9749 slots to fill with 4 options each, hence the total number is $4^{9749} \approx 3.04 \times 10^{5869}$.

Prob. 9. This problem can best be tackled by imagining that there are only 3 cities, A, B and C to be toured. If City A is chosen first, there are 2 alternatives, B and C or C and B. The other options are B,A,C and B,C,A; C,A,B and C,B,A. (We are actually using enumeration here to understand the problem). Thus there are a total of 6 options, which is really 3!. Thus for 5 cities, the number of different routes are 5!. This is equal to 120.

Prob. 11. Seven books are to be arranged in a shelf. Remember that the order is important. Thus similar to Ex 8 above, the seven books can be arranged in 7! different ways, which works out to 5040.

Prob. 13. Out of 26 letters, four are to be chosen at a time. Remember that the order is very important, since each order stands for a different word. This is a clear case for using permutation. Hence the number of words that can formed in this case is $P(26, 4)$. This is equal to $26!/22!$, which is $23 \cdot 24 \cdot 25 \cdot 26$. So the number of words is a phenomenal 3,58,800.

Prob. 15. Three awards are to be distributed. Since the awards are not identical, but different, this is a clear case in which order is important. Using principles of permutation illustrated in Section 12.1.2, the number of ways in which three awards can be presented to the class of 15 students is $P(15, 3)$. The result is $(13 \cdot 14 \cdot 15)$ which is 2730.

Prob. 17. We have the following: 6 ways to choose the first person, 5 ways to choose the second person, and so on, giving $6 \cdot 5 \cdot 4 \cdot 3 \cdot 2 \cdot 1 = 6!$ ways.

Prob. 19. The order in which the candies are chosen is not important. Hence this is a case for using Combination, as explained in Example 12.1.3. The answer is $C(10, 3)$, which is 120.

Prob. 21. Since no specific tasks are assigned to the members of the committee, this is a case for using Combinations. The three people can be selected out of ten in $C(10, 3)$ ways. This works out to $10! \div (3! \cdot 7!) = 8 \cdot 9 \cdot 10 \div (2 \cdot 3) = 120$ ways.

Prob. 23. Four balls can be selected out of fifteen balls in $C(15, 4)$ ways. Applying the formula for combinations, $15! \div (4! \cdot 11!) = 12 \cdot 13 \cdot 14 \cdot 15 \div (2 \cdot 3 \cdot 4) = 1365$ ways.

Prob. 25. If all the letters were different, they could have been arranged in 9! ways. Since there are four A's and five B's some of the words would be indistinguishable. The number of different words that can be formed is $9! \div (4! \cdot 5!) = 6 \cdot 7 \cdot 8 \cdot 9 \div (2 \cdot 3 \cdot 4) = 126$ ways.

Prob. 27. Since the order is important, we have $\begin{pmatrix} 1000 \\ 20 \end{pmatrix}$

Prob. 29.

(a) The two red balls can be picked out of the five red balls in $C(5, 2)$ ways. This works out to $5! \div (3! \cdot 2!) = 10$ ways. The two blue balls can be picked up from the four

blue balls in $C(4, 2)$ ways. This works out to $4! \div (2! \cdot 2!) = 6$ ways. One red ball and

one blue ball can be picked in $C(5, 1) \cdot C(4, 1)$ ways. The first term works out to

$5! \div 4! = 5$ ways. The second term works out to $4!/3! = 4$ ways. Hence the product is

20 ways.

(b) The totals of all three combinations works out to $10 + 6 + 20 = 36$ ways. Now the

way in which two balls can be picked from nine balls in $C(9, 2)$ ways. This works out

$9! \div (2! \cdot 7!) = 8 \cdot 9 \div 2 = 36$ ways. This total agrees with the sum obtained in part A.

Prob. 31. If all the letters were different, they could have been arranged in 14! ways. But

some of the words formed will be indistinguishable from each other since the letters repeat.

Hence the total number of different words that can be formed is

$14! \div (5! \cdot 3! \cdot 6!) = 7 \cdot 8 \cdot 9 \cdot 10 \cdot 11 \cdot 12 \cdot 13 \cdot 14/(2 \cdot 3 \cdot 4 \cdot 5 \cdot 2 \cdot 3) = 168168$ ways.

Prob. 33. S= a,b,c. The subsets of this given set are

1) ϕ the null set. 2) a 3) b 4) c 5) a,b 6) b,c 7) a,c 8) a,b,c

The total number of subsets is therefore given by $2^3 = 8$.

Prob. 35. Let the positions on the bench be numbered 1,2,3,4. If Peter sits on the

extreme ends ie. 1 or 4, Melissa would have to be on positions 2 and 3 respectively. Brian

and Hillary can sit on any of the remaining two positions. So the number of ways this is

possible is $2 \cdot 1 \cdot 2 \cdot 1 = 4$ ways. If Peter sits on positions 2 or 3, Melissa can sit on either

side of him. Brian and Hillary can sit on the other two benches. So the number of ways is

$2 \cdot 2 \cdot 2 \cdot 1 = 8$ ways, so the total is $8 + 4 = 12$ ways.

Prob. 37. The total number of ways in which 3 people can be selected from a group of 7

is $C(7, 3) = 7! \div (4! \cdot 3!) = 5 \cdot 6 \cdot 7 \div (2 \cdot 3) = 35$ ways. Now let us count the number of

committees in which the two people who do not want to be together are included. If the

two people appear together in a committee, the third person can be selected in $C(5, 1) = 5$

ways. Hence the two people who do not want to serve together will appear in 5 ways.

Subtracting 5 from $35 = 30$ ways is the total number of ways in which 3 people can be

selected out of 7 if two of them do not want to serve together.

Prob. 39. The mandatory perennial plant can be selected in $C(3, 1) = 3$ ways. Out of the

remaining 6 plants, 2 of them can be selected in $C(6, 2) = 15$ ways. Hence the total number of ways is $3 \cdot 15 = 45$ ways.

Prob. 41. We have the first group: $\begin{pmatrix} 60 \\ 20 \end{pmatrix}$, second group $\begin{pmatrix} 40 \\ 20 \end{pmatrix}$, third group $\begin{pmatrix} 20 \\ 20 \end{pmatrix}$

to get: $\begin{pmatrix} 60 \\ 70 \end{pmatrix} \begin{pmatrix} 40 \\ 20 \end{pmatrix} \begin{pmatrix} 20 \\ 20 \end{pmatrix}$.

Prob. 43. $(x+y)^4 = (x+y)^2(x+y)^2 = (x^2 + 2xy + y^2)(x^2 + 2xy + y^2) =$
$x^4 + 4(x^3)y + 6(x^2)(y^2) + 4x(y^3) + y^4$

Prob. 45. The four red cards can be drawn in $C(26,4) = 14950$ ways. The five black cards can be drawn in $C(26,5) = 65780$ ways. Therefore the total number of ways $=$ $C(26,4) \cdot C(26,5) = 983,411,000$ ways!.

Prob. 47. Under the Section Examples, the method of calculating the number of *pairs* in Poker is discussed. Now the question is about the number of ways in which exactly two pairs can be picked.

Step 1: Assign the value to the first pair (13 ways)

Step 2: Their suits can be chosen in $C(4,2)$ ways.

Step 3: Assign the value to the second pair (12 ways)

Step 4: Their suits can be chosen in $C(4,2)$ ways)

Step 5: The one remaining card should have a value which is different from the values of the two pairs. Thus this one can be chosen in $C(44,1)$ ways. Looking it another way, the last card can be chosen in $4 * C(11,1)$ ways.

Step 6. Combining all the above, the number of ways of picking exactly two pairs are:

$13 \cdot C(4,2) \cdot 12 \cdot C(4,2) \cdot C(44,1) = 247104$ ways.

Prob. 49. There are 13 ways in which 4 cards of the same value can be picked, since there are 13 values attributed to the 13 cards of each suit. The one remaining card of the *hand* can be picked in $C(48,1)$ ways. Hence the total number of ways *four of a kind* can be picked is $13 * 48$

Prob. 51. We have 3 ways for the first, 2 for second, 1 for third, giving $3 \cdot 2 \cdot 1 = 3!$

12.2   What is Probability?

Prob. 1. The sample space for the experiment of tossing a coin three times is:

$$\Omega = \{HHH, HHT, HTH, HTT, THH, THT, TTH, TTT\}$$

where H stands for Heads, T stands for tails. The Sample Space consists of a total of 8 elements.

Prob. 3. The Sample Space of selecting two balls without replacement is $\Omega = \{1,2; 1,3;$ 1,4; 1,5; 2,1; 2,3; 2,4; 2,5; 3,1; 3,2; 3,4; 3,5; 4,1; 4,2; 4,3; 4,5; 5,1; 5,2; 5,3; 5,4\}$ where 1,2,3,4,5 are the numbers on the balls. The Sample Space thus contains 20 elements.

Prob. 5. $\Omega = \{1, 2, 3, 4, 5, 6\}$ A= $\{1,3,5\}$, B= $\{1,2,3\}$ A$\cup$ B = $\{1,2,3,5\}$ A$\cap$B = $\{1,3\}$

Prob. 7. $(A\cup B)^c = \{1,2,3,5\}^c = \{4,6\}$

Prob. 9. Since $\Pr(1) + \Pr(2) + \Pr(3) + \Pr(4) + \Pr(5) = 1$,

$$\Pr(5) = 1 - \Sigma_{i=1}^{i=4} Pr(i) = 1 - 0.1 - 0.2 - 0.05 - 0.05 = 0.6$$

Prob. 11. $\Pr(A)^c = \Pr(2) + \Pr(4) = 0.25$

Prob. 13. Since $P(A) = 0.7$, $P(1) = 0.1$, we see $P(4) = 0.2$. Then $P(3) = P(B) - P(4) = 0.3$.

Prob. 15. We observe that we want $P(1, 2, 4) = 0.1 + 0.4 + 0.2 = 0.7$.

Prob. 17. $\Pr(A\cap B) = 0.1$; $\Pr(A) = 0.4$; $\Pr(A^c \cap B^c) = 0.2$. According to De Morgan's law, $(A^c \cap B^c) = (A\cup B)^c$ Therefore $(A\cup B)^c = 0.2$. Therefore $(A\cup B) = 1 - 0.2 = 0.8$ $\Pr(A\cup B) = \Pr(A) + \Pr(B) - \Pr(A\cap B)$ Therefore $0.8 = 0.4 + \Pr(B) - 0.1$ $\Pr(B) = 0.5$

Prob. 19. The Sample Space $\Omega = (HH, HT, TH, TT)$. Since all events are equally possible (the coins are fair coins) and three of the events are favorable to obtaining at least one head, the probability of obtaining at least one head = 3/4

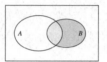

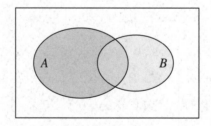

Prob. 21. We want the probability of HT, TH, HH: $\frac{3}{4}$.

Prob. 23. We have $\frac{4!}{2!2!}$ ways of having exactly 2 heads. But then, $P(2 \text{ he3ads})$

$= \frac{4!}{4}/2^4 = \frac{3}{8}$.

Prob. 25. The Probability of obtaining at least one four in a throw = 1 - Probability of obtaining no 4's on either throw is: $1 - (5 \div 6) \times (5 \div 6) = 1 - 25 \div 36 = 11 \div 36$.

Prob. 27. The Sample Space in the experiment in which two fair dice are thrown one after the other is identically the same as that in which one fair die is thrown twice, as in Problem 2. Thus the Sample Space has 36 elements which are enumerated in Problem 2(and also in Problem 22 below). All outcomes are equally probable. Out of those outcomes, {2,1; 3,1; 3,2; 4,1; 4,2; 4,3; 5,1; 5,2; 5,3; 5,4; 6,1; 6,2; 6,3; 6,4; 6,5} = 15 elements are favorable to the probability that the first number is greater than the second number. Hence the answer is $15 \div 36 = 5/12$

Prob. 29. The four possible outcomes of a cross between Cc and Cc is $\Omega = $ (CC, Cc, cC and cc). Each has probability of $1 \div 4$ since all outcomes are equally probable. Only the last case produces white flowers. Hence the probability of obtaining white flowers is $1/4$.

Prob. 31. The possible outcomes of a cross between Aa and Aa are $\Omega = $ {AA, Aa, aA, aa}. Since Aa = aA, the outcomes are = {AA, Aa, Aa, aa} all outcomes are equally probable. So the probability of obtaining Aa is $(1 + 1) \div 4 = 0.5$

Prob. 33. Let a boy be represented by B and girl as G. The possible outcomes are $\Omega = $ {BBB, BBG, BGB, BGG, GBB, GBG, GGB, GGG} All outcomes are equally probable since the sex ratio is 1:1. So the probability of obtaining all girls is $1 \div 8$

Prob. 35. We want P(no girls) $+P$(1 girl) $+P$ (2 girls)$= \frac{1}{16} + \frac{4}{16} + \frac{6}{16} = \frac{11}{16}$.

Prob. 37. Let the woman's chromosomes be: $x1f, x2f$. Men's chromosomes: $x1m, x2m$. Then the child is $\{x1fx1m, x1fx2m, x2fx1m, x2fx2m\}$ hence we get $1/4$.

Prob. 39. We have: $\frac{3}{5} \cdot \frac{2}{4} + \frac{2}{5} \cdot \frac{3}{4} = \frac{12}{20} = \frac{3}{5}$.

Prob. 41. The probability that the first card is a spade is $13 \div 52$. The probability that the second card is a spade $= 12 \div 51$. So the probability that both cards are spades $= (13 \div 52) \cdot (12 \div 51) = 1 \div 17$.

Prob. 43. The total number of ways in which 3 balls can be picked from 12 balls is $C(12,4) = 220$ ways. Out of these ways, let us count the number of favorable outcomes. If the balls are of different color, 1 ball has to be green, 1 ball has to be blue and 1 ball has to be red. 1 green ball can be picked from 4 green balls in $C(4,1) = 4$ ways. 1 blue ball can be picked from 6 blue balls in $C(6,1) = 6$ ways. 1 red ball can be picked from 2 red balls in $C(2,1) = 2$ ways. Hence the total number of ways of picking 1 green, 1 blue and 1 red balls $= 4 \cdot 6 \cdot 2 = 48$ ways. So the probability of picking all balls of different colors is $48 \div 220 = 0.218181818$

Prob. 45. The probability of obtaining at least one ace $= 1$ - Probability of obtaining no aces at all. Let us calculate the probability of obtaining no aces.

The probability that the first card is not an ace is $48 \div 52 = 0.923076923$

The probability that the second card is not an ace is $47 \div 51 = 0.921568627$

The probability that the third card is not an ace is $46 \div 50 = 0.92$

The probability that the fourth card is not an ace is $45 \div 49 = 0.918367346$

So the probability of obtaining no aces is
$(0.923076923)(0.921568627)(0.92)(0.918367346) = 0.718736725$.

So the probability of obtaining at least one ace $= 1$ - $718736725 = 0.281263274$.

Prob. 47. Let us calculate the probability of getting all red cards. The probability that the first card is red is $26 \div 52$. Since there are 26 red cards out of 52. The probability that the second card is red is $25/51$. Since there are 25 red cards out of the remaining 51 cards. The probability that the third card is red is $24 \div 50$. Since there are 24 red cards out of the remaining 50 cards. We continue this logic to all the thirteen cards.

So the probability of all cards being red $= (26 \div 52)(25 \div 51)(24/50)(23/49)(22/48)(21 \div 47)(20/46)(19 \div 45)(18/44)(17 \div 43)(16 \div 42)(15 \div 41)(14 \div 40) = 0.001722867$

Prob. 49. We first assign a value to the first pair. There are 13 values. Each pair can be picked in $C(4,2)$ ways. We then assign a value to the second pair. There are 12 values. Each pair can be picked in $C(4,2)$ ways. The remaining card can be picked in $C(44,1)$ or $4 \cdot C(11,1)$ ways. The five cards can be selected in $C(52,5)$ ways.

Combining the above steps, the probability of obtaining exactly two pairs is:
$(13 \cdot C(4,2) \cdot 12 \cdot C(12,2) \cdot C(44,1)) \div C(52,5) = 22464 \div 2598960$. This works out to a probability of 0.008643457

Prob. 51. a) $\dfrac{\dbinom{N-10}{7}\dbinom{100}{3}}{\dbinom{N}{10}}$

b) We get $N = 333$.

12.3 Conditional Probability and Independence

Prob. 1. $P(B/A) = 13/51$ (since there are 13 spades out of the remaining 51 cards)

Prob. 3. $\frac{13}{50}$, as there are 13 clubs left.

Prob. 5. We define two events

A = (probability that first ball is blue)

B = (probability that the second ball is green)

$P(A) = 5/11$ (there are 5 blue balls out of 11 balls) It is given that the event A has occurred. We have to compute $P(B/A)$ Now $P(B/A) = 6/10 = 3/5$ (since there are 6 green balls out of the remaining 10 balls)

Prob. 7. We define two events

A = (probability that the younger child is a girl)

B = (probability that the older child is a girl)

$P(A) = 1/2$ (assuming a 1:1 sex ratio)

It is given that the younger child is a girl.

$P(B) = P(A \cap B)/P(A)$ and since the events are independent, it is equal to

$((1/2) \cdot (1/2))/(1/2) = 1/2$

Prob. 9. The sample space of numbers add up to 7 is $\Omega = (1, 6; 2, 5; 3, 4; 4, 3; 5, 2; 6, 1)$ and the probability that the first number is a 4 is therefore $1/6$

Prob. 11. The sample space $\Omega = (HHH, HHT, HTH, HTT, THH, THT, TTH)$ since at least one head has occurred. (The only element left out is TTT). Hence the probability that the first coin is heads $= 4/7$.

Prob. 13. We have: $P(A|B) = \frac{P(A \cap B)}{P(B)} = \frac{\frac{1}{16}}{\frac{4}{16}} = \frac{1}{4}$, where $A =$ four heads, $B =$ first and third are heads.

Prob. 15. Let A = (test result is negative)

Let B1 = person is infected

Let B2 = person is not infected

$P(B1) = 1/100 = 0.01$

$P(B2) = 99/100 = 0.99$

$P(A/B1) = 1 - 0.9 = 0.1$

$P(A/B2) = 1 - 0.15 = 0.85$

$P(A) = P(A/B1) \cdot P(B1) + P(A/B2) \cdot P(B2)$

$P(A) = 0.1 \cdot 0.01 + 0.85 \cdot 0.99 = 0.001 + 0.8415 = 0.8425$

Prob. 17. Let A = test is negative, $B1$ = person infected, B_2 = person not infected, then:

$$P(B_1) = \frac{4}{100}, \quad P(B_2) = \frac{96}{100}.$$

$$P(A|B_1) = 0.07, \quad P(A|B_2) = 0.81$$

So, $P(A) = P(A|B_1) \cdot P(B_1) + P(A|B_2) \cdot P(B_2) = 0.7804$

$$0.07 \cdot \frac{4}{100} + 0.81 \cdot \frac{96}{100}$$

Prob. 19. Bag 1: 3 blue balls, Bag 2: 4 green balls, Bag 3: 2 blue, 1 green.

Let $P(A)$ = probability of picking a blue ball.

$P(Bi)$ = probability of picking i-th bag, $i = 1, \ldots, 3$

$P(A|B_1) = 1$, $P(A|B_2) = 0$, $P(A|B_3) = \frac{2}{3}$, $P(B_1) = P(B_2) = P(B_3) = \frac{1}{3}$. So,

$$P(A) = P(A|B_1) \cdot P(B_1) + P(A|B_2) \cdot P(B_2) + P(QA|B_3) \cdot P(B_3) = \frac{5}{9}.$$

Prob. 21. Let $P(A)$ = probability that the first card is an ace.

Let $P(B)$ = probability that the second card is an ace.

$P(A) = 4/52 = 1/13$ since there are 4 aces out of 52 cards.

Now we compute $P(B)$.

(Now the second card can be an ace if the first card is not an ace and the second card is an ace OR if both the cards are aces)

Therefore $P(B) = 48/52 \cdot 4/51 + 4/52 \cdot 3/51 = 1/13$

Hence the two probabilities are equal.

Prob. 23. 40 percent are of genotype CC

60 percent are of genotype Cc

White flowering tea plant is of genotype cc

Let $P(A) =$ probability of obtaining white flowers

Let $P(Bi) =$ probability of picking red flowering pea plant

Therefore $P(B1) = 0.4$ and $P(B2) = 0.6$

Now cross between CC and cc results in red plants because the only

possible result of such a cross is Cc.

Therefore $P(A/B1) = 0$

Cross between Cc and cc results in either (Cc, cc)

Therefore $P(A/B2) = 1/2$

$P(A) = P(A/B1) \cdot P(B1) + P(A/B2) \cdot P(B2)$

$P(A) = 0 + (1/2 \cdot 0.6) = 0.3$

Prob. 25. Let $P(A)$ be probability of picking heads

Let (PBi) be probability of picking i-th coin, $i = 1, 2$

$P(B1) = P(B2) = \frac{1}{2}$ since there are equal chances of picking either

coin

$P(A/B1) = 1/2$ since the first coin is fair

$P(A/B2) = 1$ since the second coin has two heads.

$P(A) = P(A/B1) \cdot P(B1) + P(A/B2) \cdot P(B2) = 1/2 \cdot 1/2$

$+1 \cdot 1/2 = 1/4 + 1/2 = 3/4$

Prob. 27. Let $P(A)$ denote probability that the card is a spade

$P(A) = 13/52$

Let $P(B)$ denote probability that the card is an ace $P(B) = 4/52$ Let $P(A \cap B) = $ P(card is ace of spade)$= 1/52$ Since $P(A) \cdot P(B) = 13/52 \cdot 4/52 = P(A \cap B)$ it follows that A and B are independent.

Prob. 29. 5 green and 6 blue balls. $P(A) = 5/11$ (since there are 5 green out of 11 balls); $P(B/A) = 4/10$ (since there will be 4 green balls left out of 10 balls (if the first ball picked is green) which is $P(A \cap B)/P(A)$

Therefore $P(A \cap B) = 4/10 \cdot 5/11 = 20/110 = 2/11$ Now the second ball can be green if the both the balls are green or if the first is blue and the second is green.

$P(B) = 5/11 \cdot 4/10 + 6/11 \cdot 5/10 = 20/110 + 30/110 = 50/110$

$P(A) \cdot P(B) = 5/11 \cdot 50/110 = 250/1210 = 25/121$

Since $P(A) \cdot P(B)$ is not equal to $P(A \cap B)$, the events are not independent.

Prob. 31. The sample space is $\Omega = (BBB, BBG, BGB, BGG, GBB, GBG, GGB, GGG)$ where B denotes a boy and G denotes a girl.

(a) A = (all children are girls)= 1/8 (element 8)

(b) B = (at least one boy) s 7/8(elements 1 through 7)

(c) C = (at least two girls) is 4/8 (elements 4,6,7,8) = $1 \div 2$

(d) D = (at most two boys) = 7/8 (elements 2 through 8)

Prob. 33. Let $P(A)$ = probability that the student will pass. Let $P(Bi)$ = probability that the student guesses the i-th question correctly. P(Bi) = 1/4 for i = 1 to 10 Since the events are independent, $P(A) = P(B1) \cdot P(B2) \cdot P(B3) \cdots P(Bn) = (1/4)^{10} = 1/1048576$ This means that, if you guess the answers, the chance of passing is less than one million!

Prob. 35. In this problem, it is more convenient to think in terms of complement of an event. Probability that an insect lives more than 5 days = 0.1 Probability that at least one insect will be alive after 5 days = 1 - (Probability that no insects will be alive after 5 days) $= 1 - 0.9^{10} = 1 - 0.348678 = 0.651321$

Prob. 37. Let A = test result is positive, B1 be the probability of infection, and B2 be the probability of non infection then B1 = 1/50, B2 = 49/50 , and $P(A/B1) = 0.95$, $P(A/B2) = 0.1$.

We are interested in $P(B1/A)$ that is the probability that the person is infected given that the result is positive. $P(Bi/A) = X/Y$ where $X = P(A/Bi) \cdot P(Bi)$ and $Y = \Sigma_{j=1}^{j=n} P(A/Bj) \cdot P(Bj)$ Hence the required probability is: $1/50 \cdot 0.95$ / $(1/50 \cdot 0.95 + 49/50 \cdot 0.10) = 0.019/ (0.019 + 0.098) = 0.019/0.117 = 0.16239$

Prob. 39. Let the probability that the fair coin was picked be P(A), P(B) denote probability that heads was got. P(B) = fair coin was picked and heads was got or two heads coin was picked.

So P(B) = 1/2·1/2 + 1/2·1 = 3/4 and P(A) = P(A∩B)/P(B) = (1/2·1/2) / (3/4) = 1/3

Prob. 41. Let E denote the probability that the woman is a carrier, then $P(E) = 1/2$. If F denotes the event that her son is healthy, $P(F/E) = 1/2$. Hence

$P(E/F) = P(E \cap F)/P(F)$ and $P(F/E)P(E)/P(F) = (1/2)(1/2)/P(F)$ and

$P(F) = (1/2)(1/2) + (1/2)1 = 3/4$ and $P(E/F) = (1/4)/(3/4) = 1/3$

Prob. 43.

(a) This problem is identical to 36 (b). The solution is not repeated.

(b) We have no other information about III 2 other than she may have inherited it from her mother. Therefore the probability is 1/2 that II 2 is a carrier.

(c) Here the problem is different. We have some more information about II 2 in that she has a healthy son. Let P(E) be the probability that II 2 is a carrier. P(E) = 1/2 Let P(F) denote the event that III 1 is healthy. We have to compute P(E/F)

$P(E/F) = P(E \cap F)/P(F) = P(F/E) \cdot P(E)/P(F)$

Now $P(F/E) = 1/2$ We need to compute P(F) as $P(F) = 1/2 \cdot 1/2 + 1/2 \cdot 1 = 3/4$

$P(E/F) = (0.5 \cdot 0.5)/(0.75) = 1/3$

12.4 Discrete Random Variables and Discrete Distributions

Prob. 1. The sample space $\Omega = (HH, HT, TH, TT)$

Outcome	Value	Outcome	Value	Outcome	Value	Outcome	Value
X(HH)	0	X(HT)	1	X(TH)	1	X(TT)	2

Probability Table					
Probability	Value	Probability	Value	Probability	Value
$P(X=0)$	1/4	$P(X=1)$	2/4	$P(X=2)$	1/4

Prob. 3. Given the sample space, we have:

$$P(X=0) = \frac{6}{36}, \quad P(X=1) = \frac{10}{36}, \quad P(X=2) = \frac{8}{36},$$

$$P(X=3) = \frac{6}{36}, \quad P(X=4) = \frac{4}{36}, \quad p(x=5) = \frac{2}{36}.$$

Prob. 5. 3 Green and 2 blue

Probability	Computation	Value
$P(X = 0)$	$(2/5) \cdot (1/4)$	$2/20$
$P(X = 1)$	$(3/5) \cdot (2/4) + (2/5) \cdot (3/4)$	$12/20$
$P(X = 2)$	$(3/5) \cdot (2/4)$	$6/20$

Prob. 7.

$$P(X = 0) = \frac{\binom{13}{0}\binom{39}{3}}{\binom{52}{3}}, \quad P(X = 1) = \frac{\binom{13}{1}\binom{39}{2}}{\binom{52}{3}},$$

$$P(X = 2) = \frac{\binom{13}{3}\binom{39}{1}}{\binom{52}{3}}, \quad P(X = 3) = \frac{\binom{13}{3}\binom{39}{0}}{\binom{52}{3}},$$

Prob. 9. Values of distributions function F(x) for given values of mass function and $P(X = x)$ are tabulated below. Graph for $F(x)$ follows.

x	P(X=x)	F(x)
-3	0.2	0.2
-1	0.3	0.5
1.5	0.4	0.9
2	0.1	1

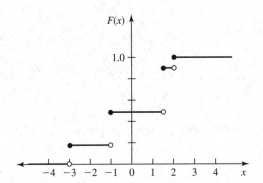

Prob. 11. This exercise is the 'converse' of the previous exercise, in the sense that, given $F(x)$, we have to find P. The values are tabulated below.

Range of x	< -2	$-2 \leq x < 0$	$0 \leq x < 1$	$1 \leq x < 2$	$x >= 2$
F(x)	0	0.2	0.3	0.7	1
P(x)	0	0.2	0.1	0.4	0.3

Prob. 13. $S = (1, 2, 3, 4, 5, 6, 7, 8, 9, 10)$ and $p(k) = k/N$

(a) By the definition of probability,

$$\Sigma_1^2 p(k) = 1$$

therefore $N = 55$

(b) The probability that $X < 8$ is given by $(1 + 2 + 3 + 4 + 5 + 6 + 7)/55 = 28/55$

Prob. 15.

(a)

No. of leaves	19	21	20	13	18	14	17	15	12	16
Relative frequency	1/25	1/25	2/25	4/25	2/25	5/25	5/25	3/25	1/25	1/25

(b)   1) Direct average can be found by first evaluating the numerator.

$$\begin{bmatrix} 19 + 21 + 20 + 13 + 18+ \\ 14 + 17 + 14 + 17 + 17+ \\ 13 + 15 + 12 + 15 + 17+ \\ 15 + 16 + 18 + 17 + 14+ \\ 14 + 14 + 13 + 20 + 13 \end{bmatrix}$$

The numerator is thus 396. Hence the direct average is $396/25 = 15.84$

2) By using relative frequency

$(19 \cdot 1 + 21 \cdot 1 + 20 \cdot 2 + 13 \cdot 4 + 18 \cdot 2 + 14 \cdot 5 + 17 \cdot 5 + 15 \cdot 3 + 12 \cdot 1 + 16 \cdot 1)/25 = $
$396/25 = 15.84$

Prob. 17.

(a) Forming the relevant table, we have:

No. of leaves	7	8	3	2	5	6	9	10	4
Relative frequency	2/25	6/25	1/25	1/25	3/25	4/25	3/25	3/25	2/25

(b)   1) Direct average can be found by first finding the numerator.

$$\begin{bmatrix} (7 + 8 + 8 + 3 + 2 + 5 + 6 + 9 + 10 + 6 + 8 + 8 + 7+ \\ 6 + 6 + 9 + 10 + 4 + 4 + 8 + 6 + 9 + 10 + 5 + 5 + 8) \end{bmatrix}$$

The total is 171. Hence the direct average is $171/25 = 6.84$

2) By using relative frequency

$$\begin{bmatrix} (7 \cdot 2 + 8 \cdot 6 + 3 \cdot 1 + 2 \cdot 1 + 5 \cdot 3+ \\ 6 \cdot 4 + 9 \cdot 3 + 10 \cdot 3 + 4 \cdot 2) \end{bmatrix}$$

The above totals to 171. Hence the direct average is: $171/25 = 6.84$

Prob. 19. The table to be used is as follows:

x	−2	−1	0	1
P(X=x)	0.1	0.4	0.3	0.2

(a) $EX = -2 \cdot 0.1 + -1 \cdot 0.4 + 0 \cdot 0.3 + 1 \cdot 0.2 = -0.4$

(b) $EX2 = (-2) \cdot (-2) \cdot 0.1 + (-1) \cdot (-1) \cdot 0.4 + (0) \cdot (0) \cdot 0.3 + (1) \cdot (1) \cdot 0.2 = 1$

(c) $E(X(X-1)) = (-2) \cdot (-3) \cdot 0.1 + (-1) \cdot (-2) \cdot 0.4 + (0) + 1 \cdot (0) \cdot 0.2 = 1$

Prob. 21. The mean μ is obtained by totaling the entries in the third row, so it is -0.1

x	-3	-1	1.5	2
$P(x = X)$	0.2	0.3	0.4	0.1
$x \cdot P(x = X)$	-0.6	-0.3	0.6	0.2

Variance

$= (-3-(-0.1))^2 \cdot (0.2) + (-1-(-0.1))^2 \cdot (0.3) + (1.5-(-0.1))^2 \cdot (0.4) + (2-(-0.1))^2 \cdot (0.1) = 3.39$

Standard Deviation $= \sqrt{\text{Variance}} = 1.841195$

Prob. 23. In this instance, all cases are equiprobable. Hence for each $x_i, i = 1, 2, 3 \cdots 10$,

$P(x_i = X) = 1/10 = 0.1$

(a) $EX = \Sigma_{i=1}^{i=10} x_i \cdot P(X) = 55(0.1) = 5.5$

(b) $var(X) = E(X - \mu)^2$ For clarity, the terms that are to be summed up to find variance are tabulated below.

x	$(x - \mu)^2 \cdot P$	Value
1	$(1 - 5.5)^2) \cdot (0.1)$	2.025
2	$(2 - 5.5)^2 \cdot (0.1)$	1.225
3	$(3 - 5.5)^2 \cdot (0.1)$	0.625
4	$(4 - 5.5)^2 \cdot (0.1)$	0.225
5	$(5 - 5.5)^2 \cdot (0.1)$	0.025
6	$(6 - 5.5)^2 \cdot (0.1)$	0.025
7	$(7 - 5.5)^2 \cdot (0.1)$	0.225
8	$(8 - 5.5)^2 \cdot (0.1)$	0.625
9	$(9 - 5.5)^2 \cdot (0.1)$	1.225
10	$(10 - 5.5)^2 \cdot (0.1)$	2.025
VARIANCE = 8.25		

Prob. 25.

(a) By definition of expectation, $E(x) = \Sigma_{\forall x} x \cdot p(x)$, we have

$$E(aX + b) = \Sigma_{\forall x}(ax + b) \cdot p(x)$$

(b) $E(aX + b) = \Sigma_{\forall x}(ax + b) \cdot p(x) = \Sigma_{\forall x}(ax) \cdot p(x) + \Sigma_{\forall x} b \cdot p(x) =$
$a \cdot \Sigma_{\forall x} x \cdot p(x) + b \cdot 1 = aE(x) + b$ (since $\Sigma_{\forall x} p(x) = 1$)

Prob. 27.

(a) $P(X = 1, Y = 0) = 0.1$

(b) $P(X = 1) = 0.1 + 0.4 = 0.5$

(c) $P(Y = 0) = 0.3 + 0.1 = 0.4$

(d) $P(Y = 0/X = 1) = (0.1)/(0.1 + 0.4) = 0.2$

Prob. 29.

(a) $E(X) = -2 \cdot 0.1 + -1 \cdot 0 + 0 \cdot 0.3 + 1 \cdot 0.4 + 2 \cdot 0.05 + 3 \cdot 0.15 = 0.75.$

$E(Y) = -2 \cdot 0.2 + -1 \cdot 0.2 + 0 + 1 \cdot 0.3 + 2 \cdot 0 + 3 \cdot 0.2 = 0.3$

(b) Since X and Y are independent, $E(X + Y) = EX + EY = 0.75 + 0.3 = 1.05.$

(c) $var(X) = E(X^2) - (EX)^2$ Let us compute $E(X^2)$:

$E(X^2) = (-2) \cdot (-2) \cdot (0.1) + 0 + 0 + (1) \cdot (1) \cdot (0.4) + 2 \cdot 2 \cdot (0.05) + 3 \cdot 3 \cdot (0.15) =$
$0.4 + 0.4 + 0.2 + 1.35 = 2.35$

$var(X) = 2.35 - 0.75 \cdot (0.75) = 1.7875$

$var(Y) = E(Y^2) - (EY)^2$

Let us compute now

$E(Y2) = (-2) \cdot (-2) \cdot (0.2) + (-1) \cdot (-1) \cdot (0.2) + 0 + 1 \cdot 1 \cdot (0.3) + 0 + 3 \cdot 3 \cdot (0.2) =$
$0.8 + 0.2 + 0.3 + 1.8 = 3.1$

$var(Y) = 3.1 - 0.3 \cdot (0.3) = 3.01$

(d) Since X and Y are independent

$var(X + Y) = var(X) + var(Y) = 1.7875 + 3.01 = 4.7975$

Prob. 31.

(a) $var(X) = E(X - EX)^2$. Lets look at $X - EX$. This value may be positive or negative. However, $(X - EX)^2$ is always positive because $(+ve) \cdot (+ve) = (+ve)$ and $(-ve) \cdot (-ve)$ is also positive, therefore var(X) is always ≥ 0

(b) Now $var(X) = EX2 - (EX)^2$, since $var(X) \geq 0, [EX2 - (EX)2] \geq 0, EX2 \geq (EX)^2$

Prob. 33. The total number of outcomes is $2^{10} = 1024$ If X is the number of heads, lets calculate the probability distribution of X.

$$
\begin{bmatrix}
X & P(x = X) & \\
0 & C(10,0)/1024 & 1/1024 \\
1 & C(10,1)/1024 & 10/1024 \\
2 & C(10,2)/1024 & 45/1024 \\
3 & C(10,3)/1024 & 120/1024 \\
4 & C(10,4)/1024 & 210/1024 \\
5 & C(10,5)/1024 & 252/1024 \\
6 & C(10,6)/1024 & 210/1024 \\
7 & C(10,7)/1024 & 120/1024 \\
8 & C(10,8)/1024 & 45/1024 \\
9 & C(10,9)/1024 & 10/1024 \\
10 & C(10,10)/1024 & 1/1024
\end{bmatrix}
$$

$P(X = 5) = 252/1024 = 0.24609375$

$P(X \geq 8) = (45 + 10 + 1)/1024 = 0.0546875$

$P(X \leq 9) = (1+10+45+120+210+252+210+120+45+10)/1024 = 1023/1024 = 0.999023$

Prob. 35. Toss a fair die 6 times. The total number of outcomes is $6^6 = 46656$

$$
\begin{bmatrix}
X & P(X) & Calculation & Result \\
1 & C(6,1) \cdot (1/6) \cdot (5/6)^5 & 3125/7776 & 0.40187772 \\
2 & C(6,2) \cdot (1/6)^2 \cdot (5/6)^4 & 9375/46656 & 0.2009387 \\
3 & C(6,3) \cdot (1/6)^3 \cdot (5/6)^3 & 2500/46656 & 0.0535836 \\
4 & C(6,4) \cdot (1/6)^4 \cdot (5/6)^2 & 375/46656 & 0.008037 \\
5 & C(6,5) \cdot (1/6)^5 \cdot (5/6) & 30/46656 & 0.000643 \\
6 & C(6,6) \cdot (1/6)^6 & 1/46656 & 0.000021433
\end{bmatrix}
$$

Prob. 37. $\frac{14}{64}$.

Prob. 39. We have:

$$
\begin{array}{ccc}
X & P(x) & Result \\
0 & (\frac{3}{5})^3 & 0.216 \\
1 & C(2,1)(\frac{2}{5})(\frac{3}{5})^2 & 0.288
\end{array}
$$

So, $P(X \leq 1) = 0.504$.

Prob. 41. 20% of all plants are infected, therefore $(100 - 20) = 80\%$ are not infected. If 20 plants are picked, Since the events are independent, the probability that none of them carried aphids is $(0.8)^{20} = 0.011529215$.

Prob. 43. In a box contains 10 apples the probability that an apple is spoiled is 0.1

(a) The expected number of spoiled apples per box is $\Sigma_{\forall x} x \cdot p(x) = \Sigma_{i=1}^{i=10} 1 \cdot (0.1) = 1$.

(b) The probability that a box picked at random from the shipment will contain no spoiled apples s $(0.9)^{10} = 0.3486$. Hence the expected number of boxes that contain no spoiled apples s 3.486

Prob. 45. The probability of guessing a question right is 1/4, so $E(X) = (1/4) \cdot 50 = 12.5$

Prob. 47. With 12 green and 24 blue balls

(a) If ten balls are taken out and six of the ten are blue. The total number of ways that ten balls can be picked from 36 balls is $C(36, 10) = 254186856$. The number of ways that 6 blue balls can be picked from 24 blue balls is $C(24, 6) = 134596$. The number of ways that 4 green balls can be picked from 12 green balls is is $C(12, 4) = 495$, therefore the probability of picking 6 blue balls $134596 \cdot 495/(254186856) = 0.26211$

(b) The probability distribution is binomial with $n = 10$, and success probability is $24/36$.

The probability $P(S10 = 6)$ is:

$C(10, 6) \cdot (24/36)^6 \cdot (12/36)^4 = (210) \cdot (2/3)^6 \cdot (1/3)^4 = 13440/59049 = 0.2276$

Prob. 51. Let's go through Example 28 where Mendel's experiments are described,

Result of Mendel's experiment		
Total No. of seeds is 40		
Characteristic of seed	Quantity	Probability of outcome shown in Col.2
Round Yellow	20	9/16
Round Green	10	3/16
Wrinkled Yellow	8	3/16
Wrinkled Green	2	1/16

This is obviously an exercise in multinomial distribution. The number of ways in which the 40 seeds can belong to the four categories mentioned is:

$$\frac{40!}{(20! \cdot 10! \cdot 8! \cdot 2!)}$$

Now the probability of each phenotype occurring from among the 16 possible genotypes is given in the Table above. Taking this into account, the probability of the outcome given in the above Table is:

$$(9/16)^{20} \cdot (3/16)^{10} \cdot (3/16)^8 \cdot (1/16)^2$$

So, finally the probability of the 'successful' outcome as desired by Mendel can be found by multiplying the two long preceding expressions with factorials and powers of numbers and so on.

Prob. 53. This is a multinomial distribution

Colour and Number of balls sampled		Probability of picking	
Color	Number	Symbol	Value
Green	N1	p1	6/24
Blue	N2	p2	8/24
Red	N3	p3	10/24

We want to find $P(N1 = 2, N2 = 2, N3 = 2)$, so

$(C(6,2) \cdot C(4,2) \cdot 1) \cdot (6/24)^2 \cdot (8/24)^2 \cdot (10/24)^2 = (15 \cdot 6 \cdot 36 \cdot 64 \cdot 100)/(24^6) = 0.108506$

Prob. 55. For convenience, let tabulate the data:

Genotype and Number			Probability	
Genotype	Symbol	Value	Symbol	Value
CC	N1	5	p1	1/4
Cc	N2	12	p2	2/4
cc	N3	6	p3	1/4

We want to find $P(N1, N2, N3)$. This is a multinomial distribution The number of

offspring is $5 + 12 + 6 = 23$ and the required probability

$C(23,5) \cdot C(18,12) \cdot (1/4)^(5) \cdot (2/4)^{12} \cdot (1/4)^6 =$

$((33649) \cdot (18564) \cdot (2^{12}))/((4^5) \cdot (4^12) \cdot (4^6)) = 0.036359$

Prob. 57. Similar to problem 56 we have:

$$P(N1, N2, N3) = C(6, 2) \cdot C(4, 3) \cdot (\frac{1}{4})^2 \cdot (\frac{2}{4})^3(\frac{1}{4}) = \frac{5}{16}.$$

Prob. 59. We construct a punnet square:

	A	a
A	AA	Aa
a	Aa	aa

Hence the probability is $(\frac{3}{4})^4$.

Prob. 61. For $k = 1$, the probability of heads appearing on the first throw is 1/2 For

$k = 2$, the probability of tails and then heads is $(1/2)^2 = 1/4$ For $k = 3$, the probability of

tails, tails and then heads is $(1/2)^3 = 1/8$

Prob. 63. $\frac{1}{8}$

Prob. 65.

$$\begin{bmatrix} k & P(X = k) \\ 1 & 1/2 = 0.5 \\ 2 & 1/2 \cdot 1/2 = 0.25 \\ 3 & (1/2)^3 = 0.125 \end{bmatrix}$$

Therefore $P(X > 3) = (1 - 0.5 - 0.25 - 0.125) = (1 - 0.875) = 0.125$

Prob. 67.

$$\begin{bmatrix} x & Computing\,P(X = x) & Final\,value\,of\,P \\ 1 & 1/2 & 0.5 \\ 2 & (1/2)^2 & 0.25 \\ 3 & (1/2)^3 & 0.125 \\ 4 & (1/2)^4 & 0.0625 \end{bmatrix}$$

Therefore $P(X \le 4) = 0.5 + 0.25 + 0.125 + 0.0625 = 0.9375$

Prob. 69.

$$\begin{bmatrix} x & P(X) \\ 1 & 1/15 \\ 2 & (14/15) \cdot (1/15) \\ 3 & (14/15)^2 \cdot (1/15) \\ 4 & (14/15)^3 \cdot (1/15) \\ \hline k & (14/15)^{(k-1)} \cdot (1/15) \\ \hline 20 & (14/15)^{19} \cdot (1/15) \end{bmatrix}$$

The Table above helps us to draw up another Table (more of the same!) in which the probability that AT LEAST the given number of draws are needed is tabulated

Number of	Probability that AT LEAST the given
draws	number of draws are needed
One	$1 - 0 = 1$
Two	$(1 - S(1))$
Three	$(1 - S(1) - S(2))$
Twenty	$(1 - S(1) - S(2) \cdots - S(19))$

Now let's compute $S(1) + (S2) + \cdots + S(19)$. This is a geometric progression with first term 1/15 and common ratio 14/15. The sum to 19 terms is given by

$(1/15)(1 - (14/15)20)/(1 - 14/15) = 0.74838$

Therefore the probability that at least 20 draws are needed $1 - 0.74838 = 0.2516$

Prob. 71. The distribution is geometric with $p = 5/30$, so $ET = 1/p = 30/5 = 6$ and $var(T) = (1 - p)/p^2 = (1 - 5/30)/(5/30)^2 = 30$

Prob. 73. (a) $(\frac{9}{10} \, {}^5 \, \frac{1}{10})$

(b) $\frac{9}{10} \cdot \frac{8}{9} \cdot \frac{7}{8} \cdot \frac{6}{7} \cdot \frac{5}{6} \cdot \frac{1}{5}$

Prob. 75.

(a) The probability that the first success occurs on the k-th trial is $(1 - p)^{(k-1)} \cdot p$

(b) $\binom{k-1}{1}p^2(1 - p)^{k-2}$

Prob. 77. We can see that it is Poisson distributed with $\lambda = 2$ $P(X = k) = e^{(-\lambda)} \cdot \lambda^k/k!$, so

$$
\begin{bmatrix}
x & P(x) \\
0 & 0.135335283 \\
1 & 0.270670566 \\
2 & 0.270670566 \cdot (1/n) \\
3 & 0.180447044
\end{bmatrix}
$$

Prob. 79. X is Poisson distributed with $\lambda = 1$

(a) $P(X \geq 2) = 1 - P(X = 0) - P(X = 1)$, since $P(X = 0) = 0.3678$

$P(X = 1) = 0.3678$, $P(X \geq 2) = 1 - 0.3678 - 0.3678 = 0.2642$

(b) $P(1 \leq X \leq 3) = P(1) + P(2) + P(3)$, since $P(X = 1) = 0.3678$ $P(X = 2) = 0.1839$,

$P(X = 3) = 0.0613$, $P(1 \leq X \leq 3) = 0.3678 + 0.1839 + 0.0613 = 0.613$

Prob. 81. X is Poisson distributed with $\lambda = 1.5$

$P(X = k) = e^{-\lambda} \cdot \lambda^k / k!$

$P(X = 0) = 0.2231$

$P(X = 1) = 0.3346$

$P(X = 2) = 0.2510$

$P(X = 3) = 0.1255$

$P(X > 3) = 1 - P(X = 0) - P(X = 1) - P(X = 2) - P(X = 3)$

$= 1 - 0.2231 - 0.3346 - 0.2510 - 0.1255 = 0.06578$

Prob. 83. X is poisson distributed with $\lambda = 2$

$P(X = k) = e^{-\lambda} \cdot \lambda^k / k!$

$P(X = 0) = 0.1353$

$P(X = 1) = 0.2706$

The probability that X is at least $2 = 1 - P(X = 0) - P(X = 1)$

$= 1 - 0.1353 - 0.2706 = 0.5940$

Prob. 85. X is Poisson distributed with $\lambda = 7$, $P(X = k) = e^{-\lambda} \cdot \lambda^k / k!$ and

$P(X = 0) = 0.000911$

Prob. 87. X is Poisson distributed with $\lambda = 0.5$ $P(X = k) = e^{-\lambda} \cdot \lambda^k / k!$ Probability that

there is at least one typo in a page = 1 - Probability that there are no typos

$= 1 - P(X = 0) = 1 - 0.6065 = 0.3934$

Prob. 89. X is poisson distributed with $\lambda = 3$, $P(X = k) = e^{-\lambda} \cdot \lambda^k / k!$,

$P(X = 0) = 0.0497$ $P(X = 1) = 0.1493$. The probability of at least two substitutions is

$1 - P(0) - P(1) = 1 - 0.0497 - 0.1493 = 0.801$

Prob. 91.

(a) Given that $\lambda = 3$, we are asked to find the probability $P(X + Y) = 2$. Let us first

tabulate the probability $P(X = k)$ for k = 0,1,2 using Poisson equation and making the appropriate substitutions.

k	$e^{-\lambda}$	λ^k	$k!$	$P(X = k)\, or\, P(Y = k)$
0	0.049787068	1	1	0.049787068
1	0.049787068	3	1	0.149361205
2	0.049787068	9	2	0.224041807

$X + Y$ will have the value 2 under anyone of the following conditions.

$(X = 0)$ AND $(Y = 2)$ OR

$(X = 1)$ AND $(Y = 1)$ OR

$(X = 2)$ AND $(Y = 0)$

It follows therefore that

$P(X + Y = 2) = P(X = 0) \cdot P(Y = 2) + P(X = 1) \cdot P(Y = 1)$

$+P(X = 2) \cdot P(Y = 0) = 2 \cdot (P(X = 0) \cdot P(X = 2) + (P(X = 1))^2 =$

$2 \cdot 0.049787068 \cdot 0.224041807 + (0.149361205)^2 = 0.044617539$

(b) Given that $X + Y = 2$, we are asked to find the probability that $X = k\, for\, k = 0, 1, 2$

Case1: Let $k = 0$. This means that $X = 0$ and $Y = 2$ The required probability for case 1 is:

Case 1 $P(X = 0), (Y = 2))/(P(X + Y) = 2)$ which is:

$0.011154385/0.044617539 = 0.2500$

Case 2:

$(P(X = 1) \cdot P(Y = 1))/(P(X + Y) = 2) = 0.022308770/0.044617539 = 0.5000$

Case 3: is identical to case 1. Notice that the probabilities of the above three cases add up to exactly unity.

Prob. 93. Using tables for Poisson with $\lambda = 4$, $\lambda = 2$, we get

$$P(X = 2 | X + Y = 3) = \frac{P(X = 2)P(X + Y = 3)}{P(X = 2, Y = 1)} = (\frac{2}{3})^2.$$

Prob. 95. Since only 1 in 1000 experiences side effect $p = 1/1000$ Since 500 is the strength of the test group, $n = 500$. We shall use Poisson approximation approach.

$\lambda = n \cdot p = 500 \cdot 1/1000 = 0.5 \; P(X = k) = e^{-\lambda} \cdot \lambda^k/k! \; P(X = 0) = 0.60653$ is the probability that none of the 500 persons experiences side effect

Prob. 97.

(a) The probability that there are no cases of Down's syndrome

$$P(X = 0) = (699/700)^{1000} = 0.2394$$

The probability that there is exactly one case of Down's syndrome

$$P(1) = C(1000, 1) \cdot (699/700)^{999} \cdot (1/700) = (1000) \cdot (0.239748) \cdot (1/700) = 0.342498.$$

Hence the probability that there is at most one case is: $P(0) + P(1) = 0.5818$

(b) We are using a Poisson approximation $\lambda = n \cdot p = 1000 \cdot 1/700 = 1.42857$

$P(X = k) = e^{-\lambda} \cdot \lambda^k/k! \; P(X = 0) = 0.2396 \; P(X = 1) = 0.3423$. Hence the probability that there is at most one case is $P(X = 0) + P(X = 1) = 0.5819$

Prob. 99. There are $P = 50$ parasitoids, vying with one another to find a host. The probability of any one of them encountering a host is $a = 0.03$ Hence, the probability that the host escapes detection is (for such cases, we take the only the first term of Poisson distribution) is $e^{-0.3 \cdot 50} = e^{-1.5} = 0.22313016$

12.5   Continuous Distributions

Prob. 1. As per Equation 12.28 in Example 1 in Section 12.5.1, any non-negative function which satisfies the condition

$$\int_{-\infty}^{\infty} f(x)dx = 1$$

is a candidate for being a density function. Carrying out the indicated non-negative test,

$$\int_{-\infty}^{\infty} (3e^{-3x})dx = \int_{0}^{\infty} (3e^{-3x})dx = \frac{3e^{-3x}}{-3} \bigg|_{0}^{\infty} = 1$$

Thus $f(x)$ is indeed a probability density function. The corresponding distribution function can be seen, from above, to be:

$$F(x) = \begin{cases} -e^{-3x} & \text{for } x \geq 0 \\ 0 & \text{for } x < 0 \end{cases}$$

Note that $F(x)$ is discontinuous at $x = 0$. Discontinuities are permitted for $F(x)$ but not for $f(x)$.

Prob. 3. Given that $f(x) = c/(1 + x^2)$, it is necessary to determine c so that $f(x)$ is a probability density function. All we need to do is to find the area under the curve for overa all real numbers. Let $x = \tan\theta$; then $dx = \sec^2\theta d\theta$

$$\int_{-\infty}^{\infty} c\frac{dx}{1 + x^2} = \int_{-\pi/2}^{\pi/2} c(\sec^2\theta)\frac{d\theta}{(1 + \tan^2\pi)} = c\pi$$

Hence, we conclude that $c = \frac{1}{\pi}$

Prob. 5.

$$\text{Given density function} \quad f(x) = \begin{cases} 2e^{-2x} & \text{for } x > 0 \\ 0 & \text{for } x \leq 0 \end{cases}$$

It is required to find Expectation and Variance.

$$EX = \int_0^{\infty} (x2e^{-2x}dx = xe^{-2x} + (0.5)e^{-2x} \Big|_0^{\infty} = \frac{1}{2}$$

The variance is found from the usual expression $(EX^2 - (EX)^2)$

$$EX^2 = \int_0^{\infty} (x^2)2e^{-2x})dx = -x^2e^{-2x} \Big|_0^{\infty} + \int_0^{\infty} 2xe^{-2x}dx = \frac{1}{2}$$

Hence the variance is $\frac{1}{2} - (\frac{1}{2})^2 = \frac{1}{4}$

Prob. 7. In this problem we are given the distribution function $F(x)$ to begin with.

$$F(x) = (1 - 1/x^2)$$

The density function is hence:

$$f(x) = \frac{2}{x^3}$$

EX can be found by a simple integration, to be

$$(-2/x) \Big|_1^{\infty} = -2$$

EX^2 can also be found, by another simple integration to be $-2\ln(x)$. Using the given lower and upper limits, we write:

$$-2\ln(x) \Big|_1^{\infty}$$

The interesting point is that at the upper limit, the value of $\ln(x)$ is unbounded. Hence Var(x) is unbounded.

Prob. 9.

$$\text{Given density function}\quad f(x) = \begin{cases} (a-1)x^{-a} & \text{for } x > 1 \\ 0 & \text{for } x \le 1 \end{cases}$$

$$\begin{aligned} EX2 &= \int_1^\infty (a-1)x^{1-a}dx \\ &= ((a-1)/(2-a))x^{2-a}\Big|_1^\infty \\ &= \text{unbounded if } a < 2 \end{aligned}$$

(a) If $a < 2$ then EX is unbounded, since x^{2-a} increases without limit.

(b)

$$EX = (a-1)/(a-2) \quad \text{if} \quad a > 2$$

Prob. 11.

(a) To prove that the curve is symmetric about $x = \mu$, we need to show that $f(x+\mu) = f(x-\mu)$. Define $K = 1/(\sigma\sqrt{2\pi}$, $f(x) = Ke^{-(x-\mu)^2/2\sigma^2}$, then we need to show that $f(\mu+x) = f(\mu-x)$. Obviously $e^{-(\mu+x-\mu)^2} = e^{-(\mu-x-\mu)^2}$ thus, it is clear that the normal distribution curve is symmetric about $x = \mu$

(b) To prove that the maximum of the curve is at $x = \mu$, we need to show that the first derivative is zero at that point. Let us also make the substitution $v = \sigma^2$

Let $f(x) = e^{-(x-\mu)^2/2v}$, we only need to prove that $df(x)/dx = 0$ and $f'(x) = -\frac{1}{v}(x-\mu)f(x)$, it is clear that at $x = \mu$ $f'(x) = 0$.

Note that we have omitted K in the above expression, as it plays no role in differentiation (or in the conclusion)

(c) Now, the problem is to establish the inflection points. At such points, the curve does not go through a maximum or a minimum though the derivative goes through zero. The curve goes upward, stops for a while as if to decrease and then changes its mind and continues its journey upwards! That is why the first derivative goes through zero, though the curve does not have a maximum at that point. So, we need to examine the second derivative at such a point. In order to reduce clutter in long algebraic expressions, let us first simplify the distribution function by making appropriate substitutions. Let us also omit the constant K as it plays no role here.

$$\text{Normal Curve} f(x) \;=\; Ke^{-(x-\mu)^2/2\sigma^2}$$

$$\text{Put} \quad (x-\mu) \;=\; z$$

$$\text{Then} \quad dx \;=\; dz$$

$$\text{Variance} \quad v \;=\; \sigma^2$$

K can be omitted for simplicity

$$\text{Incorporating above changes} \quad f(z) \;=\; e^{-z^2/2v}$$

$$\text{Differentiating once} \quad f'(z) \;=\; (-z/v)f(z)$$

$$\text{Differentiating again} \quad f''(z) \;=\; (-1/v)(zf'(z) + f(z))$$

$$\text{Simplifying} \quad f''(z) \;=\; (f(z)/v)(1 - z^2/v)$$

$$\text{It is clear that} \quad f''(z) \;=\; 0 \quad \text{at} \quad z = \pm\sigma$$

$$\text{Thus, inflection occurs at} \quad x \;=\; \mu \pm \sigma$$

Note that f(z) is obtainable from f(x) by a mere translation by μ on the x- axis. Hence the shape of the curve is preserved

(d) Graph of f(x) for $\mu = 2$ and $\sigma = 1$:

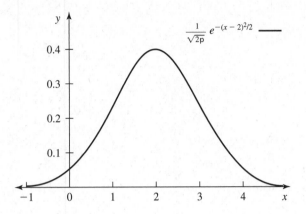

Prob. 13. Given that

$$\mu = 12.8 \quad \text{and} \quad \sigma = 2.7$$

we can evaluate the following:

$$\mu - 2\sigma = 7.4 \quad \text{and} \quad \mu + 2\sigma = 18.2$$

From the graph of normal distribution, it is known that 95 percent of the area falls between

$$\mu - 2\sigma \quad \text{and } \mu + 2\sigma$$

i.e. between 7.4 and 18.2 respectively. Hence

$$P(X \in [7.4, 18.2] = 0.95)$$

In a similar manner, it is possible to find an interval such that 99 percent falls within this interval. From the graph of normal distribution it is known that 99 percent of the area falls between

$$\mu - 3\sigma \quad \text{and} \quad \mu + 3\sigma$$

This works out to 4.7 and 20.9 respectively. Hence $P(X \in [4.7, 20.9] = 0.99)$.

Prob. 15. In this problem, we are asked to determine the fraction of population that will fall in the interval $P(\mu \leq X < \infty)$. Since the normal distribution curve is symmetric about μ, the area to the right of μ is one-half the total area. Hence the answer to the question is $\frac{1}{2}$.

Prob. 17. The range mentioned in this problem, which is similar to the previous one is: $-\infty$ and $(\mu + 3\sigma)$.

$$Area \Big|_{-\infty}^{\mu} = \frac{1}{2} \quad \text{Known fact} \tag{12.5.1}$$

$$Area \Big|_{\mu-3\sigma}^{\mu+3\sigma} = 0.995 \quad \text{Known from Appendix B} \tag{12.5.2}$$

$$\text{Hence} \quad Area \Big|_{\mu}^{\mu+3\sigma} = (0.995/2) = 0.495 \tag{12.5.3}$$

$$\text{Adding (1) and (3) above} \quad Area \Big|_{-\infty}^{\mu+3\sigma} = (0.5 + 0.495) = 0.995 \tag{12.5.4}$$

$$\tag{12.5.5}$$

Hence the fraction of population that will fall in the range given in this problem is 0.995

Prob. 19. In this problem, both the lower and upper limits of X are to the left of μ The range given is $(-\infty)$ to $\mu - 2\sigma$. The fraction of population in this range is: $(1.0 - 0.95)/2 = 0.025$

Prob. 21. For $\mu = 3$ and $\sigma = 2$

(a) The probability $P(X \le 4)$ is to be determined. The normalized random variable $Z = (X - \mu)/\sigma$ is $(4 - 3)/2 = 0.5$. From Appendix B, the area under the standard normalised curve to the left of $Z = 0.5$ is 0.6915. Hence $P(X \le 4) = 0.6915$. In other words, there is a 69.15 percent probability that the random variable is less than or equal to 4. Since parts (b), (c), (c) of this Problem are similar to (a) let us consider tabulating the key results.

(b)

$$\text{Range of X} = P(2 \le X \le 4)$$

$$\text{Corresponding range of Z} = P(-0.5 \le Z \le 0.5)$$

$$\text{Normalized curve} \quad Area \Big|_{-\infty}^{0.5} = 0.6915$$

$$\text{Normalized curve} \quad Area \Big|_{-\infty}^{0} = 0.5$$

$$\text{Hence} \quad Area \Big|_{0}^{0.5} = 0.1915$$

$$\text{Hence} \quad Area \Big|_{-0.5}^{0.5} \quad = \quad 2(0.1915) = 0.383$$

In other words the probability of the random variable being in the range (2 to 4) is 38.3 percent.

(c) Though this is similar to (b) above, the important difference is that the range in question is above a certain value. Hence the area determination has to be done carefully.

$$
\begin{aligned}
\text{Range of X} \quad &= \quad P(X > 5) \\
\text{Corresponding range of Z} \quad &= \quad P(Z > 1) \\
\text{Normalized curve} \quad Area \Big|_{-\infty}^{1} \quad &= \quad 0.8413 \\
\text{Normalized curve} \quad Area \Big|_{-\infty}^{\infty} \quad &= \quad 1.0 \\
\text{Hence} \quad Area \Big|_{1}^{\infty} \quad &= \quad 1.0 - 0.8413 = 0.1587 \\
\text{Area above given X} \quad Area \Big|_{5}^{\infty} \quad &= \quad 0.1587
\end{aligned}
$$

To put the final result in words, the probability of the random variable X being more than 5 is 15.87 percent.

(d) In this case also, area determination has to done in a range which is negative. So the Table in Appendix B has to be used with caution!

Description	Equation
Range of X	$P(X \leq 0)$
Corresponding Range of Z	$P(Z \leq -1.5)$
Area in Normalized Curve	$Area \left. \right|_{-\infty}^{1.5} = 0.9332$
Using symmetry	$Area \left. \right|_{-1.5}^{\infty} = 0.9332$
So, area we are looking for	$Area \left. \right|_{-\infty}^{-1.5} = 1 - 0.9332 = 0.0668$

Prob. 23. Here, we are asked to find the value of x, given its probability. So, we will have to find the range of absissa of normal curve given the value of the enclosed area. For (a) to (d) below, $\mu = 1$ and $\sigma = 2$

(a) Given that $P(X \leq x) = 0.9$, we are asked to find value of x.

$$\text{From Appendix B} \quad Area \left. \right|_{\infty}^{1.28} = 0.8997 \tag{12.5.6}$$

$$\text{Similarly } Area \left. \right|_{\infty}^{1.29} = 0.9015 \tag{12.5.7}$$

$$\text{Find } X = \mu + Z\sigma = 1 + (1.28)2 = 3.56 \tag{12.5.8}$$

$$\tag{12.5.9}$$

The value of Z that will give the required area of 0.9 (instead of 0.8997) is taken to be 1.28 Hence $P(X \leq 3.56) = 0.9$

(b)

$$\text{Given that} \quad P(X > x) = 0.4$$

$$\text{It follows that} \quad P(X \leq x) = 0.6$$

$$\text{from Appendix B} \quad Area \left. \right|_{-\infty}^{0.25} = 0.5987$$

$$\text{from Appendix B} \quad Area \left. \right|_{-\infty}^{0.26} = 0.6026$$

Hence, we can interpolate and say that the area to the left of the vertical line at $z = 0.255$ will be 0.6. $X = \mu + Z\sigma = 1 + 2(0.255) = 1.51$ and $P(X > 1.51) = 0.4$

(c) Note that, in working out this part, we can benefit from the result obtained above in part (b). The area in the range $1.51 < x < \infty$ has been derived to be 0.4. The area in the range $1 < x < \infty$ is known to be 0.5. Hence the area in the range $1 < x < 1.51$ is 0.1 By symmetry, the area in the range 0.49 to 1 is also 0.1. Thus the area in the range $-\infty$ to 0.49 is 0.4. Thus, $P(X \leq 0.49) = 0.4$.

(d) A new element added in this question is modulus.

$$P(|X - 1| < x) = 0.5$$

As before, $\mu = 1$ and $\sigma = 2$ First, let us take a close look at the standard normalised distribution curve. The mean is zero. The value of Z increases from zero to infinity in the positive direction and decreases from zero to infinity in the negative direction. If you consider $|Z|$, it increases in both directions.

$$\text{Area in the region} \quad -\infty \leq z < 0.675 \ = \ 0.75$$
$$\text{Area in the region} \quad -\infty \leq z < 0 \ = \ 0.5$$
$$\text{Hence, area in the interval} \quad 0 \leq z < 0.675 \ = \ 0.25$$
$$\text{By symmetry, area in} \quad 0 \leq z < -0.675 \ = \ 0.25$$
$$\text{Thus, area in the interval} \quad 0.675 \leq z \leq 0.675 \ = \ 0.5$$
$$\text{So we can conclude that} \quad |z| \leq 0.675 \ = \ 0.5$$

From the Z world, we can go over to the X world by a linear transformation. The shape of the curve, the area under given intervals, etc., will not change. However, the numerical values of X range will be different from those of Z.

$$\text{Converting z1} \quad X_1 \ = \ 1 - 2 * 0.675 = -0.3$$
$$\text{Converting z2} \quad X_2 \ = \ 1 + 2 * 0.675 = 2.3$$

it is easily verified that $P(|X - 1| \leq 2.3) = 0.5$.

Prob. 25.

(a)

Given data	$\mu = 500$; $\sigma = 100$
Question	Find $P(X > 700)$
Let us normalise x to z	$z = (700 - 500)/100 = 2.0$
Area under $-\infty < z < 2.0$	found from Tables as 0.9772
Area beyond 2.0	$(1.0 - 0.9772) = 0.0228$
Probability that score exceeds 700	Answer is 2.28 percent
Conclusion	SAT is a tough exam

(b) To find the answer, we first note, from the standard normal curve, that the area under $-\infty < z < 1.28$ is 0.8997 which is almost equal to 0.9. This means that 10 percent of students scored higher than what corresponds to a value of z of 1.28. Using the conversion formula, $x = \mu + z\sigma = 500 + 1.28100 = 628$ To conclude, 10 percent of students obtained a score of 628 or higher.

Prob. 27. Tabulating the data and computations as usual

Value of z corresponding to 5000 gm	$(5000 - 3720)/527 = 2.49$
Area for range $-\infty < z < 2.49$	from std. tables 0.9925
Area for range $z > 2.49$	$(1 - 0.9925) = 0.0075$
Percentage of animals having a weight exceeding 5000 gm	0.75
QED	

Prob. 29.

$$
\begin{aligned}
\text{Given data} &= \mu = 2 \\
\text{Given data} &= \sigma = 1 \\
\text{To find} &= P(0 \leq X \leq 3) \\
\text{Converting to Z} &= (3 - 2)/1.0 = 1.0 \\
\text{Hence } Area \Big|_1^\infty &= (1 - 0.8413) = 0.1587
\end{aligned}
$$

Prob. 31.

$$f(z) = (1/\sqrt{2\pi})e^{-z^2/2}$$

$$\int_{-\infty}^{\infty} f(z)dz = 1$$

$$\int_{-\infty}^{\infty} zf(z)dz = 0$$

$$\int_{-\infty}^{\infty} z^2f(z)dz = 1$$

Now we proceed to find $E(X)$. In all equations below, the limits of integration are $-\infty$ to ∞

$$\text{Define } K = (1/\sqrt{2\pi})$$

$$f(x) = Ke^{-(x-\mu)^2}/2\sigma^2$$

$$\int xf(x)dx = K\int xe^{-(x-\mu)^2}/(2\sigma^2)$$

$$\text{Substitute } x = \mu + \sigma z$$

$$\text{Hence } dx = \sigma dz$$

$$\int xf(x)dx = K\mu\int e^{-\sigma^2 z^2}dz + K\sigma\int ze^{-\sigma^2 z^2}dz$$

$$\text{since } \int zf(z)dz = 0$$

$$\int xf(x)dx = K\mu\int e^{-\sigma^2}z^2$$

$$= K\mu/K$$

$$= \mu$$

Thus EX value is proved to be equal to μ

Prob. 33. It is required to find $E|X|$, given that X is standard normally distributed. It means that mean is zero and standard deviation is one.

Let us split the interval of integration into two parts: $-\infty$ to zero and zero to $+\infty$. Thus $EX = (EX1 + EX2)$, where EX1 and EX2 are defined and derived below. Also note that in the first interval, where x is negative, $|x|$ can be replaced by -x.

$$\text{std. normal fn} \quad f(x) = (1/\sqrt{2\pi})e^{-x^2/2}$$

$$\begin{aligned}
\text{substitute} \quad K &= (1/\sqrt{2\pi}) \\
\text{split integration into two parts} \quad EX &= EX1 + EX2 \\
EX1 &= \int_{-\infty}^{0} |x|e^{-x^2/2}dx \\
\text{When x is -ve} \quad |x| &= -x \\
\text{Hence} \quad EX1 &= K\int_{0}^{-\infty} xe^{-x^2/2}dx \\
\text{Due to symmetry} \quad EX1 &= K\int_{0}^{\infty} xe^{-x^2/2}dx \\
EX1 &= K \\
\text{By virtue of symmetry} \quad EX2 &= K \\
\text{Final answer} \quad EX &= 2K = \sqrt{2}/\pi
\end{aligned}$$

Prob. 35. Maximum score in the examination is given to be 100 marks.
$\mu = 74$and $\sigma = 11$

(a) Finding the percentage of students scoring above 90:

$$\begin{aligned}
\text{Converting to Z} \quad Z &= (90 - 74)/11 = 1.455 \\
Area\Big|_{\infty}^{1.455} &= 0.9272 \\
Area\Big|_{\infty}^{\infty} &= 1.0 \quad \text{Known Property} \\
Area\Big|_{1.455}^{\infty} &= 0.0728
\end{aligned}$$

So, the percentage of students scoring above 90 is 7.28%

(b) It is required to find the percentage of students who score in the range $60 \leq X \leq 80$

$$\begin{aligned}
Z1 &= (60 - 74)/11 = -1.723 \\
Z2 &= (80 - 74)/11 = 0.5455
\end{aligned}$$

$$Area \Big|_{-\infty}^{0.5455} = 0.7071 \quad \text{from Appendix B}$$

$$Area \Big|_{-\infty}^{1.273} = 0.898 \quad \text{from Appendix B}$$

$$Area \Big|_{\infty}^{-1.273} = 0.898 \quad \text{By symmetry}$$

$$Area \Big|_{-\infty}^{-1.273} = 0.102$$

$$Area \Big|_{-1.273}^{0.5455} = (0.707 - 0.102) = 0.605$$

(c) Finding the minimum score of highest 10 percent of class.

$$Area \Big|_{-\infty}^{?} = 0.9 \quad \text{to find the upper limit}$$

$$Area \Big|_{-\infty}^{0.816} = 0.9 \quad \text{from Appendix B}$$

$$X = \mu + \sigma Z$$

$$X = 74 + 110.816$$

$$= 82.98 \quad \text{lowest mark of the upper 10\% students}$$

Prob. 37. The occurrences with first digit after decimal being 3 are: 0.31 to 0.39; 1.31 to 1.39 2.31 to 2.39 3.31 to 3.39 There are four occurrences each of width 0.1. Thus the total is 0.4. It is one-tenth of the total width of four. Hence the probability of the first digit after decimal being 3 is $(0.4/4) = 10$ percent.

Prob. 39. For a uniform distribution, the expressions for mean and variance are:

$$\text{Mean of uniform distribution} \quad \mu = (a + b)/2$$

$$\text{Variance} \quad \sigma^2 = (b - a)^2/12$$

$$\text{Given that} \quad \mu = 4; \sigma^2 = 3$$

$$\text{It can be easily shown that} \quad a = 1; b = 7$$

$$\text{It is also possible that} \quad a = 7; b = 1$$

Prob. 41. $E(Y) = 3/4)$

Prob. 43. Computer programs are available for conducting simulation of probability trials. For example, the famous Bernoulli trial can be conducted for the case where $P(X = H) \leq 0.6$. In the 10 trials below, the computer-generated random numbers are normally distributed. The outcomes of the trials in terms of Heads or Tails are listed below:

- 0.1905 H

- 0.4285 H

- 0.9963 T

- 0.1666 H

- 0.2223 H

- 0.6885 T

- 0.0489 H

- 0.3567 H

- 0.0719 H

- 0.8661 T

Prob. 45. (a) $(1 - x)^n$

(b) Use L'Hopital Rule on the exponential limit.

Prob. 47. A probability density function f(x) is given:

$$f(x) = \begin{cases} \lambda e^{-\lambda x} & \text{for } x > 0 \\ 0 & \text{for } x \leq 0 \end{cases}$$

If you want to verify that f(x) given here is indeed a proper candidate for a distribution function, go to Problem 1.

$$\text{By definition} \quad EX \; = \; \int_0^\infty x\lambda e^{-\lambda x} dx$$

$$\text{Substitute} \quad u = x$$

$$\text{Also substitute} \quad dv = e^{-\lambda x}dx$$

$$\text{Hence} \quad v = e^{-\lambda x}/(-\lambda)$$

$$\text{Substituting} \quad EX = xe^{-\lambda x}\Big|_0^\infty + e^{-\lambda x}/-\lambda \Big|_0^\infty$$

$$\text{Thus} \quad EX = (1/(\lambda))$$

Prob. 49. The reliability index of many industrial products follows an exponential distribution. The average life of a battery is given to be three months. This means that $\lambda = 1/3$.

$$\text{Given density function} \quad f(x) = \begin{cases} (1/3)e^{-x/3} & \text{for } x > 0 \\ 0 & \text{for } x \le 0 \end{cases}$$

The probability that the battery will last for more than four months can be found by calculating the area under the distribution curve beyond the value of four.

$$\text{Area under curve} \quad Area = \int_4^\infty \frac{1}{3}e^{-x/3}dx$$

$$= -e^{-x/3}\Big|_4^\infty$$

$$= 0.2636$$

So, the probability is 26% percent that the battery will last longer than four months.

Prob. 51. The lifetime of a radioactive atom is assumed to be exponentially distributed with an average lifetime of 27 days.

(a) It is required to find the probability that the atom will not decay during the first 20 days of starting to observe the atom.

$$\text{Given density function} \quad f(x) = \begin{cases} (1/27)e^{-x/27} & \text{for } x > 0 \\ 0 & \text{for } x \le 0 \end{cases}$$

The probability that the atom will not decay during the first twenty days can be found by calculating the area under the distribution curve beyond the value of twenty

$$\text{Area under curve} \quad Area = \int_2^{0\infty} \frac{1}{27}e^{-x/27}dx$$

$$= -e^{-x/27} \Big|_2^{0\infty}$$

$$= 0.47676$$

So, the probability is 45 percent that the atom will not decay during the first twenty days.

(b) If the atom is to last for the second twenty days, it should definitely not have decayed during the first twenty days. Also the value of λ is constant and such a case is called a non-aging process. Thus the answer to part (b) is that the probability is 45 percent that the atom will not decay during the second twenty days.

Prob. 53. a) $P(X = 0) = e^{-4}4^0/0! = e^{-4}$.

b) $P(X = 2) = e^{-4}4^2/2! = 8e^{-4}$.

c) Since they are independent:

$$P(X = 2, Y = 0) + P(X = 1, Y = 1) + P(X = 0, Y = 2)$$

$$= 8 \cdot e^{-4} \cdot e^{-4} + 4e^{-4} \cdot 4 \cdot e^{-4} + e^{-4} \cdot 8e^{-4} = 32e^{-8}.$$

d) $P(Y|X + Y = 2) = \frac{1}{4}$.

Prob. 55. We have $\frac{1}{4}$ of hour $= 15$ minutes

Prob. 57. (a) We have $F(X) = \begin{cases} 1 - e^{-\frac{x}{3}}, & x > 0 \\ 0, & x \le 0 \end{cases}$

$P(\text{fails within 2 years}) = F(2) = 0.51$

$P(\text{fails after 2 years}) = 1 - F(2) = 0.49$

(b) We need $P(N(5) = 1) = \frac{5}{3}e^{-\frac{5}{3}}$.

Prob. 59. The lifetime of a technical device is known to be exponentially distributed with a mean life of five years.

(a) It is required to find the probability that the device fails after three years.

$$\text{Given density function} \quad f(x) = \begin{cases} (1/5)e^{-x/5} & \text{for } x > 0 \\ 0 & \text{for } x \le 0 \end{cases}$$

From the density function, we evaluate thd distribution function by integration. As the result has been discussed earlier, we write:

$$\text{Distribution function} \quad F(x) = \begin{cases} (1 - e^{-x/5} & \text{for } x > 0 \\ 0 & \text{for } x \le 0 \end{cases}$$

Knowing the distribution function, it is easy to evaluate failure data.

$$\text{Distribution function} \quad F(x) = 1 - e^{-0.2x}$$

$$\text{P(Device fails within 3 years)} = F(3)$$

$$\text{P(Device fails after 3 years)} = 1 - F(3) = 0.549$$

Thus there is a 55 percent chance that the device will last longer than 3 years.

(b) First let us find the probability that it works for at least six years.

$$\text{Distribution function} \quad F(x) = (1 - e^{-0.2x})$$

$$\text{P(Device fails within 6 years)} = F(6)$$

$$\text{P(Device has worked for 6 years)} = 1 - F(6) = e^{-1.2}0.3012$$

$$\text{P(Device has worked for 7 years)} = 1 - F(7) = e^{-1.4}0.2466$$

The probability that the device works for six years is 30.12 percent. The probability that the device works for seven years would obviously be less and is thus equal to 24.66 percent. Now, it is clear that:

$$T(\text{works for six years}) \subset T(\text{works for seven years})$$

Thus, the answer to the question in part (b) is 24.66 percent.

Prob. 61. A technical device, the lifetime of which follows exponential distribution, has the parameter $\lambda = 0.2$ per year.

(a) Expected life time of the above device is five years.

(b)

$$\text{Median lifetime} \quad x_m \stackrel{\text{def}}{=} P(X > x_m)$$

The median lifetime is defined as age x_m, at which P(not having failed by age x_m) is 0.5.

$$\text{By definition} \quad S(x_m) \;=\; e^{-\lambda x_m} = 0.5$$

$$\text{From given data} \quad e^{-\frac{x}{5}} \;=\; 0.5$$

$$\text{From given data} \quad x \;=\; 0.6935 = 3.465$$

Thus the median lifetime is 3years, 5 months and about 17 days

Prob. 63. In this problem, the hazard rate function is not a constant as in the previous two examples, but is itself a function of time. This is the key difference between non-ageing entities and ageing ones.

(a) In this part, it is required to find survival chance for the specified period.

$$\text{Given hazard rate function} \quad \lambda(x) \;=\; 0.3 + 0.1e^{0.01x}$$

$$\text{P(It lives for more than five days)} \quad P(T > 5) \;=\; S(5)$$

$$\text{Survival function} \quad S(x) \;=\; e^{-\int_0^x \lambda(u)\,du}$$

$$\text{Doing step by step} \quad 0.1\int_0^5 e^{0.01u}\,du \;=\; 10(e^{0.05} - 1) = 0.5127$$

$$\text{Doing step by step} \quad 0.3\int_0^5 0.3\,du \;=\; 1.5$$

$$\text{Add the two numbers} \quad 0.5127 + 1.5 \;=\; 2.0127$$

$$\text{Now find the survival function} \quad S(5) \;=\; e^{-2.0127} = 0.13363$$

Thus the chances of the organism living for 5 days and more is 13.363 percent.

(b) In this part, it is required to find the probability that the organism survives for a period of between 7 and 10 days.

$$\text{Hazard rate function (data)} \quad \lambda(x) \;=\; 0.3 + 0.1e^{0.01x}$$

$$\text{P(It lives for more than 10 days)} \quad P(T > 5) \;=\; S(5)$$

$$\text{Survival function} \quad S(x) \;=\; e^{-\int_0^x \lambda(u)du}$$

$$\text{evaluate integral} \quad 0.1\int_0^1 0e^{0.01u}du \;=\; 10(e^{0.1} - 1) = 1.0517$$

$$\text{this part of integral is simple} \quad 0.3\int_0^1 00.3du \;=\; 3$$

$$\text{Add the above two numbers} \quad 1.0517 + 3 \;=\; 4.0517$$

$$\text{Now find the survival function} \quad S(10) \;=\; e^{-4.0517} = 0.01739$$

For a moment, compare this value of 1.739 percent for S(10) with that of 13.363 percent obtained for S(5). Why is $S(10) < S(5)$? Another interesting point should be mentioned at this stage. If one were to calculate S(15) and compare S(5) - S(10) with S(10) - S(15), it will be found that the two are not equal. That is the indication that the process here is an ageing process.

Evaluating the survival function for 7 days.

$$\text{Data given for hazard rate function} \quad \lambda(x) \;=\; 0.3 + 0.1e^{0.01}x$$

$$\text{P(It lives for more than seven days)} \quad P(T > 5) \;=\; S(5)$$

$$\text{Survival function by definition} \quad S(x) \;=\; e^{-\int_0^x \lambda(u)du}$$

$$\text{first part of integral} \quad 0.1\int_0^7 e^{0.01u}du \;=\; 10(e^{0.07} - 1) = 0.0725$$

$$\text{Doing step by step} \quad 0.3\int_0^7 du \;=\; 2.1$$

$$\text{Add the two numbers} \quad 0.0725 + 2.1 \;=\; 2.1725$$

$$\text{Now find the survival function} \quad S(7) \;=\; e^{-2.1725} = 0.1139$$

$$\text{P(organism has a life between 7 to 10 days)} \;=\; S(7) - S(10)$$

$$=\; 0.1139 - 0.01739$$

$$=\; 0.0965$$

Thus the chances of finding an organism alive between 7 and 10 days is 9.65 percent.

Prob. 65. Age is defined as the number of years at which the probability of the device/organism not having failed is 0.5. It will be seen below that the relevant equation cannot be solved in a closed form and hence an iterative numerical solution is needed.

$$\text{Hazard rate function is} \quad \lambda(x) \;=\; 1.2 + 0.3e^{0.5x}$$

$$\text{Survival function is} \quad S(x) \;=\; e^{-\int_0^x \lambda(u)du}$$

$$\text{Transcendental equation in xm} \quad Sx_m \;=\; e^{-\int_0^{x_m} \lambda(u)du} = 0.5$$

$$\text{exp part of integral} \quad 0.3\int_0^{x_m} e^{0.5u}du \;=\; 0.6(e^{0.5x_m} - 1)$$

$$\text{simpler part of integral} \quad 1.2\int_0^{x_m} du \;=\; 1.2x_m$$

$$\text{For simplicity, let} \quad y(x_m) \;=\; 1.2x_m + 0.6(e^{0.5x_m} - 1)$$

$$\text{equation to be solved} \quad e^{-y(x_m)} \;=\; 0.5$$

$$\text{Taking Ln on both sides} \quad -y(x_m) \;=\; \ln(0.5) = -0.6931$$

$$\text{equation for age} \quad 1.2x_m + 0.6(e^{0.5x_m} - 1) \;=\; 0.6931$$

$$\text{equation simplified} \quad x_m + 0.5e^{0.5x_m} \;=\; 1.0776$$

$$\text{exp upto 2 terms} \quad x_m + 0.5(1 + 0.5x_m + 0.5x_m^2/2 \;=\; 1.0776$$

$$\text{simplifying} \quad x_m^2 + 20x_m - 9.248 \;=\; 0$$

$$\text{solution} \quad x_m \;=\; 0.45$$

$$\text{verifying} \quad 0.45 + 0.5e^{0.50.45} \;=\; -0.0014$$

Thus, the mean life (age) is 0.45

Prob. 67. This and the next few problems are on the use of Weibull law by which the hazard rate function increases according to a power law. Such law applies in general to technical devices.

(a) It is required to find the probability that the organism with known hazard rate function will survive for longer than 50 days.

$$\text{Hazard rate function} \quad \lambda(x) \;=\; 210^{-5}x^{1.5}$$

$$P(T > 50 days) \quad = \quad S(50)$$

$$\text{Survival function} \quad S(50) \quad = \quad e^{-\int_0^5 0\lambda(u)du}$$

$$\text{Substituting} \quad S(50) \quad = \quad \int_0^5 0(210^{-5}u^{1.5}du)$$

$$\text{Exponent} \quad = \quad 0.810^{-5}50^{2.5}$$

$$= \quad 0.14142$$

$$\text{Hence} \quad S(50) \quad = \quad e^{-0.14142} = 0.86812$$

Thus, the probability of the organism surviving for longer than 50 days is 86.812 percent.

(b) It is required to find the probability that an organism will survive for a period between 50 and 70 days. On the same lines as before, we shall find the value of S(70)

$$\text{Hazard rate function} \quad \lambda(x) \quad = \quad 210^{-5}x^{1.5}$$

$$P(T > 70 days) \quad = \quad S(70)$$

$$\text{Survival function} \quad S(70) \quad = \quad e^{-\int_0^7 0\lambda(u)du}$$

$$\text{Substituting} \quad S(70) \quad = \quad \int_0^7 0(210^{-5}u^{1.5}du)$$

$$\text{Exponent} \quad = \quad 0.810^{-5}70^{2.5} = 0.32797$$

$$\text{Hence} \quad S(50) \quad = \quad e^{-0.32797} = 0.7204$$

$$\text{P(50 to 70 days)} \quad S(50) - S(70) \quad = \quad 0.86812 - 0.7204$$

$$= \quad 0.1477$$

The chance of an organism being alive during the period 50 to 70 days is 14.77 percent.

Prob. 69. In this and the next problem we find the median life for organisms governed by Weibull law.

$$\text{Hazard rate function is} \quad \lambda(x) \quad = \quad 410^{-5}x^{22}$$

$$\text{Survival function is} \quad S(x) \;=\; e^{-\int_0^x \lambda(u)du}$$

$$\text{equation in xm} \quad Sx_m \;=\; e^{-\int_0^{x_m} \lambda(u)du} = 0.5$$

$$\text{Integral in the exponent} \quad 4 10^{-5}(x_m)^{23}/23$$

$$\text{To solve} \quad 4 10^{-5}(x_m)^{23}/23 \;=\; -Ln0.5$$

$$\text{solving} \quad (x_m)^{23} \;=\; 3.986$$

$$\text{median life} \quad x_m \;=\; 1.062$$

Prob. 71.

$$\text{Hazard rate function is} \quad = \quad \lambda(x)$$

$$\text{Survival function is} \quad S(x) \;=\; e^{-\int_0^x \lambda(u)du}$$

$$\text{Survival function is} \quad S(x+1) \;=\; e^{-\int_0^{x+1} \lambda(u)du}$$

$$\text{Shorter Survival} \quad S(x) \;=\; e^{-\int_0^x \lambda(u)du}$$

$$\text{Ln} \quad LnS(x+1) \;=\; -\int_0^{x+1} \lambda(u)du$$

$$\text{Ln} \quad LnS(x) \;=\; -\int_0^x \lambda(u)du$$

$$\text{Subtracting} \quad LnS(x+1) - LnS(x) \;=\; -\int_x^{x+1} \lambda(u)du$$

$$\text{QED} \quad LnS(x+1)/LnS(x) \;=\; -\int_x^{x+1} \lambda(u)du$$

12.6   Limit Theorems

Prob. 1. As X is exponentially distributed with parameter $\lambda = \frac{1}{2}$, the density function is:

$$f(x) = \frac{1}{2}e^{\frac{-x}{2}}$$

So we can find Probability P as:

$$P(X \geq 3) = \int_3^\infty \frac{1}{2}e^{\frac{-x}{2}} = e^{\frac{-x}{2}}\Big|_3^\infty = e^{\frac{-3}{2}}$$

Now computing EX we have:

$$EX = \int_0^\infty x f(x) dx = \int_0^\infty \frac{x}{2} e^{\frac{-x}{2}} dx = 2$$

Thus, according to Markov's inequality:

$$P(X \geq 3) \leq \frac{EX}{3} = \frac{2}{3}$$

and the exact value is much lower.

Prob. 3. We will split the summation in twi parts, before and after a and then estimate one of them:

$$EX = \sum_x x P(X = x) = \sum_{x < a} x P(X = x) + \sum_{x \geq a} x P(X = x)$$

we can only split the sums because they are both nonnegative and convergent, and since each of the summands is a nonnegative number,

$$EX \geq \sum_{x \geq a} x P(X = x)$$

and observe that the sum extends over values of $x \geq a$, so

$$EX \geq \sum_{x \geq a} x P(X = x) \geq \sum_x a P(X = x) = a \sum_x P(X = x) = a P(X \geq a)$$

dividing by a, the Markov inequality follows.

Prob. 5. Exact: $P(|X| \geq 1) = 0$; Chebyshev's inequality: $P(|X| \geq 1) \leq \frac{1}{3}$

Prob. 7. Using Chebyshev's inequality

$$P(|X - \mu| \geq c) \leq \frac{\sigma^2}{c^2}$$

Thus,

$$P(|X - 10| \geq 5) \leq \frac{9}{5^2} = 0.36$$

Prob. 9. $\frac{1}{n} \sum_{i=1}^n X_i$ converges to 3/2 as $n \to \infty$

Prob. 11. Since $E|X_i| = \infty$, we cannot apply the law of large numbers as stated in Section 12.6.

Prob. 13. The sample size should be at least 380.

Prob. 15. 0.1587

Prob. 17. (a) 0.0023 (b) 0.83

Prob. 19. (a) 0.1192 (b) 0.182 (c) 0.1521

Prob. 21. (a) -11.2 (b) 0.579

Prob. 23. 69

Prob. 25. 385

Prob. 27. (a) 0.3660 (b) 0.3679 (c) 0.0582

Prob. 29. (a) 0.1849 (b) 0.1755 (c) 0.1896

Prob. 31. Likely not.

Prob. 33. (a) 0.6065, 0.3033, 0.0758 (b) 0.8461

Prob. 35. 0.1429

Prob. 37. 0.0485

12.7 Statistical Tools

Prob. 1. Re-arranging the sample in increasing order to find median, we will get:

$$0, 3, 6, 12, 13, 17, 18, 21, 25, 47$$

So the median is

$$\frac{13 + 17}{2} = 15$$

$$\sum_{k=1}^{10} x_k = 162 \quad \text{and} \quad \sum_{k=1}^{10} x_k{}^2 = 4246$$

Hence,

$$\bar{X}_n = \frac{162}{10} \approx 16.2$$

and

$$S_n^2 = \frac{1}{n-1}\left[\sum_{k-1}^{n} x_k{}^2 - \frac{1}{n}(\sum_{k-1}^{n} x_k)^2\right] = \frac{1}{9}(4246 - \frac{162^2}{10}) \approx 180.2$$

Prob. 3. Re-arranging gives: 29, 28, 30, 34, 35, 36, 38, 42, 45, 45.

median: $\frac{35+36}{2} = 35.5$

mean: 36

variance: 43.1

Prob. 5.

$$\sum_{k=1}^{7} x_k f_k = 3830 \quad \text{and} \quad \sum_{k=1}^{7} x_k^2 f_k = 46782$$

Hence in order to find the mean,

$$\bar{X}_n = \frac{3830}{321} \approx 11.93$$

and

$$S_n^2 = \frac{1}{n-1}\left[\sum_{k=1}^{7} x_k^2 f_k - \frac{1}{n}\left(\sum_{k=1}^{7} x_k f_k\right)^2\right] = \frac{1}{320}\left(46782 - \frac{3830^2}{321}\right) \approx 3.389$$

Prob. 7. Finding out the frequency of each clutch size:

Relative Frequency	Frequency
0.05	15
0.09	27
0.12	36
0.19	57
0.23	69
0.12	36
0.13	39
0.07	21

$$\sum_{k=1}^{8} x_k f_k = 1707 \quad \text{and} \quad \sum_{k=1}^{8} x_k^2 f_k = 10749$$

Hence in order to find the mean,

$$\bar{X}_n = \frac{1707}{300} = 5.69$$

and

$$S_n^2 = \frac{1}{n-1}\left[\sum_{k-1}^{l} x_k^2 f_k - \frac{1}{n}(\sum_{k-1}^{l} x_k f_k)^2\right] = \frac{1}{299}\left(10749 - \frac{1707^2}{300}\right) \approx 3.465$$

Prob. 9. We have:

sample	sample mean
(1,1)	1.0
(1,6)	3.5
(1,8)	4.5
(6,1)	3.5
(6,6)	6.0
(6,8)	7.0
(8,1)	4.5
(8,6)	7.0
(8,8)	8.0

Prob. 11.

$$\sum_{k=1}^{n}(X_k-\bar{X}) = \sum_{k=1}^{n}\left(X_k - \frac{1}{n}\sum_{k=1}^{n}X_k\right) = \sum_{k=1}^{n}X_k - \sum_{k=1}^{n}\left(\frac{1}{n}\sum_{k=1}^{n}X_k\right) = \sum_{k=1}^{n}X_k - n\frac{1}{n}\sum_{k=1}^{n}X_k = 0$$

Prob. 13. Grouping the summation by the elements of the same value we have:

$$\bar{X} = \frac{1}{n}\sum_{k=1}^{n}X_k = \frac{1}{n}\sum_{k=1}^{l}X_k f_k$$

Where l is number of X_k, i.e. the total number of distinct samples.

Prob. 15.

(a) Since x is exponential, we have $\bar{x} = \frac{1}{\lambda} = \frac{1}{3}$, variance: $\frac{1}{n\lambda^2} = \frac{1}{450}$.

(b) Since we have 1000 samples, the histograms will be approximately normal.

Prob. 17. The three samples are given below in the table:

Number of sample	Sample$_1$	Number of sample	Sample$_1$
1	0.01	6	.46
2	.1	7	.55
3	.19	8	.64
4	.28	9	.73
5	.37	10	.82

Number of sample	Sample$_2$	Number of sample	Sample$_2$
1	.03	6	.48
2	.12	7	.57
3	.21	8	.66
4	.30	9	.75
5	.39	10	.84

Number of sample	Sample$_3$	Number of sample	Sample$_3$
1	.05	6	.45
2	.13	7	.53
3	.21	8	.61
4	.29	9	.69
5	.37	10	.77

(a) For $Sample_1$:

$$\sum_{k=1}^{10} x_k = 4.15 \quad \text{and} \quad \sum_{k=1}^{10} x_k{}^2 = 2.2309$$

Hence,

$$\bar{X}_n = \frac{4.15}{10} \approx 0.415$$

and

$$S_n^2 = \frac{1}{n-1}\left[\sum_{k-1}^{n} x_k{}^2 - \frac{1}{n}(\sum_{k-1}^{n} x_k)^2\right] = \frac{1}{9}(2.3095 - \frac{4.15^2}{10}) \approx 0.06525$$

For $Sample_2$:

$$\sum_{k=1}^{10} x_k = 4.35 \quad \text{and} \quad \sum_{k=1}^{10} x_k{}^2 = 2.5605$$

Hence,

$$\bar{X}_n = \frac{4.35}{10} \approx 0.435$$

and

$$S_n^2 = \frac{1}{n-1}\left[\sum_{k-1}^{n} x_k{}^2 - \frac{1}{n}(\sum_{k-1}^{n} x_k)^2\right] = \frac{1}{9}(2.5605 - \frac{4.35^2}{10}) \approx 0.07425$$

For *Sample₃*:

$$\sum_{k=1}^{10} x_k = 3.81 \quad \text{and} \quad \sum_{k=1}^{10} {x_k}^2 = 2.209$$

Hence,

$$\bar{X}_n = \frac{3.81}{10} \approx 0.381$$

and

$$S_n^2 = \frac{1}{n-1}\left[\sum_{k-1}^{n} {x_k}^2 - \frac{1}{n}(\sum_{k-1}^{n} x_k)^2\right] = \frac{1}{9}(2.209 - \frac{3.81^2}{10}) \approx 0.08415$$

(b) Combining the three samples:

$$\sum_{k=1}^{30} x_k = 12.31 \quad \text{and} \quad \sum_{k=1}^{30} {x_k}^2 = 7.0004$$

Hence,

$$\bar{X}_n = \frac{12.31}{30} \approx 0.41033$$

and

$$S_n^2 = \frac{1}{n-1}\left[\sum_{k-1}^{n} {x_k}^2 - \frac{1}{n}(\sum_{k-1}^{n} x_k)^2\right] = \frac{1}{29}(7.0004 - \frac{12.31^2}{30}) \approx 0.06721$$

(c) True value of mean will be $\mu = 0.5$ and variance can be computed to $\sigma^2 = 1/12$, as we expect the combined sample, because of being larger, provides a smaller variance, since $var(\bar{X}_n) = \sigma^2/n$.

Prob. 19.

$$\sum_{k=1}^{10} x_k = 162 \quad \text{and} \quad \sum_{k=1}^{10} {x_k}^2 = 4246$$

Hence,

$$\bar{X}_n = \frac{162}{10} \approx 16.2$$

and

$$S_n^2 = \frac{1}{n-1}\left[\sum_{k-1}^{n} {x_k}^2 - \frac{1}{n}(\sum_{k-1}^{n} x_k)^2\right] = \frac{1}{9}(4246 - \frac{162^2}{10}) \approx 180.18$$

For sample error:

$$S_{\bar{X}} = \frac{S_n}{\sqrt{n}} \approx 4.245$$

Prob. 21. Since X_k is from a normal distribution then

$$P(-1.96 \leq Z \leq 1.96) = 0.95$$

and the confidence interval is given by:

$$[\bar{X}_n - 1.96\frac{\sigma}{\sqrt{n}}, \bar{X}_n + 1.96\frac{\sigma}{\sqrt{n}}]$$

Now using the fact that the $\bar{X}_n = 0.061$ and the $\sigma = 0.7427$ for this sample, the confidence interval is given by $[-0.3993, 0.5213]$.

Prob. 23. Use your calculator and prepare a data set for in-class discussion.

Prob. 25. The sample mean is

$$\bar{X}_n = \frac{117}{162} \approx 0.722$$

The estimate for the probability of germination success of seeds is therefore $\hat{p} = 0.722$. The standard error is

$$S.E. = \sqrt{\frac{\hat{p}(1-\hat{p})}{n}} = \sqrt{\frac{0.722 \times 0.288}{162}} \approx 0.03583$$

We can thus report the result as 0.722 ± 0.03583. Since n is large, we find a confidence interval as

$$[0.722 - (1.96)(0.03583), 0.722 + (1.96)(0.03583)] = [0.6518, 0.7922]$$

Prob. 27. To facilitate the computation, we construct the following table:

x_k	y_k	$x_k - \bar{x}$	$y_k - \bar{y}$	$(y_k - \bar{x})(x_k - \bar{x})$
-3	-6.3	-2.5	-4.4167	11.04175
-2	-5.6	-1.5	-3.7167	5.57505
-1	-3.3	-0.5	-1.4167	0.70835
0	0.1	0.5	1.9833	0.99165
1	1.7	1.5	3.58333	5.37495
2	2.1	2.5	3.9833	9.95825
$\bar{x} = -0.5$	$\bar{y} = -1.8833$	$\sum(x_k - \bar{x})^2$ $= 17.5$	$\sum(y_k - \bar{y})^2$ $= 67.96833$	$\sum(x_k - \bar{x})(y_k - \bar{y})$ $=33.65$

Now,

$$\hat{b} \;=\; \frac{33.65}{17.5} = 1.92286$$
$$\hat{a} \;=\; -1.8833 - (1.92286)(-0.5) = -0.92187$$

Hence, the linear regresion line is given by

$$y = 1.92x - 0.92$$

The coefficient of determination r^2 is given by

$$r^2 = \frac{[\sum(x_k - \bar{x})(y_k - \bar{y})]^2}{\sum(x_k - \bar{x})^2 \sum(y_k - \bar{y})^2}$$

Substituting the values we will find

$$r^2 = \frac{(33.65)^2}{(17.5)(67.96)} = 0.9521$$

Prob. 29. We know that

$$\bar{y} - a - b\bar{x} = 0$$

Where

$$\bar{y} = \frac{1}{n}\sum y_k \qquad \text{and} \qquad \bar{x} = \frac{1}{n}\sum x_k$$

So the sum of the residuals about the linear regression line is equal to zero.

Prob. 31. To facilitate the computation, we construct the following table:

x_k	y_k	$x_k - \bar{x}$	$y_k - \bar{y}$	$(y_k - \bar{x})(x_k - \bar{x})$
69	15	-10.8	-2	21.6
70	15	-9.8	-2	19.6
72	16	-7.8	-1	7.8
75	16	-4.8	-1	4.8
81	17	1.2	0	0
82	17	2.2	0	0
83	16	3.2	-1	-3.2
84	18	4.2	1	4.2
89	20	9.2	3	27.6
93	20	13.2	3	39.6
$\bar{x} = 79.8$	$\bar{y} = 17$	$\sum (x_k - \bar{x})^2$ $= 589.6$	$\sum (y_k - \bar{y})^2$ $= 30$	$\sum (x_k - \bar{x})(y_k - \bar{y})$ $= 122$

Now,

$$\hat{b} = \frac{122}{589.6} = 0.20692$$

$$\hat{a} = 17 - (0.20692)(79.8) = 0.487788$$

Hence, the linear regresion line is given by

$$y = 0.20x + 0.48$$

The coefficient of determination r^2 is given by

$$r^2 = \frac{[\sum (x_k - \bar{x})(y_k - \bar{y})]^2}{\sum (x_k - \bar{x})^2 \sum (y_k - \bar{y})^2}$$

Substituting the values we will find

$$r^2 = \frac{(122)^2}{(589.6)(30)} = 0.841$$

12.9   Review Problems

Prob. 1.

(a) There are 25 students and the total number of days in a year are 365. So the first person have a choice of having birthday on any of the days. Therefore,

$$p_1 = \frac{365}{365}$$

The second person has a choice of only 364 days to have a birhtday, so the probability for 2 people is,

$$p_2 = \frac{365}{365} \frac{364}{365}$$

similarly, for the 25 people,

$$p_{25} = \frac{365}{365} \frac{364}{365} \frac{363}{365} \cdots \frac{341}{365} = \frac{365!}{340!\, 365^{25}} = 0.431$$

(b) We can write the general equation as:

$$p_n = \frac{365!}{(365 - n)!\, 365^n}$$

then it is easy to see that

$$\begin{aligned}
p_{n+1} &= \frac{365!}{(365 - (n+1))!\, 365^{n+1}} \\
&= \frac{365!\, (365 - n)}{(365 - n)(365 - (n+1))!\, 365^n 365} \\
&= \frac{365!\, (365 - n)}{(365 - n)!\, 365^n 365} \\
&= p_n \frac{365 - n}{365}
\end{aligned}$$

Prob. 3. There are $15 \times 14 \times 13$ ways to choose three plants (considering the order) for the first plot, and considering that choices of order (ABC, ACB, BAC, ...) does not matter, we can divide the total by 6, and the choices for the first plot are $15 \times 14 \times 13/6$, and proceed in this way for the rest of the plots, the total will be $15!/6^5 = 168,168,000$.

Prob. 5.

(a) We know that 0.42 of the seeds of a certain plant germinates, so the expected number of germinating seeds in a sample of ten seeds is:

$$\texttt{\# of germinating seeds} = 0.42 \times \texttt{\# of seeds} = 0.42 \times 10 = 4.2 \approx 4$$

(b) The probability that none of the seeds will grow in a plant are:

$$p = (1 - 0.42)^{10} = (0.58)^{10} \approx 0.0043$$

(c) On 5 pots, it will be: $5 \times 0.0043 = 0.0215$.

(d) The probability of number of pots having no seeds is: $P = 1 - 0.0043^5$.

Prob. 7.

(a The chance of passing in the first trial is 0.2, and chance of failing is 0.8, So the the probability of passing in the second trial after failing in the first trial is:

$$p = 0.8 \times 0.2 = 0.16$$

(b) The chance of passing in the first trial is 0.2, and simmilarly is the chance of passing in the second trial. So the probability of passing in the second trial is: 0.2.

Prob. 9. (a) $\mu = 170$, $\sigma = 6.098$ (b) 0.4013

Prob. 11. A white flowering plant is of genotype cc. If the red-flowering parent plant is of genotype CC is crossed with a white-flowering plant, then all offspring are of genotype Cc and therefore produce red flowers. If the red-flowering parent plants is of genotype Cc is crossed with a white-flowering plant, then with probability 0.5 an offspring is of genotype Cc (and therefore red flowering) and with a probability 0.5 of genotype cc (and therefore white flowering). Assume that percentage of red-flowering parent plants of genotype Cc is x, so the percentage of red flowering parent plants of genotype CC would be $1 - x$. Therefore, the percentage red-flowering parent plants would be:

$$P(\texttt{red offspring}) = 0.9 = x(0.5) + (1 - x)(1) = 1 - 0.5x$$

which solves to $x = 0.2$, so the percentage of red-flowering parent plants would be 20 percent.

Prob. 13.

(a) $V = \frac{1}{n} \sum_{k=1}^{n}(X_k - \bar{X}_n)^2 = \frac{n-1}{n} \frac{1}{n-1} \sum_{k=1}^{n}(X_k - \bar{X}_n)^2 = \frac{n-1}{n} S^2$

(b) $EV = E(\frac{n-1}{n} S^2) = \frac{n-1}{n} ES^2 = \frac{n-1}{n} \sigma^2$

Prob. 15. Answers may vary.

(a) Since the generated distributions will vary, we can have:

$(1, 3.5), (2, 4.8), (3, 6.9), (4, 9.1), (5, 10.95)$

(b) Following Example 11, Section 12.7 we can determine the least square line by minimizing the sum of the deviations:

$$h(a, b) = \sum_{k=1}^{5}[y_k - (a + bx_k)]^2.$$